Control of Nonlinear Systems

The book "*Control of Nonlinear Systems–Stability and Performance*" fills a crucial gap in the field of nonlinear control systems by providing a comprehensive yet accessible treatment of the subject. Unlike many existing texts that are either too complex for beginners or omit essential topics, this book strikes the right balance of mathematical rigor and practicality.

The main objective of the book is to simplify and unify the existing techniques for designing and analyzing control systems for nonlinear systems. It aims to alleviate confusion and difficulty in understanding these methods, making it an invaluable resource for students, researchers, and practitioners in the field.

By presenting the material in a tutorial manner, the book enhances the reader's understanding of the design and analysis of a wide range of control methods for nonlinear systems. The emphasis on stability and performance highlights the practical relevance of the concepts discussed in the book.

Overall, "*Control of Nonlinear Systems–Stability and Performance*" is a valuable contribution to the field of nonlinear control systems. Its emphasis on practical applications and its accessible presentation make it an indispensable resource for engineers seeking to enhance their knowledge and skills in this important area of control theory.

Automation and Control Engineering
Series Editors - Frank L. Lewis, Shuzhi Sam Ge, and Stjepan Bogdan

Optimal and Robust Scheduling for Networked Control Systems
Stefano Longo, Tingli Su, Guido Herrmann, and Phil Barber

Electric and Plug-in Hybrid Vehicle Networks
Optimization and Control
Emanuele Crisostomi, Robert Shorten, Sonja Stüdli, and Fabian Wirth

Adaptive and Fault-Tolerant Control of Underactuated Nonlinear Systems
Jiangshuai Huang, Yong-Duan Song

Discrete-Time Recurrent Neural Control
Analysis and Application
Edgar N. Sánchez

Control of Nonlinear Systems via PI, PD and PID
Stability and Performance
Yong-Duan Song

Multi-Agent Systems
Platoon Control and Non-Fragile Quantized Consensus
Xiang-Gui Guo, Jian-Liang Wang, Fang Liao, Rodney Swee Huat Teo

Classical Feedback Control with Nonlinear Multi-Loop Systems
With MATLAB® and Simulink®, Third Edition
Boris J. Lurie, Paul Enright

Motion Control of Functionally Related Systems
Tarik Uzunović and Asif Sabanović

Intelligent Fault Diagnosis and Accommodation Control
Sunan Huang, Kok Kiong Tan, Poi Voon Er, Tong Heng Lee

Nonlinear Pinning Control of Complex Dynamical Networks
Edgar N. Sanchez, Carlos J. Vega, Oscar J. Suarez and Guanrong Chen

Adaptive Control of Dynamic Systems with Uncertainty and Quantization
Jing Zhou, Lantao Xing and Changyun Wen

Robust Formation Control for Multiple Unmanned Aerial Vehicles
Hao Liu, Deyuan Liu, Yan Wan, Frank L. Lewis, Kimon P. Valavanis

Variable Gain Control and Its Applications in Energy Conversion
Chenghui Zhang, Le Chang, Cheng Fu

Manoeuvrable Formation Control in Constrained Space
Dongyu Li, Xiaomei Liu, Qinglei Hu, and Shuzhi Sam Ge

Distributed Adaptive Consensus Control of Uncertain Multi-agent Systems
Wei Wang, Jiang Long, Jiangshuai Huang, Changyun Wen

Control of Nonlinear Systems
Yongduan Song, Kai Zhao, and Hefu Ye

For more information about this series, please visit: https://www.crcpress.com/
Automation-and-Control-Engineering/book-series/CRCAUTCONENG

Control of Nonlinear Systems

Systems

Stability and Performance

Yongduan Song, Kai Zhao, and Hefu Ye

CRC Press
Taylor & Francis Group
Boca Raton London New York

CRC Press is an imprint of the
Taylor & Francis Group, an **informa** business

First edition published 2025
by CRC Press
2385 NW Executive Center Drive, Suite 320, Boca Raton FL 33431

and by CRC Press
4 Park Square, Milton Park, Abingdon, Oxon, OX14 4RN

CRC Press is an imprint of Taylor & Francis Group, LLC

© 2025 Yongduan Song, Kai Zhao, and Hefu Ye

ISBN: 978-1-032-75527-4 (hbk)
ISBN: 978-1-032-75528-1 (pbk)
ISBN: 978-1-003-47436-4 (ebk)

DOI: 10.1201/9781003474364

Typeset in Nimbus Roman
by KnowledgeWorks Global Ltd.

Publisher's note: This book has been prepared from camera-ready copy provided by the authors.

To my family for all the love, support and understanding.
Yongduan Song

To my family, for all the support, love and kindness.
Kai Zhao

To my family, for all the support, love and kindness.
Hefu Ye

Contents

SECTION I Fundamentals of Nonlinear Systems

SECTION II Control of Nonlinear Systems

SECTION III   Performance/Prescribed-time Control

Preface

Motivation: Most practical engineering systems exhibit complex dynamics, high nonlinearities, and strong couplings while operating in challenging and dynamic environments. The control problem for such systems becomes intricate, emphasizing the theoretical and practical significance of developing advanced control schemes for nonlinear systems with varying uncertainties. Additionally, the need for these systems to operate reliably and meet specified performance criteria, such as convergence rate, time, and overshoot, poses substantial challenges in control design and stability analysis, attracting increasing research interest within the control community.

Numerous textbooks and monographs address nonlinear control, with or without considerations for prescribed performance. However, the abundance of seemingly unrelated designs and analytical methods in literature often overwhelms researchers and engineers, leading to confusion. The primary objective of this book is to streamline and simplify the understanding of a wide array of control methods for nonlinear systems. The book, a culmination of several years of research, does not aim to introduce new results but rather to consolidate, unify, and present existing techniques in a tutorial fashion.

Scope: The book primarily focuses on introducing typical and popular control methods for different types of nonlinear systems amidst various uncertainties, whether parametric/non-parametric or vanishing/non-vanishing, and under known or unknown control gains (both the signs and the magnitudes). It covers essential stability theorems and concepts like asymptotic stability, exponential stability, uniformly ultimately bounded (UUB) stability, and finite-time stability. Additionally, it delves into control design techniques and tools such as the backstepping technique for strict-feedback nonlinear systems, tuning function method to address over-parametrization issues, core function technique for non-parametric uncertainties, Young's inequality for establishing stability analysis, and the Nussbaum gain technique for handling unknown control directions and magnitudes. Furthermore, the book explores advanced control schemes like adaptive/robust adaptive control, fractional power state feedback-based finite-time control, regular state feedback-based prescribed-time control, among others.

Organizational Structure: The book is structured into three parts:

Part I - Fundamentals of Nonlinear Systems: Chapters 1 to 3 introduce nonlinear system concepts, classical examples, special forms, and the fundamentals of the Lyapunov Theorem for autonomous or non-autonomous systems, along with essential design steps and tools.

Part II - Control of Nonlinear Systems: Chapters 4 to 7 focus on control design for systems facing uncertainties under known or unknown control directions, emphasizing the steady-state performance of closed-loop systems. This section covers various methods for first-order and normal-form nonlinear systems, adaptive control of uncertain strict-feedback systems, and control methods for multiple-input and multiple-output (MIMO) systems.

Part III - Performance/Prescribed-time Control: Chapters 8 to 13 delve into performance and prescribed-time control algorithms for nonlinear systems under uncertainties, discussing accelerated adaptive control, scaling function transformation-based control methods, finite-time control criteria, and prescribed-time control schemes for both SISO and MIMO systems.

Target Audience: The book caters to researchers, engineers, and students with backgrounds in electrical engineering, mechanical engineering, or applied mathematics, as well as academics working on nonlinear systems, adaptive control, prescribed performance control, and finite-time control. It serves as introductory material for fresh graduates entering the control domain and as a reference textbook for postgraduate students interested in research. The book also appeals to professionals in engineering fields, including control engineers, mechanical engineers, and aerospace engineers, due to its structure-specific methods. Physicists may also find the content relevant for its unique approach to control systems.

Acknowledgments: The authors would like to express their sincere gratitude to the following individuals for their inspiring early work on the control of nonlinear systems: Petar V. Kokotovic, Petros A. Ioannou, Miroslav Krstic, Jean-Jacques Slotine, Frank L. Lewis, Jing Sun, Changyun Wen, Andrey Polyakov, Sanjay P. Bhat, Charalampos P. Bechlioulis, Xudong Ye, and Zhiyong Chen. The authors would also like to thank their editors, Sam Ge and Frank L. Lewis, for their interest in the project and their professionalism. Additionally, the authors acknowledge Elsevier and Wiley for granting permission to reuse materials from publications copyrighted by these publishers. Finally, the authors gratefully acknowledge the National Natural Science Foundation of China (No. 61933012), the National Key Research and Development Program of China (No. 2022YFB4701400/4701401), the Chongqing Top-Notch Young Talents Project (No. cstc2024ycjh-bgzxm0085), the International Cooperation Program (No. 61860206008), the Institute of Artificial Intelligence, and the School of Automation at Chongqing University for their generous support in bringing this book to fruition.

Chongqing, China   *Yongduan Song*
Chongqing, China   *Kai Zhao*
Chongqing, China   *Hefu Ye*

Authors

Yongduan Song received the Ph.D. degree in electrical and computer engineering from Tennessee Technological University, Cookeville, TN, USA, in 1992. He held a tenured Full Professor with North Carolina A&T State University, Greensboro, NC, USA, from 1993 to 2008, and a Langley Distinguished Professor with the National Institute of Aerospace, Hampton, VA, USA, from 2005 to 2008. He is currently the Dean of the Institute of Artificial Intelligence, Chongqing University, Chongqing, China. He was one of the six Langley Distinguished Professors with the National Institute of Aerospace (NIA), Hampton, VA, USA, and the Founding Director of Cooperative Systems with NIA. His current research interests include intelligent systems, guidance navigation and control, bio-inspired adaptive, and cooperative systems.

Prof. Song was a recipient of several competitive research awards from the National Science Foundation, the National Aeronautics and Space Administration, the U.S. Air Force Office, the U.S. Army Research Office, and the U.S. Naval Research Office. He has served/been serving as an Associate Editor for several prestigious international journals, including the IEEE Transactions on Automatic Control, IEEE Transactions on Neural Networks and Learning Systems, IEEE Transactions on Intelligent Transportation Systems, and IEEE Transactions on Systems, Man, and Cybernetics: Systems. He is currently the Editor-in-Chief for the IEEE Transactions on Neural Networks and Learning Systems.

Kai Zhao received the Ph.D. degree in control theory and control engineering from Chongqing University, Chongqing, China, in 2019. He held research positions with the Department of Computer and Information Science at the University of Macau, and the Department of Electrical and Computer Engineering at the National University of Singapore. Since November 2023 he has been a full professor at the School of Automation, Chongqing University, China.

His research interests include adaptive control, constrained control, intelligent control systems, and robotic control and applications. Dr. Zhao was a recipient of Excellent Doctoral Dissertation Award of Chongqing Municipality in 2020, Excellent Doctoral Dissertation (Nomination) Award of Chinese Association of Automation (CAA) in 2020, and Outstanding Reviewer Award of IEEE Transactions on Cybernetics in 2022.

Hefu Ye received the B.Eng. degree from the School of Information Science and Engineering, Harbin Institute of Technology, Harbin, China, in 2019. He is currently pursuing the Ph.D. degree with the School of Automation, Chongqing University, Chongqing, China. During Oct. 2022 and Oct. 2023, he was a Joint Ph.D. student at the School of Electrical and Electronic Engineering, Nanyang Technological University, Singapore.

His research interests include robotic systems, robust adaptive control, prescribed performance control, and prescribed-time control. Mr. Ye is an active Reviewer for many international journals, including the IEEE Transactions on Automatic Control, IEEE Transactions on Neural Networks and Learning Systems, and IEEE Transactions on Systems, Man, and Cybernetics: Systems.

Acronyms

Abbreviations and Notations

UAS	uniformly asymptotically stable
UGAS	uniformly globally asymptotically stable
US	uniformly stable
UUB	uniformly ultimately bounded
Log	logarithm
MIMO	multiple-input and multiple-output
SISO	single-input and single-output
MPC	model predictive control
LTI	linear time-invariant
TSM	terminal sliding mode
PPB	prescribed performance bound
NN	neural network
RBFNN	radial basis function neural network
DOF	degree of freedom
AAC	accelerated adaptive control
TAC	traditional adaptive control
PT-ISS	Prescribed-time Input-to-state Stable
PT-ISS+C	Prescribed-time Input-to-state Stable+Convergent
ISS	Input-to-state Stable
$\mathbb{R}$	set of real numbers
$\mathbb{R}_+$	non-negative set of real numbers
$\mathbb{R}^n$	n-dimensional real Euclidean space
$\mathbb{R}^{n \times m}$	set of $n \times m$ real matrix
$|\bullet|$	absolute value of a real number
$\|\bullet\|_1$	1-norm of matrix
$\|\bullet\|_2 \ (\|\bullet\|)$	2-norm of matrix
$\|\bullet\|_\infty \ (\|\bullet\|)$	∞-norm of matrix
$\mathrm{sgn}(\bullet)$	sign function
$\tanh(\bullet)$	hyperbolic tangent function
$\tan(\bullet)$	tangent function
max	maximum
min	minimum
diag	diagonal
$I_{n \times n}$	$n \times n$ identity matrix
$A^\top$	transpose of matrix A
A^{-1}	inverse of square matrix A
$\lambda_{\max}(\bullet)$	maximum eigenvalue of real square matrix $\bullet$
$\lambda_{\min}(\bullet)$	minimum eigenvalue of real square matrix $\bullet$
$\mathrm{diag}\{a_1, \cdots, a_n\}$	diagonal matrix with $a_1, \cdots, a_n$ on the diagonal
sup	supremum, the least upper bound

inf	infimum, the greatest lower bound
$\forall$	for all
$\in$	belongs to
$\rightarrow$	tends to
$\Rightarrow$	implies
$A \subset B$	a set A is the subset of a set B
$A \subsetneq B$	a set A is the proper subset of a set B
$< (>)$	less (greater) than
$\leq (\geq)$	less (greater) than or equal to
$\ll (\gg)$	much less (greater) than
Σ	summation
$f : A \rightarrow B$	a function f mapping a set A into a set B
$\dot{y}$	the first derivative of y with respect to time
$\ddot{y}$	the second derivative of y with respect to time
$y^{(i)}$	the i-th derivative of y with respect to time
$\mathscr{C}^i$	a set of functions that have continuous derivatives up to the order i
w.r.t.	with respect to
$\mathrm{sat}(\cdot)$	the saturation function
$P > 0$	a positive definite matrix P
$P \geq 0$	a positive semi-definite matrix P

Section I

Fundamentals of Nonlinear Systems

1 Introduction

Many practical engineering applications are essentially highly nonlinear systems, e.g., robotic systems [1, 2, 3, 4], high-speed trains [5, 6], and aircraft [7, 8]. Therefore, the control of nonlinear systems has received much attention in recent decades, and numerous research results have been obtained [9, 10, 11, 12, 13, 14].

In this chapter, we first introduce the state space description of nonlinear systems, autonomous/non-autonomous systems, and nonlinear control systems in Section 1.1. Then, in Section 1.2, we give some typical practical nonlinear examples. To facilitate the descriptions later, we introduce some special kinds of nonlinear systems in Section 1.3. Finally, in Section 1.4, we close this chapter with the organization of this book.

1.1 NONLINEAR SYSTEMS

The state space representation of a nonlinear system is described by the following equation [14]:

$$\dot{x}(t) = f(x(t), t), \; x(t_0) = x_0, \; t \geq t_0 \geq 0, \tag{1.1}$$

where $t \in \mathbb{R}_+$ is the time variable with $t_0 \geq 0$ being the initial time, $x \in \mathbb{R}^n$ is the state vector with x_0 being the initial state, and $f : \mathbb{R}^n \times [t_0, \infty) \to \mathbb{R}^n$ is locally Lipschitz in x and piecewise continuous in t. For convenience, the equation (1.1) can often be simplified as $\dot{x} = f(x, t)$.

A special class of nonlinear systems is linear systems. The dynamics of linear systems are of the form:

$$\dot{x} = A(t)x, \tag{1.2}$$

where $x \in \mathbb{R}^n$ is the system state, and $A(t)$ is an $n \times n$ matrix.

1.1.1 AUTONOMOUS/NON-AUTONOMOUS SYSTEMS

Linear systems are classified as either time-varying or time-invariant, depending on whether the system matrix A [as shown in (1.2)] varies with time or not. In nonlinear systems, these adjectives are traditionally replaced by "autonomous" and "non-autonomous", respectively.

Now, we give the definitions of autonomous and non-autonomous systems.

Definition 1.1 *[14] The nonlinear system (1.1) is said to be autonomous if the system equation does not depend explicitly on time t, i.e., if the system's state equation can be rewritten as:*

$$\dot{x} = f(x).$$

Otherwise, the system is called the non-autonomous system.

DOI: 10.1201/9781003474364-1

3

Obviously, LTI systems are autonomous, and linear time-varying systems are non-autonomous. Strictly speaking, all physical systems are non-autonomous because none of their dynamic characteristics is strictly time-invariant. The concept of an autonomous system is an idealized notion, like the concept of an LTI system. In practice, however, some system properties may change slowly, and sometimes we can neglect their time variation without causing any practical meaning error.

The fundamental difference between autonomous and non-autonomous systems lies in the fact that the state trajectory of an autonomous system only depends on $(t - t_0)$ but is independent of the initial time t_0, while the state trajectory of a non-autonomous system generally relies on both time variable t and the initial time t_0. Now we give an intuitive explanation from the following example.

Example 1.1 *[1] For the following autonomous system:*

$$\dot{x} = -\lambda x, \ x \in \mathbb{R}, \ x(t_0) = x_0, \ t \geq t_0 \geq 0,$$

where λ is a positive constant, $x(t_0)$ is the initial state, and t_0 is the initial time. Its solution is:

$$x(t) = x(t_0)\exp(-\lambda(t - t_0)), \tag{1.3}$$

with $\exp(\bullet)$ being an exponential function of $\bullet$. However, for the following non-autonomous system:

$$\dot{x} = -\frac{x}{1+t}, \ x \in \mathbb{R}, \ x(t_0) = x_0, \ t \geq t_0 \geq 0,$$

it can be verified that for any initial state $x(t_0)$ with any initial time $t_0 \geq 0$, the solution is:

$$x(t) = x(t_0)\frac{1+t_0}{1+t}. \tag{1.4}$$

From (1.3) and (1.4), it is easy to see the differences between the autonomous system and the non-autonomous system.

Such differences may lead to the different descriptions of basic concepts and Lyapunov theorems for the autonomous system and the non-autonomous system (see Chapter 2).

1.1.2   AUTONOMOUS/NON-AUTONOMOUS NONLINEAR CONTROL SYSTEM

A general non-autonomous nonlinear control system is described by the following equation [15]:

$$\begin{cases} \dot{x} = f(x,u,t), \ x(t_0) = x_0, \ t \geq t_0 \geq 0, \\ y = h(x,u,t), \\ y_m = h_m(x,u,t), \end{cases} \tag{1.5}$$

where $x \in \mathbb{R}^n$ is the system state with initial state x_0, $u \in \mathbb{R}^m$ is the control input, $y \in \mathbb{R}^p$ is the system output, $y_m \in \mathbb{R}^p$ is the measurement output, $f : \mathbb{R}^n \times \mathbb{R}^m \times [t_0, \infty) \rightarrow$

$\mathbb{R}^n$, $h : \mathbb{R}^n \times \mathbb{R}^m \times [t_0, \infty) \to \mathbb{R}^p$, and $h_m : \mathbb{R}^n \times \mathbb{R}^m \times [t_0, \infty) \to \mathbb{R}^p$ are nonlinear functions. Furthermore, in this book we normally assume that the measurement output is equal to the system output, namely, $y_m = y$.

A special case of (1.5) where t does not appear explicitly in the right-hand side of these equations is as follows:

$$\begin{cases} \dot{x} = f(x, u), \; x(t_0) = x_0, \; t \geq t_0 \geq 0, \\ y = h(x, u), \\ y_m = h_m(x, u). \end{cases} \tag{1.6}$$

We call (1.6) an autonomous nonlinear control system.

A special case of the autonomous system (1.6) is:

$$\begin{cases} \dot{x} = f(x) + g(x)u, \; x(t_0) = x_0, \; t \geq t_0 \geq 0, \\ y = h(x), \\ y_m = h_m(x), \end{cases} \tag{1.7}$$

where $f : \mathbb{R}^n \to \mathbb{R}^n$, $g : \mathbb{R}^n \to \mathbb{R}^{n \times m}$, $h : \mathbb{R}^n \to \mathbb{R}^p$, and $h_m : \mathbb{R}^n \to \mathbb{R}^p$ are nonlinear functions. We call (1.7) an affine nonlinear control system.

Furthermore, it must be emphasized that a control system is composed of a controller and a plant (including a sensor and actuator dynamics), the non-autonomous nature of a control system may be due to a time-variation either in the plant or in the control law. Specifically, a time-invariant plant with dynamics:

$$\dot{x} = f(x, u)$$

may lead to a non-autonomous closed-loop system if a controller dependent on time is chosen, i.e., if $u = g(x, t)$. For example, the closed-loop system of the simple plant $\dot{x} = -x + u$ can be nonlinear and non-autonomous by choosing u to be nonlinear and time-varying (e.g., $u = -x^2 \sin(t)$).

1.2 NONLINEAR EXAMPLES

1.2.1 PENDULUM

Consider the simple pendulum shown in Fig. 1.1, where l represents the length of the pendulum rod and the mass of the pendulum is m. Assuming that the rod is rigid and has zero mass, θ represents the angle between the rod and the vertical axis passing through the center point. The pendulum swings freely in the vertical plane and moves on a circle with radius l. According to Newton's second law of motion, the motion equation of the pendulum can be expressed as [14]:

$$ml^2\ddot{\theta} + mgl\sin(\theta) + b\dot{\theta} = 0,$$

where g is the gravity of the pendulum, and b is the friction coefficient.

In order to obtain the state model, taking the state variables $x_1 = \theta$ and $x_2 = \dot{\theta}$, then the state equation is:

$$\begin{cases} \dot{x}_1 = x_2, \\ \dot{x}_2 = -\frac{g}{l}\sin(x_1) - \frac{b}{ml^2}x_2. \end{cases}$$

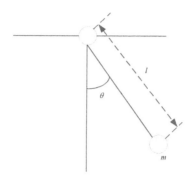

FIGURE 1.1 Pendulum.

Let $\dot{x}_1 = \dot{x}_2 = 0$, getting that the equilibrium points are located at $(n\pi, 0)$, $n = 0, \pm 1, \pm 2, \cdots$. From the physical description of the pendulum, it has only two equilibrium points $(0,0)$ and $(\pi, 0)$, and other equilibrium points coincide with these two equilibrium points. Importantly enough, several unrelated physical systems are modeled by equations similar to the pendulum equation, such as the model of the Josephson junction circuit and phase-locked loop [14].

1.2.2 MASS-DAMPING-SPRING SYSTEM

In the mass-damping-spring system shown in Fig. 1.2, object m sliding on a horizontal surface and connected to a vertical surface by a spring is subjected to an external force. Define the displacement of the object from the reference point as x, according to Newton's second law of motion, one obtains [14]:

$$m\ddot{x} = F - c\dot{x} - F_f, \tag{1.8}$$

where $-c\dot{x}$ is the friction resistance, the negative sign means the direction of resistance is opposite to the direction of speed, F_f is the restoring force of the spring,

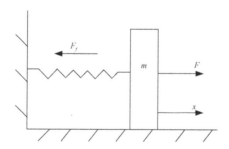

FIGURE 1.2 Mass-damping-spring system.

which depends only on x, i.e., $F_f = g(x)$. When the displacement is relatively small, it can be modeled by a linear function $g(x) = kx$, where k is the elastic coefficient. When x is large, $g(x)$ has a nonlinear relationship with x. For the softening spring, $g(x) = k\left(1 - a^2x^2\right)x$, and for the hardening spring, $g(x) = k\left(1 + a^2x^2\right)x$.

Under a period of the external force $F = A\cos(\omega t)$, considering the hardening spring, (1.8) can be written as:

$$m\ddot{x} + c\dot{x} + kx + ka^2x^3 = A\cos(\omega t). \tag{1.9}$$

This is a classic example of studying nonlinear systems with periodic excitations.

For a linear spring in the presence of the static friction force, coulomb friction force as well as linear viscous friction force, when the external force is zero, we have:

$$m\ddot{x} + c\dot{x} + kx + \gamma(x, \dot{x}) = 0, \tag{1.10}$$

where

$$\gamma(x, \dot{x}) = \begin{cases} \mu_k mg\,\mathrm{sgn}(\dot{x}), & \text{for } |\dot{x}| > 0, \\ -kx, & \text{for } \dot{x} = 0 \text{ and } |x| \le \mu_s mg/k, \\ -\mu_s mg\,\mathrm{sgn}(x), & \text{for } \dot{x} = 0 \text{ and } |x| > \mu_s mg/k, \end{cases}$$

in which μ_k is the kinetic friction coefficient, and μ_s is the static friction coefficient.

Let $x_1 = x$ and $x_2 = \dot{x}$, the state equation (1.10) is:

$$\begin{cases} \dot{x}_1 = x_2, \\ \dot{x}_2 = -\frac{k}{m}x_1 - \frac{c}{m}x_2 - \frac{\gamma(x,\dot{x})}{m}. \end{cases}$$

1.2.3 ROBOTIC SYSTEMS

The dynamic model of m-joint robotic systems is described by [1, 16, 17]:

$$H(q, p)\ddot{q} + N_g(q, \dot{q}, p)\dot{q} + G_g(q, p) + \tau_d(\dot{q}, p) = u, \tag{1.11}$$

where $q = [q_1, \cdots, q_m]^\top \in \mathbb{R}^m$ denotes the position, $H(q, p) \in \mathbb{R}^{m \times m}$ denotes the inertia matrix, $p \in \mathbb{R}^l$ denotes the unknown parameter vector, $N_g(q, \dot{q}, p) \in \mathbb{R}^{m \times m}$ denotes the centripetal-Coriolis matrix, $G_g(q, p) \in \mathbb{R}^m$ represents the gravitation vector, $\tau_d(\dot{q}, p) \in \mathbb{R}^m$ denotes the frictional and disturbing force, and $u \in \mathbb{R}^m$ is the control input.

For the above considered robotic systems, it has the following properties [1, 3, 4]:

Property 1.1 *The inertia matrix $H(\cdot)$ is positive definite and symmetric, so is $G(\cdot) = H^{-1}$. Moreover, there exist positive constants $\underline{\lambda}$ and $\bar{\lambda}$ such that:*

$$\underline{\lambda} \le \|G\| \le \bar{\lambda}, \ \underline{\lambda}\|\mathbf{x}\|^2 \le \mathbf{x}^\top G\mathbf{x} \le \bar{\lambda}\|\mathbf{x}\|^2, \ \forall \mathbf{x} \in \mathbb{R}^m.$$

Property 1.2 *For any $q \in \mathbb{R}^m$ and $\dot{q} \in \mathbb{R}^m$, there exist positive constants γ_H, γ_N, and γ_G such that:*

$$\|H\| \le \gamma_H, \ \|N_g\| \le \gamma_N\|\dot{q}\|, \ \|G_g\| \le \gamma_G.$$

1.3 SPECIAL FORMS OF NONLINEAR SYSTEMS

Unlike linear systems, nonlinear systems have various forms. This section mainly introduces two kinds of typical nonlinear systems, including normal-form nonlinear systems and strict-/pure-feedback nonlinear systems.

1.3.1 NORMAL-FORM NONLINEAR SYSTEMS

The n-order normal-form nonlinear system is given as [18]:

$$\begin{cases} \dot{x}_k = x_{k+1}, \quad k = 1, 2, \cdots, n-1, \\ \dot{x}_n = f(\bar{x}_n, p) + g(\bar{x}_n) u, \\ y = x_1, \end{cases} \tag{1.12}$$

where $\bar{x}_i = [x_1, \cdots, x_i]^\top \in \mathbb{R}^i \, (i = 1, \cdots, n)$ is the system state, $u \in \mathbb{R}$ and $y \in \mathbb{R}$ are the control input and system output, respectively, $g : \mathbb{R}^n \to \mathbb{R}$ denotes the control gain, $f : \mathbb{R}^n \times \mathbb{R}^r \to \mathbb{R}$ denotes a nonlinear function, and $p \in \mathbb{R}^r$ represents a constant parameter vector. Many practical engineering systems can be described by such a form, i.e., inverted pendulum model and single-joint robot system model.

1.3.2 STRICT-/PURE-FEEDBACK NONLINEAR SYSTEMS

A nonlinear system with the following form is called the strict-feedback nonlinear system [9]:

$$\begin{cases} \dot{x}_k = f_k(\bar{x}_k) + g_k(\bar{x}_k) x_{k+1}, \quad k = 1, 2, \cdots, n-1, \\ \dot{x}_n = f_n(\bar{x}_n) + g_n(\bar{x}_n) u, \\ y = x_1, \end{cases} \tag{1.13}$$

where $x_1 \in \mathbb{R}, \cdots, x_n \in \mathbb{R}$ are the system states with $\bar{x}_k = [x_1, \cdots, x_k]^\top \in \mathbb{R}^k \, (k = 1, \cdots, n)$, $f_k : \mathbb{R}^k \to \mathbb{R}$ denotes a nonlinear function, $g_k : \mathbb{R}^k \to \mathbb{R}$ is the control coefficient, $u \in \mathbb{R}$ and $y \in \mathbb{R}$ are the control input and system output, respectively.

It is worth mentioning that if $f_1(x_1) = \cdots = f_{n-1}(\bar{x}_{n-1}) = 0$ and $g_1(x_1) = \cdots = g_{n-1}(\bar{x}_{n-1}) = 1$, then (1.13) can be simplified to the normal-form nonlinear system (1.12), which indicates that such a normal-form nonlinear system is a special case of the strict-feedback nonlinear system (1.13).

Remark 1.1 *The key to analyzing and identifying the forms of nonlinear systems is to observe $f_k(\cdot)$:*

(1) If $f_1(x_1) = \cdots = f_{n-1}(\bar{x}_{n-1}) = 0$ and $g_1(x_1) = \cdots = g_{n-1}(\bar{x}_{n-1}) = 1$, then (1.13) can be simplified to the normal-form nonlinear system;
(2) If the nonlinear function $f_k(\cdot)$ contains the states $x_1, \cdots, x_k$, then (1.13) belongs to the strict-feedback nonlinear system;
(3) If the nonlinear function $f_k(\cdot)$ includes not only the states $x_1, \cdots, x_k$, but also the state x_{k+1}, that is, $f_k(\cdot) = f_k(\bar{x}_k, x_{k+1})$, then (1.13) becomes the following

pure-feedback nonlinear system:

$$\begin{cases} x_k = f_k(\bar{x}_{k+1}), \ k = 1, 2, \cdots, n-1, \\ \dot{x}_n = f_n(\bar{x}_n, u), \\ y = x_1, \end{cases} \tag{1.14}$$

which is more general than the strict-feedback form. Since the control design of pure-feedback systems is similar to that of strict-feedback systems, we will focus in this book on the introduction of strict-feedback nonlinear systems. Interested readers can refer to related references [19] and [20] to learn more about the design procedures for pure-feedback nonlinear systems;

(4) *If the nonlinear function $f_k(\cdot)$ includes not only the states $x_1, \cdots, x_k$ but also the other states x_j ($j \geq k+2$), then (1.13) is called a nonstrict-feedback (or nonpure-feedback) nonlinear system. For example:*

$$\begin{cases} \dot{x}_1 = f_1(\bar{x}_3), \\ \dot{x}_2 = f_2(\bar{x}_5, x_6), \\ \dot{x}_k = f_k(\bar{x}_k, \cdots, x_n), \ k = 3, 4, \cdots, n-1, \\ \dot{x}_n = f_n(\bar{x}_n), \\ y = x_1. \end{cases} \tag{1.15}$$

The control design of (1.15) is beyond the scope of this book, interested readers may refer to the relevant literature.

1.4 ORGANIZATION OF THIS BOOK

In this book, we focus on introducing the control problems of various types of nonlinear systems. For such problems, it mainly contains the following three aspects: the considered plant, the control objectives, and the control method.

The controlled plant mainly involves first-order systems and high-order systems, SISO systems and MIMO systems, normal-form systems and strict-feedback systems. For the control objectives, the primary objective is to ensure the stability of closed-loop systems by applying the Lyapunov theorems. In addition to the stability requirement, prescribed transient responses including the convergence time, convergence rate as well as convergence accuracy are also crucial in many practical applications. To this end, we illustrate some typical and useful control methods, for example, prescribed performance control, finite-time control, and prescribed-time control.

This book consists of three parts with a total of 13 chapters. The first part, ranging from Chapters 1 to 3, is an introduction to the concepts of nonlinear systems, the Lyapunov theorem, and some typical design methods or tools. The second part consists of Chapters 4 to 7, which are mainly concerned with control methods for nonlinear systems. The third part, consisting of Chapters 8 to 13, focuses on the performance/prescribed-time control of nonlinear systems. Fig. 1.3 shows the outline scheme for the contents of this book.

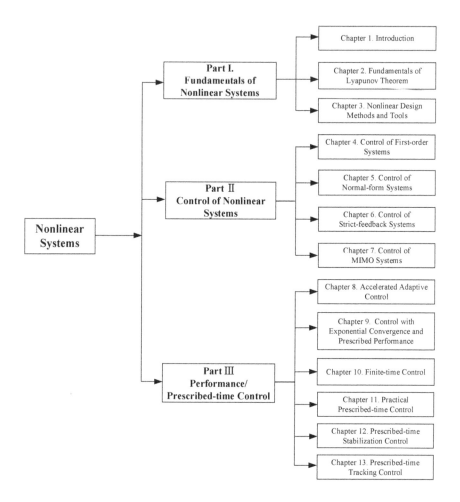

FIGURE 1.3   The diagram of content arrangement.

2 Fundamentals of Lyapunov Theorem

Given a control system, the first and most important question about its various properties is whether it is stable because an unstable control system is normally useless and potentially dangerous. Therefore, every control system, whether linear or nonlinear, involves a stability problem that should be carefully studied.

The most useful and general approach for studying the stability of nonlinear control systems is the theory introduced in the late 19th century by the Russian mathematician–*Aleksandr Mikhailovich Lyapunov*. Lyapunov's work-The General Problem of Motion Stability-includes two methods for stability analysis (i.e., the linearization method and direct method) and was first published in 1892. The linearization method draws conclusions about, for a nonlinear system, local stability around an equilibrium point from the stability properties of its linear approximation. The direct method is not restricted to local motion and determines the stability properties of a nonlinear system by constructing a scalar energy-like function for the considered system and examining the corresponding time variation. The linearization method and direct method constitute the so-called Lyapunov stability theorem.

In this chapter, we focus on introducing the basic knowledge and stability theorems of nonlinear systems. This chapter is organized as follows. In Section 2.1, we present the definitions and properties of various norms and functions that are used in the remainder of this book. The concepts about nonlinear autonomous systems, equilibrium points, and system stability are given in Section 2.2; Lyapunov's linearization method and direct method are introduced in Section 2.3 and Section 2.4, respectively. In Section 2.5, the invariant set theorem is introduced; the stability concept of non-autonomous systems is shown in Section 2.6 and the corresponding stability analysis is given in Section 2.7; Barbalat's Lemma is introduced in Section 2.8. Finally, some useful inequalities and theorems are given in Section 2.9.

2.1 PRELIMINARIES

2.1.1 NORMS AND $\mathscr{L}_P$ SPACES

For many of the arguments for scalar equations to be extended and remain valid for vector equations, we need an analog for vectors of the absolute value of a scalar. This is provided by the norm of a vector.

Definition 2.1 *The norm $\|x\|$ of a vector x is a real-valued function with the following properties:*

(i) $\|x\| \geq 0$ with $\|x\| = 0$ if and only if $x = 0$;

DOI: 10.1201/9781003474364-2

(ii) $\|\alpha x\| = |\alpha| \|x\|$ *for any scalar α;*
(iii) $\|x+y\| \leq \|x\| + \|y\|$ *(triangle inequality).*

The norm $\|x\|$ of a vector x can be thought of as the size or length of the vector x. Similarly, $\|x - y\|$ can be thought of as the distance between the vectors x and y.

An $m \times n$ matrix A represents a linear mapping from n-dimensional space $\mathbb{R}^n$ into m-dimensional space $\mathbb{R}^m$. We define the induced norm of A as follows:

Definition 2.2 *Let $\| \bullet \|$ be a given vector norm. Then for each matrix $A \in \mathbb{R}^{m \times n}$, the quantity $\|A\|$ defined by*

$$\|A\| \overset{\Delta}{=} \sup_{\substack{x \neq 0 \\ x \in \mathbb{R}^n}} \frac{\|Ax\|}{\|x\|} = \sup_{\|x\| \leq 1} \|Ax\| = \sup_{\|x\|=1} \|Ax\|$$

is called the induced (matrix) norm of A corresponding to the vector norm $\| \bullet \|$.

The induced matrix norm satisfies three properties in Definition 2.1. Some of the properties of the induced norm that we will often use in this book are summarized as follows:

(i) $\|Ax\| \leq \|A\| \|x\|, \forall x \in \mathbb{R}^n$;
(ii) $\|A+B\| \leq \|A\| + \|B\|$;
(iii) $\|AB\| \leq \|A\| \|B\|$.

where A and B are arbitrary matrices of compatible dimensions. Table 2.1 shows some of the most commonly used norms.

Example 2.1 *(i) Let $x = [1, 2, -10, 0]^\top$. Using Table 2.1, we have:*

$$\|x\|_\infty = 10, \quad \|x\|_1 = 13, \quad \|x\|_2 = \sqrt{105}.$$

(ii) Let

$$A = \begin{bmatrix} 0 & 5 \\ 1 & 0 \\ 0 & -10 \end{bmatrix}, \quad B = \begin{bmatrix} -1 & 5 \\ 0 & 2 \end{bmatrix}.$$

Table 2.1
Commonly Used Norms

Norms on $\mathbb{R}^n$	Induced Norms on $\mathbb{R}^{m \times n}$
$\|x\|_\infty = \max_i |x_i|$ (infinity norm)	$\|A\|_\infty = \max_i \sum_j |a_{ij}|$ (row sum)
$\|x\|_1 = \sum_i |x_i|$	$\|A\|_1 = \max_j \sum_i |a_{ij}|$ (column sum)
$\|x\|_2 = \left(\sum_i |x_i|^2 \right)^{1/2}$	$\|A\|_2 = \left[\lambda_{\max} \left(A^\top A \right) \right]^{(1/2)}$, where $\lambda_{\max}(M)$
(Euclidean norm)	is the maximum eigenvalue of M

Using Table 2.1, we have:

$$\|A\|_1 = 15, \quad \|A\|_2 = 11.18, \quad \|A\|_\infty = 10$$
$$\|B\|_1 = 7, \quad \|B\|_2 = 5.465, \quad \|B\|_\infty = 6$$
$$\|AB\|_1 = 35, \quad \|AB\|_2 = 22.91, \quad \|AB\|_\infty = 20$$

which can be used to verify property (iii) of the induced norm.

For functions of time, we define the $\mathscr{L}_p$ norm

$$\|x\|_p \triangleq \left(\int_0^\infty \|x(\tau)\|^p d\tau \right)^{1/p}$$

for $p \in [1, \infty)$ and say that $x \in \mathscr{L}_p$ when $\|x\|_p$ exists (i.e., when $\|x\|_p$ is finite). The $\mathscr{L}_\infty$ norm is defined as:

$$\|x\|_\infty \triangleq \sup_{t \geq 0} \|x(t)\|$$

and we say that $x \in \mathscr{L}_\infty$ when $\|x\|_\infty$ exists.

In the above $\mathscr{L}_p$, $\mathscr{L}_\infty$ norm definitions, $x(t)$ can be a scalar or a vector function. If x is a scalar function, then $\| \bullet \|$ denotes the absolute value. If x is a vector function in $\mathbb{R}^n$, $\| \bullet \|$ denotes any norm in $\mathbb{R}^n$.

In the remaining chapters of this book, we adopt the following notation regarding norms unless stated otherwise. We will drop the subscript 2 from $\| \bullet \|_2$ when dealing with the Euclidean norm, the induced Euclidean norm, and the $\mathscr{L}_2$ norm. If $x : \mathbb{R}_+ \to \mathbb{R}^n$, then $\|x(t)\|$ represents the vector norm in $\mathbb{R}^n$ at each time t, $\|x\|_p$ represents the $\mathscr{L}_p$ norm of the function $\|x(t)\|$. If $A \in \mathbb{R}^{m \times n}$, then $\|A\|_i$ represents the induced matrix norm corresponding to the vector norm $\| \bullet \|_i$. If $A : \mathbb{R}_+ \to \mathbb{R}^{m \times n}$ has elements that are functions of time t, then $\|A(t)\|_i$ represents the induced matrix norm corresponding to the vector norm $\| \bullet \|_i$ at time t.

2.1.2 POSITIVE DEFINITE MATRICES

A square matrix $A \in \mathbb{R}^{n \times n}$ is called symmetric if $A = A^\top$. A symmetric matrix A is called positive semi-definite if for every $x \in \mathbb{R}^n$, $x^\top A x \geq 0$ and positive definite if $x^\top A x > 0$, $\forall x \in \mathbb{R}^n$ with $\|x\| \neq 0$. It is called negative semi-definite (negative definite) if $-A$ is positive semi-definite (positive definite). The definition of a positive definite matrix can be generalized to non-symmetric matrices. In this book, we will always assume that the matrix is symmetric when we consider positive or negative definite or semi-definite properties. We write $A \geq 0$ if A is positive semi-definite, and $A > 0$ if A is positive definite. We write $A \geq B$ and $A > B$ if $A - B \geq 0$ and $A - B > 0$, respectively.

A symmetric matrix $A \in \mathbb{R}^{n \times n}$ is positive definite if and only if any one of the following conditions holds:

(i) $\lambda_i(A) > 0$, $i = 1, 2, \cdots, n$, where $\lambda_i(A)$ denotes the i-th eigenvalue of A;

(*ii*)   there exists a nonsingular matrix A_1 such that $A = A_1 A_1^\top$;
(*iii*) every principal minor of A is positive;
(*iv*) $x^\top A x \geq \alpha \|x\|^2$ for some $\alpha > 0$ and $\forall x \in \mathbb{R}^n$.

The decomposition $A = A_1 A_1^\top$ in (*ii*) is unique when A_1 is also symmetric. In this case, A_1 is positive definite, it has the same eigenvectors as A, and its eigenvalues are equal to the square roots of the corresponding eigenvalues of A. We specify this unique decomposition of A by denoting A_1 as $A^{\frac{1}{2}}$, i.e., $A = A^{\frac{1}{2}} A^{\frac{\top}{2}}$ where $A^{\frac{1}{2}}$ is a positive definite matrix and $A^{\frac{\top}{2}}$ denotes the transpose of $A^{\frac{1}{2}}$.

A symmetric matrix $A \in \mathbb{R}^{n \times n}$ has n orthogonal eigenvectors and can be decomposed as:

$$A = U^\top \Lambda U \tag{2.1}$$

where U is a unitary (orthogonal) matrix (i.e., $U^\top U = I$) with the eigenvectors of A, and Λ is a diagonal matrix composed of the eigenvalues of A. Using (2.1), it follows that if $A \geq 0$, then for any vector $x \in \mathbb{R}^n$,

$$\lambda_{\min}(A)\|x\|^2 \leq x^\top A x \leq \lambda_{\max}(A)\|x\|^2.$$

Furthermore, if $A \geq 0$ then $\|A\|_2 = \lambda_{\max}(A)$, and if $A > 0$ we also have $\|A^{-1}\|_2 = \frac{1}{\lambda_{\min}(A)}$, where $\lambda_{\max}(A)$ and $\lambda_{\min}(A)$ are the maximum and minimum eigenvalues of A, respectively. Moreover, we should note that if $A > 0$ and $B \geq 0$, then $A + B > 0$, but it is not true in general that $AB \geq 0$.

2.2   STABILITY CONCEPTS OF AUTONOMOUS SYSTEMS

It is possible for a system trajectory to correspond to only a single point. Such a point is called an equilibrium point. In fact, many stability problems are naturally formulated w.r.t. the equilibrium points. Therefore, before presenting the concepts of stability, we introduce the concept of autonomous systems again.

Consider the autonomous system:

$$\dot{x} = f(x), \quad x(t_0) = x_0, \tag{2.2}$$

where $f : \mathbb{D} \to \mathbb{R}^n$ is a local Lipschitz function, and $\mathbb{D} \subset \mathbb{R}^n$ is a domain that contains the equilibrium point. Noting that the state trajectory of an autonomous system only depends on $(t - t_0)$ but is independent of the initial time t_0, we define that $t_0 = 0$ in the rest of this section.

2.2.1   EQUILIBRIUM POINTS

Definition 2.3 *A state x^* is an equilibrium state (or equilibrium point) of the autonomous system (2.2) if the state x^* satisfies:*

$$\dot{x}^* = f(x^*) \equiv 0. \tag{2.3}$$

The equilibrium points can be found by solving the nonlinear algebraic equation (2.3), where the equilibrium points may not be unique.

For a linear time-invariant system $\dot{x} = Ax$. If A is nonsingular, the system has a single equilibrium point. If A is singular, there are multiple solutions that satisfy $Ax = 0$, and the system has multiple equilibrium points. Furthermore, a nonlinear system can have several (or infinitely many) isolated equilibrium points.

In linear system analysis and design, for notational and analytical simplicity, we often transform the linear system equations in such a way that the equilibrium point is the origin of the state space. We can do the same thing for the nonlinear systems (2.2) about a specific equilibrium point.

Let us say that the equilibrium point of interest is x^*. By introducing a new variable:

$$y = x - x^*,$$

and substituting $x = y + x^*$ into equation (2.2), a new set of equations on the variable y are obtained:

$$\dot{y} = f(y + x^*). \tag{2.4}$$

One can easily verify that there is a one-to-one correspondence between the solutions of (2.2) and those of (2.4). In addition, $y = 0$, the solution corresponding to $x = x^*$, is an equilibrium point of (2.4). Therefore, instead of studying the behavior of the equation (2.2) in the neighborhood of x^*, one can equivalently study the behavior of the equation (2.4) in the neighborhood of the region.

2.2.2   STABILITY AND INSTABILITY

Since nonlinear systems may have much more complex behavior than linear systems, the notion of stability is not enough to describe the essential features of their motion. A number of stability concepts, such as asymptotic stability, exponential stability, and global asymptotic stability, are given in the following. For convenience, only the case where the equilibrium point is the origin is analyzed. For the situation where the equilibrium point is not the origin, the equilibrium point can be transformed into the origin.

A few simplifying notations are defined at this point. Let $\mathbb{B}_\varepsilon$ denote the spherical region (or ball) defined by $\|x\| < \varepsilon$ in state space, and $\mathbb{S}_\varepsilon$ the sphere itself, defined by $\|x\| = \varepsilon$. Let us first introduce the basic concepts of stability and instability for autonomous systems (2.2).

Definition 2.4 *The equilibrium point $x = 0$ of (2.2) is stable if, for each $\varepsilon > 0$, there is a $\delta = \delta(\varepsilon) > 0$ such that:*

$$\|x(0)\| < \delta \Rightarrow \|x(t)\| < \varepsilon, \quad \forall t \geq 0.$$

Otherwise, the equilibrium point is unstable.

Essentially, stability (also called stability in the sense of Lyapunov or Lyapunov stability) means that the system trajectory can be kept arbitrarily close to the origin by starting sufficiently close to it. In other words, the definition means that the origin

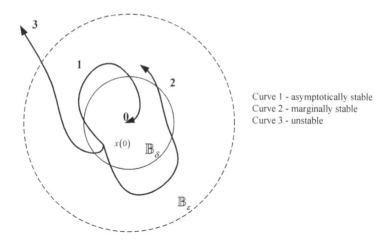

FIGURE 2.1 Concepts of stability.

is stable, if, given that we do not want the state trajectory $x(t)$ to get out of a ball of arbitrarily specified radius ε, a value $\delta(\varepsilon)$ can be found such that starting the state from within the ball $\mathbb{B}_\delta$ at time 0 guarantees that the state will stay within the ball $\mathbb{B}_\varepsilon$ thereafter. The geometrical implication of stability is indicated by the curve 2 in Fig. 2.1.

Using the mathematical symbols, Definition 2.4 can be written as: $\forall \varepsilon > 0,\ \exists \delta > 0,$ $\|x(0)\| < \delta \Rightarrow \|x(t)\| < \varepsilon$ for $\forall t \geq 0$, or

$$\forall \varepsilon > 0,\ \exists \delta > 0,\ x(0) \in \mathbb{B}_\delta \Rightarrow x(t) \in \mathbb{B}_\varepsilon,\ \forall t \geq 0.$$

Conversely, an equilibrium point $x = 0$ is unstable if there exists at least one ball $\mathbb{B}_\varepsilon$, such that for every $\delta > 0$, no matter how small, it is always possible for the system trajectory to start somewhere within the ball $\mathbb{B}_\delta$ and eventually leave the ball $\mathbb{B}_\varepsilon$ (see Fig. 2.1). Instability of an equilibrium point is typically undesirable because it often leads the system into limited cycles or results in damage to the involved mechanical or electrical components.

In the following, we will illustrate the concepts of asymptotic stability and exponential stability.

2.2.3 ASYMPTOTIC STABILITY AND EXPONENTIAL STABILITY

In many engineering applications, Lyapunov stability is not enough. For example, when a satellite's attitude is disturbed from its nominal position, we not only want the satellite to maintain its attitude in a range determined by the magnitude of the disturbance, i.e., Lyapunov stability, but also require that the attitude gradually goes back to its original value. This type of engineering requirement is captured by the concept of asymptotic stability.

Definition 2.5 *An equilibrium point $x = 0$ of (2.2) is asymptotically stable if it is stable and there is a $\delta > 0$ such that $x(t) \to 0$ as $t \to \infty$ for all $\|x(0)\| < \delta$.*

Asymptotic stability means that the equilibrium is stable, and in addition, states that started close to 0 actually converge to 0 as time t goes to infinity. Fig. 2.1 shows that the system trajectory 1 starting from within the ball $\mathbb{B}_\delta$ converges to the origin. The ball $\mathbb{B}_\delta$ is called a domain of attraction of the equilibrium point (where the domain of attraction of the equilibrium point refers to the largest such region, i.e., to the set of all points such that trajectories initiated at these points eventually converge to the origin). An equilibrium point, which is Lyapunov stable but not asymptotically stable, is called marginally stable.

However, in many engineering applications, it is still not sufficient to know that a system will converge to the equilibrium point after infinite time. There is a need to estimate how fast the system trajectory approaches. The concept of exponential stability can be used for this purpose.

Definition 2.6 *An equilibrium point $x = 0$ of (2.2) is exponentially stable if there exists an $\alpha > 0$, and for every $\varepsilon > 0$ there exists a $\delta = \delta(\varepsilon) > 0$ such that:*

$$\|x(t)\| \leq \varepsilon \exp(-\alpha t), \quad \forall t \geq 0, \tag{2.5}$$

for $\|x(0)\| < \delta$.

(2.5) means that the state vector of an exponentially stable system converges to the origin faster than the exponential function $\exp(-\alpha t)$, the positive number α is often called the rate of exponential convergence. For instance, for the following system:

$$\dot{x} = -\left(1 + \sin^2(x)\right) x,$$

its solution is:

$$x(t) = x(0) \exp\left(-\int_0^t \left[1 + \sin^2(x(\tau))\right] d\tau\right).$$

Therefore we have:

$$|x(t)| \leq |x(0)| \exp(-t),$$

which shows that the system state is exponentially convergent to $x = 0$ with a rate $\alpha = 1$.

Note that exponential stability implies asymptotic stability. But asymptotic stability does not guarantee exponential stability, as can be seen from the following system:

$$\dot{x} = -x^2, \quad x(0) = 1, \tag{2.6}$$

whose solution is $x = \frac{1}{1+t}$, a function slower than any exponential function $\exp(-\alpha t)$ with $\alpha > 0$.

2.2.4 LOCAL AND GLOBAL STABILITY

The above definitions are formulated to characterize the local behavior of systems, i.e., how the state evolves after starting near the equilibrium point. Local properties tell little about how the system will behave when the initial state is some distance away from the equilibrium, then the global concept is required for this purpose.

Definition 2.7 *If asymptotic (or exponential) stability holds for any initial states, then the equilibrium point is said to be asymptotically (or exponentially) stable in the large. It is also called globally asymptotically (or exponentially) stable.*

The simple system in (2.6) is globally asymptotically stable, as can be seen from its solution. Furthermore, for linear systems, no matter it is time-invariant or time-varying, continuous-time or discrete-time, based on the superposition principle, if the equilibrium point of linear systems, $x = 0$, is asymptotically stable, the asymptotic stability is always global.

2.3 LYAPUNOV'S LINEARIZATION METHOD

The Lyapunov's linearization method (or the Lyapunov's first method) is concerned with the local stability of a nonlinear system. It is a formalization of the intuition that a nonlinear system should behave similarly to its linearized approximation for small-range motions. Because all physical systems are inherently nonlinear, Lyapunov's linearization method serves as the fundamental justification for using linear control techniques in practice, i.e., it shows that stable design by linear control guarantees the stability of the original physical system locally.

Consider the autonomous system (2.2), and assume that $f(x) : \mathbb{D} \to \mathbb{R}^n$ is continuously differentiable and $\mathbb{D}$ is a neighborhood of the origin. Then, according to the Taylor expansion, the system dynamics can be written as:

$$\dot{x} = \left(\frac{\partial f}{\partial x} \right)_{x=0} x + f_{h.o.t.}(x), \tag{2.7}$$

where $f_{h.o.t.}$ stands for the high-order term in x.

Note that the above Taylor expansion starts directly with the first-order term, due to the fact that $f(0) = 0$, since the origin is an equilibrium point. Let us use the constant matrix A to denote the Jacobian matrix of f w.r.t. x at $x = 0$ (an $n \times n$ matrix of elements $\partial f_i / \partial x_i$):

$$A = \left(\frac{\partial f}{\partial x} \right)_{x=0}.$$

Then, the system

$$\dot{x} = Ax \tag{2.8}$$

is called the linearization (or linear approximation) of the original nonlinear system (2.2) at the equilibrium point.

In practice, finding a system's linearization is often most easily done simply by neglecting any term of order higher than 1 in the dynamics, as we now illustrate.

Example 2.2 *Consider the system:*

$$\begin{cases} \dot{x}_1 = x_2^2 + x_1 \cos(x_2), \\ \dot{x}_2 = x_2 + (x_1 + 1)x_1 + x_1 \sin(x_2). \end{cases}$$

Its linearized approximation about $x = [x_1, x_2]^\top = 0$ *is:*

$$\dot{x}_1 \approx \left(\frac{\partial \left(x_2^2 + x_1 \cos(x_2) \right)}{\partial x_1} \right)_{x_1=0, x_2=0} x_1 + \left(\frac{\partial \left(x_2^2 + x_1 \cos(x_2) \right)}{\partial x_2} \right)_{x_1=0, x_2=0} x_2$$

$$= (\cos(x_2))_{x_1=0, x_2=0} x_1 + (2x_2 - x_1 \sin(x_2))_{x_1=0, x_2=0} x_2$$

$$= x_1.$$

By using similar calculation, we have:

$$\dot{x}_2 \approx \left(\frac{\partial \left(x_2 + (x_1 + 1)x_1 + x_1 \sin(x_2) \right)}{\partial x_1} \right)_{x_1=0, x_2=0} x_1$$

$$+ \left(\frac{\partial \left(x_2 + (x_1 + 1)x_1 + x_1 \sin(x_2) \right)}{\partial x_2} \right)_{x_1=0, x_2=0} x_2$$

$$= (2x_1 + \sin(x_2) + 1)_{x_1=0, x_2=0} x_1 + (1 + x_1 \cos(x_2))_{x_1=0, x_2=0} x_2$$

$$= x_1 + x_2.$$

Its linearized system can be written as:

$$\dot{x} = \begin{pmatrix} 1 & 0 \\ 1 & 1 \end{pmatrix} x, \quad A = \begin{pmatrix} 1 & 0 \\ 1 & 1 \end{pmatrix}.$$

The following result takes the relationship between the stability of the linear system (2.8) and that of the original nonlinear system (2.2).

Theorem 2.1 *Lyapunov's linearization method:*

(i)   *if the linearized system is strictly stable (i.e., if all eigenvalues of A are strictly in the left-half complex plane), then the equilibrium point $x = 0$ of (2.2) is asymptotically stable;*

(ii) *if the linearized system is unstable (i.e., if at least one eigenvalue of A is strictly in the right-half complex plane), then the equilibrium point $x = 0$ of (2.2) is unstable;*

(iii) *if the linearized system is marginally stable (i.e., all eigenvalues of A are in the left-half complex plane, but at least one of them is on the imaginary axis), then one cannot conclude anything from the linear approximation (the equilibrium point may be stable, asymptotically stable, or unstable for the nonlinear system).*

The above theorem shows that if the linearized system is strictly stable or strictly unstable, then, since the approximation is valid "not too far" from the equilibrium, the original nonlinear system itself is locally stable or locally unstable. However, if the linearized system is marginally stable, the high-order term $f_{h.o.t.}$ in (2.7) can have a decisive effect on whether the system is stable or unstable.

Lyapunov's linearization theorem shows that linear control design is a matter of consistency: one must design a controller such that the system remains in its "linear range". It also stresses major limitations of linear design: How large is the linear range? What is the extent of stability (How large is δ in Definition 2.4)? These questions motivate a deeper approach to the nonlinear control problem, Lyapunov's direct method.

2.4 LYAPUNOV'S DIRECT METHOD

The basic philosophy of Lyapunov's direct method is the mathematical extension of a fundamental physical observation: if the total energy of a mechanical (or electrical) system is continuously dissipated, then the system, whether linear or nonlinear, must eventually settle down to an equilibrium point. Thus, we may conclude the stability of a system by examining the variation of a single scalar function.

Specifically, let us consider the mass-damping-spring system in Fig. 2.2, whose dynamic equation is:

$$m\ddot{x} + b\dot{x}|\dot{x}| + k_0 x + k_1 x^3 = 0, \tag{2.9}$$

where $b\dot{x}|\dot{x}|$ represents the nonlinear dissipation or damping, and $k_0 x + k_1 x^3$ represents a nonlinear spring term. Assume that the mass is pulled away from the natural length of the spring by a large distance, and then released. Will the resulting motion be stable? It is difficult to answer this question using the definitions of stability and linearization method, so we consider the Lyapunov's direct method in the following.

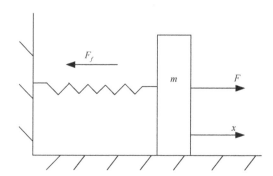

FIGURE 2.2 Mass-damping-spring system.

The total mechanical energy of the system is the sum of its kinetic energy and its potential energy:

$$V(x) = \frac{1}{2}m\dot{x}^2 + \int_0^x \left(k_0 x + k_1 x^3\right) dx = \frac{1}{2}m\dot{x}^2 + \frac{1}{2}k_0 x^2 + \frac{1}{4}k_1 x^4. \qquad (2.10)$$

Comparing the definitions of stability and total mechanical energy (2.10), one can easily see some relations between them:

(i)   zero energy corresponds to the equilibrium point $(x = 0, \dot{x} = 0)$;
(ii) asymptotically stable implies the convergence of the mechanical energy to zero;
(iii) instability is related to the growth of mechanical energy.

These relations indicate that the stability properties of the system can be characterized by the variation of the mechanical energy of the system. By differentiating (2.10) and using (2.9), the variation of energy is:

$$\dot{V}(x) = m\dot{x}\ddot{x} + \left(k_0 x + k_1 x^3\right)\dot{x} = \dot{x}\left(-b\dot{x}|\dot{x}|\right) = -b|\dot{x}|^3. \qquad (2.11)$$

Equation (2.11) implies that the energy of the system, starting from some initial values, is continuously dissipated by the damper until the mass settles down, i.e., until $\dot{x} = 0$. Physically, the mass must finally settle down at the natural length of the spring, because it is subjected to a non-zero spring force at any position other than the natural length. Thus, by examining the derivative of V along the trajectories of the system, it is possible to determine the stability of the equilibrium point.

The Lyapunov's direct method is based on a generalization of the concepts from the above mass-damping-spring system to more complex systems. Faced with a set of nonlinear differential equations, the basic procedure of the Lyapunov's direct method is to generate a scalar "energy-like" function for the dynamic system, and then examine the time variation of the corresponding scalar function. In this way, conclusions may be drawn on the stability of the set of differential equations without using the difficult stability definitions or requiring explicit knowledge of solutions.

The definitions of positive definite functions and Lyapunov functions are imposed below.

2.4.1   POSITIVE DEFINITE FUNCTIONS AND LYAPUNOV FUNCTIONS

The energy function in (2.10) has two properties. The first is a property of the function itself: it is strictly positive unless both state variables x and $\dot{x}$ are zero. The second is a property associated with the dynamics (2.9): the function is monotonically decreasing when the variables x and $\dot{x}$ vary according to (2.9). In the Lyapunov's direct method, the first property is formalized by the notion of positive definite functions, and the second is formalized by the so-called Lyapunov functions.

Let us discuss positive definite functions first.

Definition 2.8 *A scalar continuous function V (x) is said to be locally positive definite if, for $x \in \mathbb{D}$ with $\mathbb{D} \subset \mathbb{R}^n$ denoting a domain that contains the origin,*

$$\begin{cases} V(0) = 0, \\ V(x) > 0 \ \ for \ \ x \in \mathbb{D} - \{0\}. \end{cases}$$

If the above property holds over the whole state space, then V (x) is said to be globally positive definite.

The above definition implies that the function V has a unique minimum at the origin. Actually, given any function having a unique minimum in a certain ball, we can construct a locally positive definite function simply by adding a constant to that function. For example, the function $V(x_1, x_2) = x_1^2 + x_2^2 - 1$ is a lower bounded function with a unique minimum at the origin, and the addition of the constant 1 makes it a positive definite function. Of course, the function shifted by a constant has the same time derivative as the original function.

Let us describe the geometrical meaning of locally positive definite functions. Consider a positive definite function $V(x)$ of two state variables x_1 and x_2. Plotted in a 3-dimensional space, $V(x)$ typically corresponds to a surface looking like an upward cup (Fig. 2.3). The lowest point of the cup is located at the origin.

A second geometrical representation can be made as follows. Taking x_1 and x_2 as Cartesian coordinates, the level curves $V(x_1, x_2) = V_\alpha$ typically represent a set of ovals surrounding the origin, with each oval corresponding to a positive value of V_α. These ovals, often called contour curves, may be thought as the sections of the cup by horizontal planes, projected on the (x_1, x_2) plane (Fig. 2.4). Note that the contour curves do not intersect, because $V(x_1, x_2)$ is uniquely defined given (x_1, x_2).

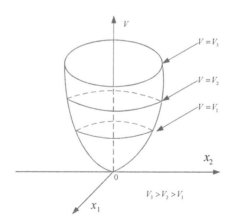

FIGURE 2.3 Typical shape of a positive definite function $V(x)$ with $x = [x_1, x_2]^\top$.

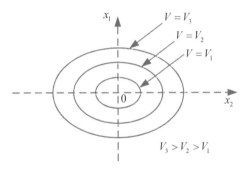

FIGURE 2.4 Interpreting positive definite functions using contour curves.

A few related concepts can be defined similarly, in a local or global sense, i.e., a function $V(x)$ is negative definite if $-V(x)$ is positive definite, $V(x)$ is positive semi-definite if $V(0) = 0$ and $V(x) \geq 0$ for $x \neq 0$, $V(x)$ is negative semi-definite if $-V(x)$ is positive semi-definite. The prefix "semi-" is used to reflect the possibility of V being equal to zero for $x \neq 0$. These concepts can be given geometrical meanings similar to the ones given for positive definite functions.

With x denoting the state of the system (2.2), a scalar function $V(x)$ actually represents an implicit function of time t. Assuming that $V(x)$ is differentiable, its derivative w.r.t. time can be found by the chain rule:

$$\dot{V} = \frac{dV(x)}{dt} = \frac{\partial V}{\partial x}\dot{x} = \frac{\partial V}{\partial x}f(x).$$

We see that because x is required to satisfy the autonomous state equation (2.2), $\dot{V}$ only depends on x. It is often referred to as "the derivative of V along the system trajectory". For the system (2.9), $\dot{V}(x)$ is computed in (2.11) and found to be negative.

Definition 2.9 *If $V : \mathbb{D} \to \mathbb{R}$ is positive definite and has continuous partial derivatives, and if its time derivative along any state trajectory of the system (2.2) is negative semi-definite, i.e., $\dot{V}(x) \leq 0$, then $V(x)$ is said to be a Lyapunov function for the system (2.2).*

2.4.2   LYAPUNOV THEOREM WITH LOCAL/GLOBAL STABILITY

The relationships between the Lyapunov functions and the stability of the system are made precise in a number of theorems in the Lyapunov's direct method. Such theorems usually have local and global versions. The local versions are concerned with stability properties in the neighborhood of the equilibrium point and usually involve a locally positive definite function.

Theorem 2.2 *[1] (Local Stability) Let $x = 0$ be an equilibrium point of (2.2) and $\mathbb{D} \subset \mathbb{R}^n$ be a domain containing $x = 0$. If there exists a scalar function $V(x)$ in $\mathbb{D}$ with continuous first partial derivatives such that:*

(i) $V(0) = 0$ and $V(x) > 0$ in $\mathbb{D} - \{0\}$ (or $V(x)$ is positive definite in $\mathbb{D}$);
(ii) $\dot{V}(x) \leq 0$ in $\mathbb{D}$ (or $\dot{V}(x)$ is negative semi-definite in $\mathbb{D}$);

then the equilibrium point $x = 0$ is locally stable. Moreover, if

$$\dot{V}(x) < 0 \ in \ \mathbb{D} - \{0\},$$

then the equilibrium point $x = 0$ is locally asymptotically stable.

It is obvious that the scalar function $V(x)$ in Theorem 2.2 satisfies the properties of Lyapunov functions. For such a Lyapunov function, the surface $V(x) = c$ with $c > 0$ is called a Lyapunov surface or a level surface. Using Lyapunov surfaces, we notice that Fig. 2.4 makes Theorem 2.2 intuitively clear. It shows Lyapunov surfaces for increasing values of c. The condition $\dot{V} \leq 0$ implies that when a trajectory crosses a Lyapunov surface $V(x) = c$, it moves inside the set $\Omega_c = \{x \in \mathbb{R}^n \,|\, V(x) \leq c\}$ and can never come out again. When $\dot{V} < 0$, the trajectory moves from one Lyapunov surface to an inner Lyapunov surface with a smaller c. As c decreases, the Lyapunov surface $V(x) = c$ shrinks to the origin, showing that the trajectory approaches the origin as time progresses. If we only know that $\dot{V} \leq 0$, we cannot make sure that the trajectory will approach the origin, but we can conclude that the origin is stable since the trajectory can be contained inside any ball $\mathbb{B}_\varepsilon$ by requiring the initial state $x(0)$ to lie inside a Lyapunov surface contained in that ball.

In applying the above theorem for the analysis of a nonlinear system, one goes through the two steps of choosing a positive definite function, and then determining its derivative along the path of the nonlinear system. The following example illustrates this procedure.

Example 2.3 A simple pendulum with viscous damping is described by:

$$\ddot{\theta} + \dot{\theta} + \sin(\theta) = 0.$$

Consider the following scalar function:

$$V(\theta, \dot{\theta}) = (1 - \cos(\theta)) + \frac{\dot{\theta}^2}{2}.$$

One easily verifies that this function is locally positive definite, which represents the total energy function, composed of the sum of the potential energy and the kinetic energy of the pendulum. Its time derivative is:

$$\dot{V}(\theta, \dot{\theta}) = \dot{\theta}\sin(\theta) + \dot{\theta}\ddot{\theta} = -\dot{\theta}^2 \leq 0.$$

Therefore, by invoking Theorem 2.2, one concludes that the origin is a stable equilibrium point. In fact, using physical insight, one easily sees the reason why $\dot{V} \leq 0$, namely that the damping term absorbs energy. Actually, $\dot{V}$ is precisely the power dissipated in the pendulum. However, with this Lyapunov function, one cannot draw conclusions on the asymptotic stability of the system, because $\dot{V}$ is only negative semi-definite.

The following example illustrates the asymptotic stability result.

Example 2.4 *Consider the following nonlinear system:*

$$\begin{cases} \dot{x}_1 = x_1\left(x_1^2 + x_2^2 - 2\right) - 4x_1 x_2^2, \\ \dot{x}_2 = 4x_1^2 x_2 + x_2\left(x_1^2 + x_2^2 - 2\right). \end{cases}$$

Given a positive definite function:

$$V(x_1, x_2) = x_1^2 + x_2^2,$$

its derivative along the system trajectory is:

$$\dot{V} = 2\left(x_1^2 + x_2^2\right)\left(x_1^2 + x_2^2 - 2\right).$$

Thus, $\dot{V}$ is locally negative definite in the region defined by $x_1^2 + x_2^2 < 2$. Therefore, the above theorem indicates that the origin is locally asymptotically stable.

The above theorem applies to the local analysis of stability. In order to assert the global asymptotic stability of a system, one might naturally expect that the domain $\mathbb{D}$ in the above local theorem has to be the whole state-space. This is indeed necessary, but it is not enough. An additional condition on the function V has to be satisfied: $V(x)$ must be radially unbounded, by which we mean that $V(x) \to \infty$ as $\|x\| \to \infty$ (in other words, as x tends to infinity in any direction). We then obtain the following powerful result.

Theorem 2.3 *(Global Stability) Let $x = 0$ be an equilibrium point of (2.2). If there exists a scalar function $V(x)$ with continuous first partial derivatives such that:*

(i) $V(0) = 0$ and $V(x) > 0$, $\forall x \neq 0$ *(or $V(x)$ is positive definite);*
(ii) $\dot{V}(x) < 0$, $\forall x \neq 0$ *(or $\dot{V}(x)$ is negative definite);*
(iii) $V(x) \to \infty$ as $\|x\| \to \infty$.

Then the equilibrium point $x = 0$ is globally asymptotically stable.

Here, we pause to stress the reason of the radial unboundedness condition. It is to ensure that the contour curves $V(x) = V_\alpha$ correspond to closed curves. If the curves are not closed, it is possible for the state trajectories to drift away from the equilibrium point, even though the state keeps going through contours corresponding to smaller and smaller V_α (see Fig. 2.5 with $\alpha = 1, 2, 3$). For example, for the positive definite function $V = \frac{x_1^2}{1+x_1^2} + x_2^2$, the curves $V = V_\alpha$ for $V_\alpha > 1$ are open curves ($V_\alpha > 1$ may correspond to the case that $|x_1| \to \infty$ and any $x_2 \neq 0$). Fig. 2.5 shows the divergence of the state while moving toward lower and lower "energy" curves.

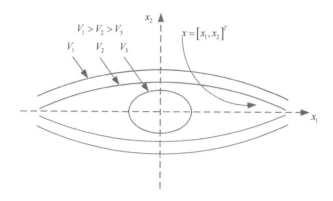

FIGURE 2.5 Motivation of the radial unboundedness condition.

Now we show the following example.

Example 2.5 *Consider the nonlinear system:*

$$\dot{x} + c(x) = 0,$$

where c is any continuous function of the same sign as its scalar argument x, i.e.,

$$xc(x) > 0 \ for \ x \neq 0.$$

Consider the Lyapunov function candidate as:

$$V = x^2.$$

This function V is radially unbounded, since it tends to infinity as $|x| \to \infty$. Its derivative is:

$$\dot{V} = 2x\dot{x} = -2xc(x).$$

Thus $\dot{V} < 0$ as long as $x \neq 0$, so that $x = 0$ is a globally asymptotically stable equilibrium point.

Many Lyapunov functions may exist for the same system. For instance, if V is a Lyapunov function for a given system, so is:

$$V_1 = \rho V^\alpha,$$

where ρ is any strictly positive constant and α is any scalar (not necessarily an integer) larger than 1. Indeed, the positive definiteness of V implies that of V_1, the positive definiteness (or positive semi-definiteness) of $-\dot{V}$ implies that of $-\dot{V}_1$. More importantly, for a given system, specific choices of Lyapunov functions may yield more precise results than others. Consider again the pendulum of Example 2.3. The following function

$$V(\theta, \dot{\theta}) = \frac{1}{2}\dot{\theta}^2 + \frac{1}{2}(\dot{\theta} + \theta)^2 + 2(1 - \cos(\theta))$$

is also a Lyapunov function for the system, because

$$\dot{V}\left(\theta,\dot{\theta}\right) = -\left(\dot{\theta}^2 + \theta\sin\left(\theta\right)\right) \leq 0.$$

However, it is interesting to note that $\dot{V}$ is actually locally negative definite for $\theta \in \left(-\frac{\pi}{2},\frac{\pi}{2}\right)$, and therefore, this modified choice of V, without obvious physical meaning, allows the asymptotic stability of the pendulum to be shown.

Along the same lines, it is important to realize that the theorems in the Lyapunov analysis are all sufficiency theorems. For a particular choice of the Lyapunov function candidate V, if the conditions on $\dot{V}$ are not met, one cannot draw any conclusion on the stability or instability of the system. The only conclusion one should draw is that a different Lyapunov function candidate should be tried.

2.5 INVARIANT SET THEOREM

The asymptotic stability of a control system is usually a very important property to be determined. However, the above theorems just described are often difficult to apply in order to assert this property. The reason is that $\dot{V}$, the derivative of the Lyapunov function candidate, is only negative semi-definite. In this kind of situation, fortunately, it is still possible to draw conclusions on asymptotic stability, with the help of the powerful invariant theorem, attributed to La Salle. This section presents the local and global versions of the invariant set theorems.

The central concept in these theorems is that of a invariant set, a generalization of the concept of equilibrium point.

Definition 2.10 *A set $\mathcal{G}$ is an invariant set for a dynamic system if every system trajectory which starts from a point in set $\mathcal{G}$ remains in $\mathcal{G}$ for all future time.*

For example, any equilibrium point is an invariant set. The attraction domain of an equilibrium point is also an invariant set. The invariant set theorems reflect the intuition that the decrease of a Lyapunov function V has to gradually vanish (i.e., $\dot{V}$ has to converge to zero) because V is lower bounded. A precise statement of this result is given as follows.

Theorem 2.4 *(Local Invariant Set Theorem) Consider an autonomous system (2.2) with a continuous function f. Let $V(x)$ be a scalar function with continuous first-order partial derivatives. Assume that:*

(i) For some $l > 0$, $\Omega_l = \{x \mid V(x) < l\}$ is a bounded region;
(ii) $\dot{V}(x) \leq 0$ holds for all $x \in \Omega_l$.

Let $\mathcal{R}$ be the set of all points within Ω_l where $\mathcal{R} = \{x \mid \dot{V}(x) = 0\}$, and $\mathcal{M}$ be the largest invariant set in $\mathcal{R}$. Then, every solution $x(t)$ originating in Ω_l tends to $\mathcal{M}$ as $t \to \infty$.

Note that in the above theorem, the word "largest" is understood in the sense of set theorem, i.e., $\mathcal{M}$ is the union of all invariant sets within $\mathcal{R}$. In particular, if the set $\mathcal{R}$ is itself invariant, then $\mathcal{M} = \mathcal{R}$.

Theorem 2.5 *(Global Invariant Set Theorem) Consider an autonomous system (2.2) with a continuous function f. Let $V(x)$ be a scalar function with continuous first-order partial derivatives. Assume that:*

(i) $V(x) \to \infty$ as $\|x\| \to \infty$;
(ii) $\dot{V}(x) \le 0$ holds over the whole state space.

Let $\mathcal{R}$ be the set of all points satisfying $\dot{V}(x) = 0$, and $\mathcal{M}$ be the largest invariant set in $\mathcal{R}$. Then all solutions globally asymptotically converge to $\mathcal{M}$ as $t \to \infty$.

Let us now illustrate applications of the invariant set theorem using the following example.

Example 2.6 *Asymptotic stability of the mass-damping-spring system.*
For the system (2.9), one can only draw the conclusion of marginal stability using the energy function (2.10) in Theorem 2.2, because $\dot{V}$ is only negative semi-definite according to (2.11). Using the invariant set theorem, however, we can show that the system is actually asymptotically stable. To do this, we only have to show that the set $\mathcal{M}$ contains only one point. The set $\mathcal{R}$ is defined by $\dot{x} = 0$, i.e., the collection of states with zero velocity or the whole horizontal axis in the phase plane $(x, \dot{x})$. Let us show that the largest invariant set $\mathcal{M}$ in this set $\mathcal{R}$ contains only the origin $x = 0$. Assume that $\mathcal{M}$ contains a non-zero position x_1, then the acceleration at this point is $\ddot{x}_1 = -(k_0/m)x_1 - (k_1/m)x_1^3 \neq 0$. This implies that the trajectory will immediately move out of the set $\mathcal{R}$ and thus also out of the set $\mathcal{M}$, which leads to a contradiction to the definition of an invariant set. Therefore, the maximum invariant set $\mathcal{M}$ only contains the origin, and the system is asymptotically stable.

2.6   STABILITY CONCEPTS OF NON-AUTONOMOUS SYSTEMS

In the previous sections, we studied the Lyapunov analysis of autonomous systems. In many practical problems, however, we encounter non-autonomous systems. For instance, a rocket taking off is a non-autonomous system because the parameters involved in its dynamic equations (such as air temperature and pressure) vary with time. Therefore, stability analysis techniques for non-autonomous systems must be developed.

2.6.1   EQUILIBRIUM POINTS

The concepts of stability for non-autonomous systems are quite similar to those of autonomous systems. However, due to the dependence of non-autonomous system behavior on the initial time t_0, the definitions of these stability concepts include t_0

explicitly. Furthermore, a new concept, uniformity, is necessary to characterize non-autonomous systems whose behavior has a certain consistency for different values of the initial time t_0. In this section, we concisely extend the stability concepts for autonomous systems to non-autonomous systems, and introduce the new concept of uniformity.

For non-autonomous systems in the following form:

$$\dot{x} = f(x,t), \quad x(t_0) = x_0, \quad \forall t \geq t_0 \geq 0, \tag{2.12}$$

where $f : \mathbb{D} \times [t_0, \infty) \rightarrow \mathbb{R}^n$ is piecewise continuous in t and locally Lipschitz in x on $[t_0, \infty) \times \mathbb{R}^n$, the equilibrium point x^* is defined by:

$$f(x^*, t) \equiv 0, \quad \forall t \geq t_0. \tag{2.13}$$

Note that this equation must be satisfied for $\forall t \geq t_0$, implying that the system should be able to stay at the point x^* all the time. For instance, one easily sees that the linear time-varying system:

$$\dot{x} = A(t)x$$

has a unique equilibrium point at the origin 0 unless A is always singular.

Example 2.7 The system $\dot{x} = -\frac{x}{1+x^2}$ has an equilibrium point at $x = 0$. The system $\dot{x} = -\frac{1}{1+t}x$ has an equilibrium point at $x = 0$. However, the system $\dot{x} = -\frac{x}{1+x^2} + b(t)$ with $b(t) \neq 0$ does not have an equilibrium point. It can be regarded as a system under the external input or disturbance $b(t)$.

2.6.2 EXTENSIONS OF THE STABILITY CONCEPTS

Let us extend the previously defined concepts of stability, instability, asymptotic stability, and exponential stability to non-autonomous systems. The key in doing so is to properly include the initial time t_0 in the definitions. For convenience, only the case where the equilibrium point is the origin is analyzed. To facilitate the description later, we rewrite the solution of (2.12), $x(t; t_0, x_0)$, as $x(t)$.

Definition 2.11 The equilibrium point $x = 0$ of (2.12) is stable if, for each $\varepsilon > 0$, and any $t_0 \geq 0$ there is a $\delta = \delta(\varepsilon, t_0) > 0$ such that:

$$\|x(t_0)\| < \delta \Rightarrow \|x(t)\| < \varepsilon, \quad \forall t \geq t_0.$$

Otherwise, the equilibrium point $x = 0$ is unstable.

Definition 2.12 The equilibrium point $x = 0$ of (2.12) is asymptotically stable if it is stable and there is a $\delta = \delta(t_0) > 0$ such that $x(t) \rightarrow 0$ as $t \rightarrow \infty$ for all $\|x(t_0)\| < \delta$.

Definition 2.13 The equilibrium point $x = 0$ of (2.12) is globally asymptotically stable if, for any $x(t_0)$, $x(t) \rightarrow 0$ as $t \rightarrow \infty$.

Definition 2.14 *The equilibrium point $x = 0$ of (2.12) is exponentially stable if there exists an $\alpha > 0$, and for every $\varepsilon > 0$ there exists a $\delta(\varepsilon) > 0$ such that:*

$$\|x(t)\| \leq \varepsilon \exp(-\alpha(t - t_0)), \quad \forall \|x(t_0)\| < \delta(\varepsilon), \quad \forall t \geq t_0.$$

Example 2.8 *Consider the first-order system $\dot{x}(t) = -a(t)x(t)$, its solution is:*

$$x(t) = x(t_0) \exp\left(-\int_{t_0}^{t} a(\tau)d\tau\right).$$

Thus, the system is stable if $a(t) \geq 0$, $\forall t \geq t_0$. It is asymptotically stable if $\int_{t_0}^{\infty} a(\tau)d\tau = +\infty$. It is exponentially stable if there exists a strictly positive constant T so that $\forall t \geq t_0$, $\int_{t}^{t+T} a(\tau)d\tau \geq \gamma$, with γ being a positive constant. Therefore,

(i)   *the system $\dot{x} = -\frac{x}{(1+t)^2}$ is stable;*
(ii) *the system $\dot{x} = -\frac{x}{1+t}$ is asymptotically stable;*
(iii) *the system $\dot{x} = -tx$ is exponentially stable.*

Furthermore, we also give the concept of "bounded solution".

Definition 2.15 *A solution $x(t)$ of (2.12) is bounded if there exists a $\beta > 0$ such that $\|x(t)\| < \beta$ for all $t \geq t_0$, where β may depend on each solution.*

2.6.3   UNIFORMITY IN STABILITY CONCEPTS

The previous concepts of Lyapunov stability for non-autonomous systems indicate the important effect of the initial time. In practice, it is usually desirable for the system to have a certain uniformity in its behavior regardless of when the operation starts. This motivates us to consider the definitions of uniform stability and uniform asymptotic stability. It is also useful to point out that because the behavior of autonomous systems is independent of the initial time, all the stability properties of an autonomous system are uniform.

Definition 2.16 *The solutions of (2.12) are uniformly bounded if for any $\alpha > 0$ and $t_0 \in \mathbb{R}_+$, there exists a $\beta = \beta(\alpha) > 0$ independent of t_0 such that if $\|x(t_0)\| < \alpha$, then $\|x(t)\| < \beta$ fro all $t \geq t_0$.*

Definition 2.17 *The solutions of (2.12) are ultimately uniformly bounded (with bounded $\mathscr{B}$) if there exists a $\mathscr{B}$ and if corresponding to any $\alpha > 0$ and $t_0 \in \mathbb{R}_+$, there exists a $T = T(a) > 0$ (independent of t_0) such that for $\|x(t_0)\| < \alpha$, $\|x(t)\| < \mathscr{B}$ holds for all $t \geq t_0 + T$.*

Definition 2.18 *The equilibrium point $x = 0$ of (2.12) is locally uniformly stable if the scalar δ in Definition 2.11 can be chosen independent of t_0, i.e., if $\delta = \delta(\varepsilon)$.*

Definition 2.19 *The equilibrium point $x = 0$ of (2.12) is locally uniformly asymptotically stable if:*

(i)   *it is uniformly stable;*
(ii) *for every $\varepsilon > 0$ and any $t_0 \in \mathbb{R}_+$, there exist a $\delta_0 > 0$ independent of t_0 and ε, and a $T(\varepsilon) > 0$ independent of t_0 such that if $\|x_0\| < \delta_0$, then $\|x(t)\| < \varepsilon$, $\forall t \geq t_0 + T(\varepsilon)$.*

Definition 2.20 *The equilibrium point $x = 0$ of (2.12) is globally uniformly asymptotically stable if:*

(i)   *it is uniformly stable;*
(ii)   *the solutions of (2.12) are uniformly bounded;*
(iii) *for any $\alpha > 0$, any $\varepsilon > 0$, and any $t_0 \in \mathbb{R}_+$, there exists $T(\varepsilon, \alpha) > 0$ independent of t_0 such that if $\|x(t_0)\| < \alpha$, then $\|x(t)\| < \varepsilon$ for all $t \geq t_0 + T(\varepsilon, \alpha)$.*

Definition 2.21 *The equilibrium point $x = 0$ of (2.12) is globally exponentially stable if there exists an $\alpha > 0$, and for any $\beta > 0$, there exists a $k(\beta) > 0$ such that:*

$$\|x(t)\| \leq k(\beta) \exp(-\alpha(t - t_0)), \quad \forall \|x(t_0)\| < \beta, \ \forall t \geq t_0.$$

By definition, uniformly asymptotic stability always implies asymptotic stability.

2.7   LYAPUNOV ANALYSIS OF NON-AUTONOMOUS SYSTEMS

We now extend the Lyapunov analysis results of autonomous systems to the stability analysis of non-autonomous systems. Although many of the ideas of autonomous systems can be similarly applied to non-autonomous cases, the conditions required in the treatment of non-autonomous systems are more complicated and more restrictive.

The basic idea of the direct method, i.e., concluding the stability of nonlinear systems using scalar Lyapunov functions, can be similarly applied to non-autonomous systems. Besides, due to the more mathematical complexity, the powerful La Salle's theorems do not apply to non-autonomous systems. This drawback will partially be compensated by Barbalat's Lemma, which will be introduced in Section 2.8.

2.7.1   TIME-VARYING POSITIVE DEFINITE FUNCTIONS

When studying non-autonomous systems using the Lyapunov's direct method, scalar functions with explicit time-dependence $V(x,t)$ may have to be used, while in autonomous system analysis time-invariant functions $V(x)$ suffice. We now introduce a simple definition of positive definiteness for such scalar functions.

Definition 2.22 *A scalar time-varying function $V : \mathbb{D} \times [t_0, \infty) \to \mathbb{R}$ is locally positive definite in $\mathbb{D} \times [t_0, \infty)$, where $\mathbb{D} \subset \mathbb{R}^n$ denotes a domain that contains the origin, if:*

(i) $V(0,t) = 0$;
(ii) *there exists a time-invariant positive definite function $V_0(x)$ such that $V(x,t) \geq V_0(x)$, $\forall t \geq t_0$.*

Thus, a time-variant function is locally positive definite if it dominates a time-invariant locally positive definite function.

Globally positive definite functions can be defined similarly if $\mathbb{D} = \mathbb{R}^n$.

Example 2.9 *The function* $V(x,t) = \frac{x^2}{1-x^2}$ *with* $x \in \mathbb{D} = \{x \in \mathbb{R} \, | \, |x| < 1\}$ *is positive definite with* $V_0(x)$ *being* $V_0(x) = \frac{x^2}{1-x^2}$, *whereas* $V(x,t) = \frac{1}{1+t}x^2$ *is not. The function* $V(x,t) = \frac{x^2}{1+x^2}$ *is positive definite for all* $x \in \mathbb{R}$ *as* V_0 *can be chosen as* $V_0 = \frac{x^2}{1+x^2}$.

Other related concepts can be defined in the same way in a local or global sense. A function $V(x,t)$ is negative definite if $-V(x,t)$ is positive definite, $V(x,t)$ is positive semi-definite if $V(x,t)$ dominates a time-invariant positive semi-definite function, $V(x,t)$ is negative semi-definite if $-V(x,t)$ is positive semi-definite.

In the Lyapunov analysis of non-autonomous systems, the concept of decrescent functions is also necessary.

Definition 2.23 *A scalar time-varying function* $V : \mathbb{D} \times [t_0, \infty) \to \mathbb{R}$ *is said to be decrescent if:*

(i) $V(0,t) = 0$;
(ii) *if there exists a time-invariant positive definite function* $V_1(x)$ *such that:*
      $V(x,t) \leq V_1(x)$, $\forall t \geq t_0$.

In other words, a scalar function $V(x,t)$ *is decrescent if it is dominated by a time-invariant positive definite function.*

Example 2.10 *A time-varying function* $V(x,t) = (1 + \sin^2(t))(x_1^2 + x_2^2)$ *is positive definite because it dominates the function* $V_0(x) = x_1^2 + x_2^2$. *This function is also decrescent because it is dominated by the function* $V_1(x) = 2(x_1^2 + x_2^2)$. *The function* $V(x,t) = \frac{1}{1+t}x^2$ *is decrescent because* $V(x,t) \leq V_1(x) = x^2$ *for* $\forall t \in \mathbb{R}_+$ *but* $V(x,t) = tx^2$ *is not.*

2.7.2   LYAPUNOV THEOREM FOR NON-AUTONOMOUS SYSTEM STABILITY

The main Lyapunov stability results for non-autonomous systems can be summarized by the following theorem.

Theorem 2.6 *(Lyapunov Theorem for Non-autonomous Systems)*

Stability: *If, in a domain* $\mathbb{D} \subset \mathbb{R}^n$ *that contains the origin, there exists a scalar function* $V : \mathbb{D} \times [t_0, \infty) \to \mathbb{R}$ *with continuous partial derivatives such that:*

(i) $V(x,t)$ *is positive definite;*
(ii) $\dot{V}(x,t)$ *is negative semi-definite.*

Then the equilibrium point $x = 0$ is stable.

Uniform stability: *If,*

(*i*) $V(x,t)$ *is positive definite;*
(*ii*) $\dot{V}(x,t)$ *is negative semi-definite;*
(*iii*) $V(x,t)$ *is decrescent.*

Then the equilibrium point $x = 0$ is uniformly stable.

Uniform asymptotic stability: *If,*

(*i*) $V(x,t)$ *is positive definite;*
(*ii*) $\dot{V}(x,t)$ *is negative definite;*
(*iii*) $V(x,t)$ *is decrescent.*

Then the equilibrium point $x = 0$ is uniformly asymptotically stable.

Global uniform asymptotic stability: *If $\mathbb{D}$ is replaced by the whole state space $\mathbb{R}^n$, and*

(*i*) $V(x,t)$ *is positive definite;*
(*ii*) $\dot{V}(x,t)$ *is negative definite;*
(*iii*) $V(x,t)$ *is decrescent;*
(*iv*) $V(x,t)$ *is radially unbounded.*

Then the equilibrium point $x = 0$ is globally uniformly asymptotically stable.

In order to understand the above theorem, we impose the definitions of positive definite functions and decrescent functions in terms of the so-called class $\mathcal{K}$ functions.

Definition 2.24 *A continuous function $\alpha : [0, p) \to [0, \infty)$ is said to belong to class $\mathcal{K}$ if it is strictly increasing and $\alpha(0) = 0$ and is said to belong to class $\mathcal{K}_\infty$ if, additionally, $p = \infty$ and $\lim_{p \to \infty} \alpha(p) \to \infty$.*

Definition 2.25 *A continuous function $\gamma : [0, p) \times [0, \infty) \to [0, \infty)$ is said to belong to class $\mathcal{K}\mathcal{L}$ if 1) for each fixed t, the mapping $\gamma(s,t)$ belongs to class $\mathcal{K}$ with respect to s; and 2) for each fixed s, the mapping $\gamma(s,t)$ is decreasing with respect to t and $\gamma(s,t) \to 0$ as $t \to \infty$.*

Example 2.11 *The function $\alpha(|x|) = \frac{x^2}{1+x^2}$ belongs to class $\mathcal{K}$ defined on $[0,\infty)$ but not to class $\mathcal{K}_\infty$. The function $\alpha(|x|) = |x|$ belongs to class $\mathcal{K}$ and class $\mathcal{K}_\infty$. The function $\gamma(|s|,t) = |s| \exp(-t)$ belongs to class $\mathcal{K}\mathcal{L}$.*

With the aid of the so-called class-$\mathcal{K}$ functions, we can now restate the above Definitions 2.22–2.23 and Theorem 2.6 as follows.

Definition 2.26 *(Definition 2.22) A scalar time-varying function $V : \mathbb{D} \times [t_0, \infty) \to \mathbb{R}$ is locally positive definite in $\mathbb{D} \times [t_0, \infty)$, where $\mathbb{D} \subset \mathbb{R}^n$ denotes a domain that contains the origin, if*

(i) $V(0,t) = 0$;
(ii) *there exists an $\alpha \in \mathcal{K}$ such that $V(x,t) \geq \alpha(|x|)$ for $\forall t \geq t_0$.*

Definition 2.27 *(Definition 2.23) A scalar time-varying function $V : \mathbb{D} \times [t_0, \infty) \to \mathbb{R}$ is said to be decrescent if*

(i) $V(0,t) = 0$;
(ii) *if there exists an $\alpha \in \mathcal{K}$ such that $V(x,t) \leq \alpha(|x|)$ for $\forall t \geq t_0$.*

Definition 2.28 *A scalar time-varying function $V : \mathbb{R}^n \times [t_0, \infty) \to \mathbb{R}$ is said to be radially unbounded if*

(i) $V(0,t) = 0$;
(ii) *if there exists an $\alpha \in \mathcal{K}_\infty$ such that $V(x,t) \geq \alpha(|x|)$ for $\forall t \geq t_0$ and $\forall x \in \mathbb{R}^n$.*

Example 2.12 *The function $V(x) = \frac{x^2}{1+x^2}$ satisfies $V(0,t) = 0$. However, because $V(x) \leq 1$, one cannot find a function $\alpha(|x|) \in \mathcal{K}_\infty$ to satisfy $V(x) \geq \alpha(|x|)$ for all $x \in \mathbb{R}$. Hence, V is not radially unbounded.*

Theorem 2.7 *(Theorem 2.6: Lyapunov Theorem for Non-autonomous Systems)*

Stability: *If, in a domain $\mathbb{D} \subset \mathbb{R}^n$ that contains the origin, there exists a scalar function $V : \mathbb{D} \times [t_0, \infty) \to \mathbb{R}$ with continuous partial derivatives and $V(0,t) = 0$ such that:*

(i)   *there exists an $\alpha \in \mathcal{K}$ such that $V(x,t) \geq \alpha(\|x\|) > 0$;*
(ii) $\dot{V}(x,t) \leq 0$.

Then the equilibrium point $x = 0$ is stable.

Uniform stability: *If,*

(i)   *there exist $\alpha, \beta \in \mathcal{K}$ such that $0 < \alpha(\|x\|) \leq V(x,t) \leq \beta(\|(x)\|)$;*
(ii) $\dot{V}(x,t) \leq 0$.

Then the equilibrium point $x = 0$ is uniformly stable.

Uniform asymptotic stability: *If,*

(i)   *there exist $\alpha, \beta \in \mathcal{K}$ such that $0 < \alpha(\|x\|) \leq V(x,t) \leq \beta(\|(x)\|)$;*
(ii) *there exists a $\gamma \in \mathcal{K}$ such that $\dot{V}(x,t) \leq -\gamma(\|x\|) < 0$.*

Then the equilibrium point $x = 0$ is uniformly asymptotically stable.

Global uniform asymptotic stability: *If $\mathbb{D}$ is replaced by the whole state space $\mathbb{R}^n$, and*

(i)   there exist $\alpha, \beta \in \mathcal{K}$ such that $0 < \alpha(\|x\|) \le V(x,t) \le \beta(\|(x)\|)$;
(ii) there exists a class $\mathcal{K}$ function γ such that $\dot{V}(x,t) \le -\gamma(\|x\|) < 0$;
(iii) $\lim_{\|x\| \to \infty} \alpha(x) = \infty$.

Then the equilibrium point $x = 0$ is globally uniformly asymptotically stable.

Theorem 2.8 *If there exists a function $V(x,t)$ defined on $\|x\| \ge R$ (where R may be large) and $t \in [t_0, \infty)$ with continuous first-order partial derivatives w.r.t. x and t and if there exist $\alpha_1, \alpha_2 \in \mathcal{K}_\infty$ such that:*

(i) $\alpha_1(\|x\|) \le V(x,t) \le \alpha_2(\|x\|)$;
(ii) $\dot{V}(x,t) \le 0$;

for all $\|x\| \ge R$ and $t \in [t_0, \infty)$, then the solutions of (2.12) are uniformly bounded. If in addition there exists $\alpha_3 \in \mathcal{K}$ and

(iii) $\dot{V}(x,t) \le -\alpha_3(\|x\|)$ *for all* $\|x\| \ge R$.

Then the solutions of (2.12) are ultimately uniformly bounded.

Example 2.13 *Consider the system:*

$$\begin{cases} \dot{x}_1 = x_2 + cx_1(x_1^2 + x_2^2), \\ \dot{x}_2 = -x_1 + cx_2(x_1^2 + x_2^2), \end{cases} \tag{2.14}$$

where c is a constant. Note that $x = [x_1, x_2]^\top = [0,0]^\top$ is the only equilibrium point. Let us choose

$$V(x) = x_1^2 + x_2^2$$

as a candidate for the Lyapunov function. $V(x)$ is positive definite, decrescent, and radially unbounded. Its time derivative along the solution of (2.14) is

$$\dot{V} = 2c(x_1^2 + x_2^2)^2.$$

If $c = 0$, then $\dot{V} = 0$, and, therefore, $x = 0$ is uniformly bounded. If $c < 0$, then $\dot{V} = -2|c|(x_1^2 + x_2^2)^2$ is negative definite, and, therefore, $x = 0$ is globally uniformly asymptotically bounded. If $c > 0$, $x = 0$ is unstable (because in this case V is strictly increasing for $\forall t \ge t_0$), and, therefore, the solutions of (2.14) are unbounded.

Example 2.14 *Consider the following system:*

$$\begin{cases} \dot{x}_1(t) = -x_1(t) - \exp(-2t)x_2(t), \\ \dot{x}_2(t) = x_1(t) - x_2(t). \end{cases}$$

Choosing the scalar function as:

$$V(x_1, x_2, t) = x_1^2 + (1 + \exp(-2t))x_2^2.$$

Since there exist class $\mathcal{K}$ functions $x_1^2 + x_2^2$ and $x_1^2 + 2x_2^2$ such that:

$$x_1^2 + x_2^2 \leq V \leq x_1^2 + 2x_2^2.$$

Thus the function $V(x,t)$ is positive definite and has an infinite upper bound. Furthermore, the derivative of $V(x,t)$ is:

$$\dot{V}(x,t) = -2[x_1^2 - x_1 x_2 + x_2^2(1 + 2\exp(-2t))],$$

then we have:

$$\dot{V} \leq -2(x_1^2 - x_1 x_2 + x_2^2) = -(x_1 - x_2)^2 - x_1^2 - x_2^2.$$

Therefore, $\dot{V}$ is negative definite. According to Theorem 2.6, the equilibrium point $x = 0$ is globally uniformly asymptotically stable.

Stability results for non-autonomous systems can be less intuitive than those for autonomous systems, and therefore, particular cases are required in applying the above theorems. For example, let us examine the statement about uniform stability of Theorem 2.6 (or Theorem 2.7) where V decrescent and $\dot{V} \leq 0$ imply $x = 0$ is uniformly stable. If we remove the restriction of V being decrescent in uniform stability, we obtain the stability, i.e., $\dot{V} \leq 0$ implies $x = 0$ is stable but not necessarily uniformly stable. Therefore, one might be tempted to expect that by removing the condition of V being decrescent in the statement of uniformly asymptotically stable, we obtain $x = 0$ is asymptotically stable, i.e., $\dot{V} < 0$ alone implies $x = 0$ is asymptotically stable. This intuitive conclusion is not true, as demonstrated by a counter example in reference [1] where a first-order differential equation and a positive definite, nondecrescent function $V(x,t)$ are used to show that $\dot{V} \leq 0$ does not imply asymptotically stable.

Furthermore, Theorem 2.6 (or Theorem 2.7) of the non-autonomous system (2.12) also holds for the autonomous system (2.2) because it is a special case of (2.12). In the case of (2.2), however, $V(x,t) = V(x)$, i.e., it does not depend explicitly on time t, and all references to the word "decrescent" and "uniform" could be deleted. This is because $V(x)$ is always decrescent and the stability of the equilibrium $x = 0$ of (2.2) implies uniformly stable.

It must be emphasized that in the previous Lyapunov theorems, the existence of Lyapunov functions is always assumed, and the objective is to deduce the stability properties of the systems from the properties of the Lyapunov functions. In view of the common difficulty in finding Lyapunov functions, one may naturally wonder whether Lyapunov functions always exist for stable systems. A number of interesting results concerning the existence of Lyapunov functions, called converse Lyapunov theorems, have been obtained in this regard. For many years, these theorems were thought to be of no practical value because, like the previously described theorems, they do not tell us how to generate Lyapunov functions for a system to be analyzed, but only represent comforting reassurances in the search for Lyapunov functions. Moreover, there are some discussions about the Lyapunov analysis of linear time-varying systems and Lyapunov's indirect method for non-autonomous systems. Interested readers may refer to references [1], [14], and [22].

2.8   BARBALAT'S LEMMA

For autonomous systems, the invariant set theorems are powerful tools to study stability, because they allow asymptotic stability conclusions to be drawn even when $\dot{V}$ is only negative semi-definite. However, the invariant set theorem is not applicable to non-autonomous systems. Therefore, the asymptotic stability analysis of non-autonomous systems is generally much harder than that of autonomous systems, since it is usually very difficult to find Lyapunov functions with a negative definite derivative. An important and simple result which partially remedies this situation is Barbalat's Lemma. Such a lemma is a purely mathematical result concerning the asymptotic properties of functions and their derivatives. When properly used for dynamic systems, particularly non-autonomous systems, it may lead to the satisfactory solution of many asymptotic stability problems.

Before discussing Barbalat's Lemma itself, let us clarify a few points concerning the asymptotic properties of functions and their derivatives. Given a differentiable function f of time t, the following three facts are important to keep in mind.

(1) $\dot{f} \to 0 \nRightarrow f$ converges.

The fact that $\dot{f}(t) \to 0$ does not imply that $f(t)$ has a limit as $t \to \infty$. Geometrically, a diminishing derivative means flatter and flatter slopes. However, it does not necessarily imply that the function approaches a limit. For example, the function $f(t) = \sin(\log(t))$ is bounded but keeps oscillating, while the derivative converges to zero as time goes to infinity, i.e.,

$$\dot{f}(t) = \frac{\cos(\log(t))}{t} \to 0 \text{ as } t \to \infty.$$

(2) f converges $\nRightarrow \dot{f} \to 0$.

The fact that $f(t)$ has a finite limit as $t \to \infty$ does not imply that $\dot{f}(t) \to 0$. For instance, while the function $f(t) = \exp(-t)\sin(\exp(2t))$ tends to zero, its derivative $\dot{f}$ is unbounded, i.e.,

$$\dot{f}(t) = -\exp(-t)\sin(\exp(2t)) + 2\exp(t)\cos(\exp(2t))$$

does not exist.

(3)   If f is lower bounded and decreasing ($\dot{f} \le 0$), then it converges to a limit.

Now, given that a function tends towards a finite limit, what additional requirement can guarantee that its derivative actually converges to zero? Barbalat's Lemma indicates that the derivative itself should have some smoothness. More precisely, we have:

Lemma 2.1 *(Barbalat's Lemma) If the differentiable function $f(t) \in \mathbb{R}$ has a finite limit as $t \to \infty$, and if it is uniformly continuous, then $\lim_{t \to \infty} \dot{f}(t) = 0$.*

Lemma 2.2 *(Barbalat's Lemma) For function $f(t) \in \mathbb{R}$, if it is uniformly continuous for $t \geq t_0$, and $\lim\limits_{t \to \infty} \int_{t_0}^{t} f(\tau)d\tau$ exists and is bounded, then $\lim\limits_{t \to \infty} f(t) = 0$.*

Barbalat's Lemma is a mathematical result about the properties of functions and their derivatives. To apply Barbalat's Lemma to the analysis of dynamic systems, one typically uses the following immediate corollary, which looks very much like an invariant set theorem in Lyapunov analysis.

Lemma 2.3 *(Lyapunov-like Lemma) If a scalar function $V(x,t)$ satisfies the following conditions:*

(*i*) *$V(x,t)$ is lower bounded;*
(*ii*) *$\dot{V}(x,t)$ is negative semi-definite;*
(*iii*) *$\dot{V}(x,t)$ is uniformly continuous in time.*

Then $\lim\limits_{t \to \infty} \dot{V}(x,t) = 0$.

Barbalat's Lemma and Lyapunov-like Lemma are very important conclusions, which provide a theoretical foundation for adaptive control.

2.9   MATHEMATICAL TOOLS

Theorem 2.9 *[23] (Young's inequality) Consider the vectors $x \in \mathbb{R}^n$ and $y \in \mathbb{R}^n$. If the positive constants p and q satisfy $\frac{1}{p} + \frac{1}{q} = 1$, the following inequality holds:*

$$x^\top y \leq \frac{\varepsilon^p}{p} \|x\|^p + \frac{1}{q\varepsilon^q} \|y\|^q$$

for any $\varepsilon > 0$. It is worth noting that if $x \in \mathbb{R}$ and $y \in \mathbb{R}$, Theorem 2.9 can be rewritten as:

$$xy \leq \frac{x^2}{2} + \frac{y^2}{2}, \quad or \quad xy \leq x^2 + \frac{y^2}{4}.$$

Theorem 2.10 *[23] For any vector $x \in \mathbb{R}^n$, the following inequality holds: $\|x+y\| \leq \|x\| + \|y\|$.*

Theorem 2.11 *[24] For any variable $\eta \in \mathbb{R}$ and function $\varepsilon : [0,\infty) \to \mathbb{R}_+$, the following inequality holds:*

$$|\eta| \leq \varepsilon(t) + \frac{\eta^2}{\sqrt{\eta^2 + \varepsilon^2(t)}}.$$

Theorem 2.12 *[25] Consider the following dynamic system:*

$$\dot{\vartheta}(t) = -\zeta\vartheta(t) + w(t),$$

where $\zeta > 0$ is a constant, $w : [0,\infty) \to \mathbb{R}_+$ is a scalar function, then for any initial condition $\vartheta(0) \geq 0$, $\vartheta(t) \geq 0$ holds $\forall t \geq 0$.

Theorem 2.13 *[26] For $\varsigma_1, \varsigma_2 \in \mathbb{R}$, let a and b be positive constants and $\overline{\vartheta}(\varsigma_1, \varsigma_2) > 0$ be a real value function. Then we have:*

$$|\varsigma_1|^a |\varsigma_2|^b \leq \frac{a}{a+b}\overline{\vartheta}|\varsigma_1|^{a+b} + \frac{b}{a+b}\overline{\vartheta}^{-\frac{a}{b}}|\varsigma_2|^{a+b}.$$

Theorem 2.14 *[27] For any $\varsigma_i \in \mathbb{R}$, $i = 1, \cdots, n$, $0 < r \leq 1$, then $\left(\sum_{i=1}^{n}|\varsigma_i|\right)^r \leq \sum_{i=1}^{n}|\varsigma_i|^r.$*

Theorem 2.15 *Let $f, V : [t_0, \infty) \to \mathbb{R}$. Then*

$$\dot{V} \leq -\alpha V + f, \quad \forall t \geq t_0 \geq 0,$$

implies that:

$$V(t) \leq \exp(-\alpha(t - t_0))V(t_0) + \int_{t_0}^{t} \exp(-\alpha(t - \tau))f(\tau)d\tau, \quad \forall t \geq t_0 \geq 0,$$

for any finite constant α.

Proof. Let $w(t) \triangleq \dot{V} + \alpha V - f$. We have $w(t) \leq 0$, and

$$\dot{V} = -\alpha V + f + w$$

implies that:

$$V(t) = \exp(-\alpha(t - t_0))V(t_0) + \int_{t_0}^{t} \exp(-\alpha(t - \tau))f(\tau)d\tau$$
$$+ \int_{t_0}^{t} \exp(-\alpha(t - \tau))w(\tau)d\tau.$$

Because $w(t) \leq 0$, $\forall t \geq t_0 \geq 0$, we have:

$$V(t) \leq \exp(-\alpha(t - t_0))V(t_0) + \int_{t_0}^{t} \exp(-\alpha(t - \tau))f(\tau)d\tau.$$

The proof is completed.

2.10 NOTES

Most materials in this chapter are standard and can be found in many textbooks on nonlinear systems and control, e.g., [1, 9, 14, 15, 21, 22]. The Lyapunov's linearization method and Lyapunov's direct method for autonomous systems are discussed firstly, which are borrowed from references [1] and [22]. The invariant set theorem is introduced for the asymptotic stability of autonomous systems, which can be found in references [1], [21], and [22]. The related definitions of non-autonomous systems and the corresponding Lyapunov theorems are based on references [1, 14, 15]. It is worth noting that the results of non-autonomous systems are quite similar to those for autonomous systems, although more involved conditions are required. The description of Barbalat's Lemma can also be found in references [1] and [22]. In addition, the examples in this chapter can be found in reference [1].

3 Nonlinear Design Methods and Tools

The complexity of nonlinear control presents a challenge that calls for the development of systematic design procedures to achieve control objectives and design specifications. It is evident that a one-size-fits-all approach is not feasible for all nonlinear systems, and designing a nonlinear feedback controller cannot rely solely on one tool. What is essential for a control engineer is a diverse set of analysis and design tools that can address a wide range of scenarios. When tackling a specific application, the engineer must utilize the most suitable tools for the task at hand.

In this chapter, we compile various nonlinear design methods and tools that hold significance and practicality for controlling nonlinear systems in real-world settings. We begin by outlining the primary control issues and qualitative specifications prevalent in the control community, such as stabilization, regulation, and tracking, in Section 3.1. The subsequent Section 3.2 delves into the steps involved in system design. While robust control is effective in dealing with uncertainties commonly found in practical engineering systems, adaptive control offers a more potent approach to enhancing system performance. Thus, we delve into the concept of adaptive control in Section 3.3 and introduce two classical adaptive control methods: model-reference adaptive control and Lyapunov-based adaptive control.

For intricate systems like strict-feedback nonlinear systems where conventional control tools may prove inadequate, we explore the backstepping method in Section 3.4. Recognizing that finite-time convergence is a crucial metric for certain plants, demanding the system state to reach zero within a finite duration, ultimately showcasing superior disturbance rejection and robustness compared to asymptotic results, we introduce various concepts and classifications of finite-time controls in Section 3.5.

3.1 CONTROL PROBLEMS WITH QUALITATIVE SPECIFICATIONS

3.1.1 CONTROL PROBLEMS

Generally, the main problems of nonlinear control systems can be divided into three categories: stabilization, regulation, and tracking.

In this section, we introduce them in the following, respectively.

1. **Stabilization**

For the problem of stabilization, a control system, called a stabilizer, is to be designed so that the system state converges to an equilibrium point, which can be

DOI: 10.1201/9781003474364-3

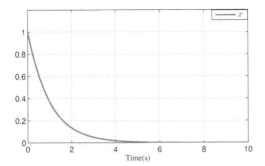

FIGURE 3.1   The schematic diagram of stabilization.

stated mathematically as:

$$\lim_{t \to T} x(t) = 0 \quad \text{or} \quad \lim_{t \to \infty} x(t) = 0,$$

where $x : [t_0, \infty) \to \mathbb{R}^n$ represents the system state and $T > 0$ denotes a positive constant, as shown in Fig. 3.1 with $n = 1$ and $t_0 = 0$.

2.   Regulation

The problem of regulation means that the system state converges to a constant, which can be stated mathematically as:

$$\lim_{t \to T} x(t) = c \quad \text{or} \quad \lim_{t \to \infty} x(t) = c,$$

where c denotes a constant, as shown in Fig. 3.2 with $n = 1$ and $t_0 = 0$. It is easily seen that if $c = 0$, the problem of regulation reduces to stabilization. Therefore, the case of stabilization is a special case of regulation.

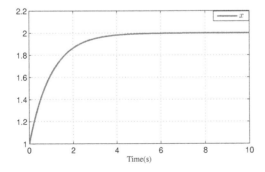

FIGURE 3.2   The schematic diagram of regulation with $c = 2$.

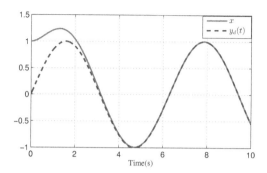

FIGURE 3.3 The schematic diagram of tracking with $y_d(t) = \sin(t)$.

3. **Tracking**

In tracking control problems, the objective is to design a tracking controller so that the system output tracks a given time-varying trajectory, which can be stated mathematically as:

$$\lim_{t \to T} x(t) = y_d(t) \quad \text{or} \quad \lim_{t \to \infty} x(t) = y_d(t),$$

where $y_d : [t_0, \infty) \to \mathbb{R}^n$ represents the time-varying signal, as shown in Fig. 3.3 with $n = 1$ and $t_0 = 0$.

It is not difficult to get the conclusion that: if $y_d(t)$ becomes a non-zero constant, the case of tracking reduces to the case of regulation; if $y_d(t)$ becomes zero, the case of tracking reduces to the case of stabilization, which implies that stabilization and regulation are special cases of tracking.

3.1.2 QUALITATIVE SPECIFICATIONS

In linear control, the desired behavior of a control system can be systematically specified, either in the time-domain or in the frequency-domain. However, systematic specification for nonlinear systems is much less obvious because the response of a nonlinear system to one command does not reflect its response to another command, and furthermore, a frequency-domain description is not possible. As a result, for nonlinear systems, one often looks for some qualitative specifications of the desired behavior to reflect the control performance. Computer simulation is an important complement to analytical tools in determining whether such specifications are met. With respect to the desired behavior of nonlinear control systems, a designer may consider the following characteristics.

(1) *Stability* must be guaranteed for the considered model, either in a local sense or global sense.

FIGURE 3.4 Plant representation.

(2) *Accuracy and speed of response* may be considered for the trajectories in the region of operation. The detailed controller design will be discussed in Chapters 8–13.
(3) *Robustness* is the sensitivity to effects which are not considered in the design, such as disturbance, measurement noise, unmodeled dynamics, etc. The system should be able to withstand these neglected effects when performing the tasks of interest.
(4) *Cost of a control system* is determined mainly by the number and type of actuators, sensors, and computers necessary to implement it. The actuators, sensors and controller complexity (affecting computing requirements) should be chosen consistently and suit the particular application.

3.2 SYSTEM DESIGN STEPS

The design of a controller capable of adjusting or modifying the behavior and response of an unknown plant to meet specific performance criteria can pose a challenging and intricate problem across various control applications. When referring to the plant, we are considering any process characterized by a defined number of inputs u and outputs y, as depicted in Figure 3.4. The inputs u of the plant undergo processing to generate multiple plant outputs y, which represent the observed output response of the plant. The primary objective of control design is to select the input u in a manner that ensures the output response y complies with predefined performance specifications.

Given that the plant process is typically intricate, incorporating diverse mechanical, electronic, hydraulic components, among others, the optimal selection of input u is often not a straightforward task. Consequently, the design of system control necessitates a series of specific steps to navigate this complexity effectively.

1. Modeling

The task of the control engineer in this step is to understand the processing mechanism of the plant, which takes a given input signal $u(t)$ and produces the output response $y(t)$, such that he or she can describe it in the form of some mathematical equations. These equations constitute the mathematical model of the plant. However, the complexity of most physical plants makes the development of such an exact model difficult or even impossible. This makes the task of modeling even more challenging, because the control engineer has to come up with a mathematical model that describes accurately the input/output behavior of the plant and yet is simple enough to be used for control design purposes. A simple model usually leads to a simple controller that is easier to understand and implement, and often more reliable for practical purposes.

A plant model may be developed by using physical laws or by processing the plant input/output data obtained by performing various experiments. Such a model, however, may still be complicated enough from the viewpoint of control design and further simplifications may be necessary. Some of the approaches often used to obtain a simplified model are:

(i)   Linearization around operating points;
(ii) Model order reduction techniques.

In approach (i) the plant is approximated by a linear model that is valid around a given operating point. Different operating points may lead to several different linear models that are used as plant models. Linearization is achieved by using Taylor expansion and approximation.

In approach (ii) small effects and phenomena outside the frequency range of interest are neglected, leading to a lower-order and simpler plant model. The reader is referred to references [28] and [29] for more details on model reduction techniques and approximations.

In general, the task of modeling involves a good understanding of the plant process and performance requirements, and may require some experience from the control engineer. The modeling of nonlinear systems is beyond the scope of this book, so we omit this step in the later chapters and assume that the model structure of nonlinear systems is known.

2. **Controller Design**

Once a model of the plant is available, one can proceed with the controller design. The controller is designed to meet the performance requirements for the plant model. If the model is a good approximation of the plant, then one would hope that the controller performance for the plant model would be close to that achieved when the same controller is applied to the plant. Noting that there are many unknown parameters and uncertain functions in the model of nonlinear systems, effective control tools must be developed to handle the corresponding detailed issues.

3. **Implementation**

In this step, the controller designed in Step 2, which is shown to meet the performance requirements for the plant model and is robust w.r.t. possible model uncertainties, is ready to be applied to the unknown plant. The implementation can be done using a digital computer, even though in some applications analog computers may be used too. Issues, such as the type of computer available, the type of interface devices between the computer and the plant, software tools, etc., need to be considered a priori. Computer speed and accuracy limitations may put constraints on the complexity of the controller that may force the control engineer to go back to Step 2 or even Step 1 to come up with a simpler controller without violating the performance requirements.

Another important aspect of implementation is the final adjustment, or as often called the tuning, of the controller to improve performance by compensating for the plant model uncertainties that are not accounted for during the design process. Tuning is often done by trial and error, and depends very much on the experience and intuition of the control engineer.

In this book, we will focus on introducing the advanced controller design of nonlinear systems, therefore, the steps of modeling and implementation are not elaborated.

3.3 ADAPTIVE CONTROL

With the development of science and technology, many kinds of advanced control schemes have emerged in the control community, for example, adaptive control [9], robust control [30], finite-time control [31], sliding mode control [13], and so on.

Many dynamic systems to be controlled have *constant or slowly-varying uncertain parameters*. For instance, robot manipulators may carry large objects with unknown inertial parameters. Power systems may be subjected to large variations in loading conditions. Fire-fighting aircraft may experience considerable mass changes as they load and unload large quantities of water. Adaptive control is an approach to the control of such systems. The basic idea in adaptive control is to estimate the uncertain plant parameters (or equivalently the corresponding controller parameters) online based on the measured system signals, and use the estimated parameters in the control input computation. An adaptive control system can thus be regarded as a control system with online parameter estimation. Since adaptive control systems, to be developed for linear plants or for nonlinear plants, are inherently nonlinear, their analysis and design are intimately connected with the Lyapunov stability theorems.

Research in adaptive control started in the early 1950s in connection with the design of autopilots for high-performance aircraft which operate at a wide range of speeds and altitudes and thus experience large parameter variations. Adaptive control was proposed as a way of automatically adjusting the controller parameters in the face of changing aircraft dynamics. Subsequently, these theoretical advances, together with the availability of cheap computation, have led to many practical applications such as robotic manipulations [32], aircraft control [33], chemical processes [34], power systems [35], ship steering [36], and bioengineering [37].

It should be emphasized that one of the reasons for the rapid growth and continuing popularity of adaptive control is its clearly defined goal: to control plants with unknown parameters. Adaptive control has been the most successful for plant models in which the unknown parameters appear linearly. Meanwhile, due to that such "linear parametrization" phenomenon widely exists in most practical systems, the development of adaptive control schemes for uncertain systems has achieved more and more attention from researchers and engineers.

In this book, to gain insights into the behavior of the adaptive control systems and also to avoid mathematical difficulties, we shall assume the unknown plant parameters are *constant* in analyzing the adaptive control designs. In practice, adaptive control systems are often used to handle time-varying unknown parameters. In order for the analysis results to be applicable to these practical cases, time-varying plant parameters must vary considerably slower than the parameter adaptation. Fortunately,

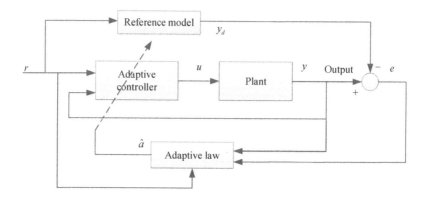

FIGURE 3.5 A model-reference adaptive control system.

this is often satisfied in practice. Note that fast parameter variations may also indicate that the modeling is inadequate, and that the dynamics causing the parameter changes should be additionally modeled.

According to different classification principles, adaptive control can be divided into different approaches, for example, "direct" and "indirect" adaptive controls [9], "Lyapunov-based" and "estimation-based" adaptive controls [9], "model-reference" and "self-tuning" adaptive controls [1, 22]. In this chapter, we mainly focus on introducing some kinds of adaptive controls.

3.3.1   MODEL-REFERENCE ADAPTIVE CONTROL

Generally, a model-reference adaptive control (MRAC) system can be schematically represented by Fig. 3.5. It is composed of four parts: a *plant* containing unknown parameters, a *reference model* for compactly specifying the desired output of the control system, an *adaptive controller* containing adjustable parameters, and an *adaptive law* for updating the adjustable parameters.

The *plant* is assumed to have a known structure, although the parameters are unknown. For linear plants, this means that the number of poles and the number of zeros are assumed to be known, but the locations of these poles and zeros are not known. For nonlinear plants, this implies that the structure of the dynamic equations is known, but some parameters are not.

A *reference model* is used to specify the ideal response of the adaptive control system to the external command. Intuitively, it provides the ideal plant response which the adaptive law should seek in adjusting the parameters. The choice of the reference model is a part of the adaptive control system design. This choice has to satisfy two requirements. On the one hand, it should reflect the performance specification in the control tasks, such as the settling time, overshoot or precision; on the other hand, this ideal behavior should be achievable for the adaptive control system, i.e., there are some inherent constraints on the structure of the reference model (e.g., the structure of the plant model is given).

The *adaptive controller* should have *perfect tracking* capacity in order to allow the possibility of tracking convergence. That is, when the plant parameters are exactly known, the corresponding controller parameters should make the plant output identical to that of the reference model. When the plant parameters are not known, the adaptive law will adjust the controller parameters so that perfect tracking is achieved.

The *adaptive law* is used to adjust the parameters in the adaptive controller. In the MRAC systems, the adaptive law searches for parameters such that the response of the plant under adaptive control becomes the same as that of the reference model, *i.e.*, the objective of the adaptation is to make the tracking error converge to zero. The main issue in adaptation design is to synthesize an adaptive law which will guarantee that the control system remains stable and the tracking error converges to zero. Many formalisms in nonlinear control can be used to this end, such as the Lyapunov theorem and passivity theory. Although the application of one formalism may be more convenient than that of another, the results are often equivalent. In later, we will introduce the Lyapunov theorem.

Example 3.1 *[1] Let us now discuss the adaptive control of the following first-order plants using the MRAC method, as an illustration of how to design and analyze an adaptive control system.*

$$\begin{cases} \dot{x} = -a_p x + b_p u, \\ y = x, \end{cases} \tag{3.1}$$

where $x \in \mathbb{R}$ is the system state, $y = x$ is the plant output, $u \in \mathbb{R}$ is its input, and $a_p \in \mathbb{R}$ and $b_p \neq 0 \in \mathbb{R}$ are constant system parameters.

In the adaptive control problem, the plant parameters a_p and b_p are unknown. Let the desired performance of the adaptive control system be specified by a first-order reference model

$$\dot{y}_d = -a_m y_d + b_m r(t), \tag{3.2}$$

where a_m and b_m are constant parameters, and $r(t)$ is a bounded external reference signal. The parameter a_m is required to be strictly positive so that the reference model is stable, and b_m is chosen strictly positive without loss of generality. The reference model can be represented by its transfer function M, $y_d = Mr$, where $M = \frac{b_m}{p + a_m}$ and p is the Laplace variable. Note that M is a strictly positive real function.

The objective of the adaptive control is to develop a controller and an adaptive law such that the error $x(t) - y_d(t)$ asymptotically converges to zero. In order to accomplish this, we have to assume the sign of the parameter b_p to be known. This is a quite mild condition, which is often satisfied in practice. The design and analysis can be divided into the following steps.

Step 1: Controller design. Choosing the adaptive controller as:

$$u = \hat{a}_r(t)r + \hat{a}_y(t)y, \tag{3.3}$$

where $\hat{a}_r$ and $\hat{a}_y$ are variable feedback gains. With this controller, the closed-loop dynamics is:

$$\dot{y} = -(a_p - \hat{a}_y b_p)y + \hat{a}_r b_p r(t). \tag{3.4}$$

The reason for the choice of controller in (3.3) is clear: it allows the possibility of perfect model matching. If the plant parameters were known, the following values of control parameters

$$a_r^* = \frac{b_m}{b_p}, \quad a_y^* = \frac{a_p - a_m}{b_p}$$

would lead to the closed-loop dynamics:

$$\dot{y} = -a_m y + b_m r,$$

which is identical to the reference model dynamics, and yields zero tracking error.

   Step 2: Adaptive law design. *We choose the adaptive law for the parameters $\hat{a}_r$ and $\hat{a}_y$. Let $e = x - y_d$ be the tracking error. The parameter errors are defined as the differences between the controller parameters provided by the adaptive law and the ideal parameters. i.e.,*

$$\tilde{\mathbf{a}}(t) = \begin{bmatrix} \tilde{a}_r \\ \tilde{a}_y \end{bmatrix} = \begin{bmatrix} \hat{a}_r - a_r^* \\ \hat{a}_y - a_y^* \end{bmatrix}. \tag{3.5}$$

The dynamics of the tracking error can be found by subtracting (3.4) from (3.2),

$$\begin{aligned}\dot{e} &= -a_m (y - y_d) + (a_m - a_p + b_p \hat{a}_y) y + (b_p \hat{a}_r - b_m) r \\ &= -a_m e + b_p (\tilde{a}_r r + \tilde{a}_y y).\end{aligned}$$

This can be conveniently represented as:

$$e = \frac{b_p}{p + a_m} (\tilde{a}_r r + \tilde{a}_y y) = \frac{1}{a_r^*} M (\tilde{a}_r r + \tilde{a}_y y), \tag{3.6}$$

with p denoting the Laplace operator. Therefore, according to Lemma 8.1 in reference [1], the adaptive laws are given as:

$$\begin{cases} \dot{\hat{a}}_r = -\operatorname{sgn}(b_p)\gamma e r, \\ \dot{\hat{a}}_y = -\operatorname{sgn}(b_p)\gamma e y, \end{cases} \tag{3.7}$$

with γ being a positive constant representing the adaptation gain.

   Step 3: Stability analysis of closed-loop systems. *The Lyapunov function candidate is selected as:*

$$V(e, \tilde{a}_r, \tilde{a}_y) = \frac{1}{2} e^2 + \frac{1}{2\gamma} |b_p| (\tilde{a}_r^2 + \tilde{a}_y^2).$$

According to the controller (3.3) and adaptive law (3.7), the derivative of V becomes:

$$\dot{V} = -a_m e^2.$$

Thus, the adaptive control system is globally stable, and the tracking error $e(t)$ converges to zero asymptotically by utilizing the Lyapunov stability theorem and Barbalat's Lemma.

Step 4: Parameter convergence analysis. In order to gain insights about the behavior of adaptive control systems, let us understand the convergence of estimated parameters. Note that the output of the stable filter in (3.6) converges to zero, thus, $\tilde{a}_r r + \tilde{a}_y y$ must converge to zero. From the adaptive law (3.7) and the tracking error convergence to zero as time goes to infinity, then the rate of the parameter estimates converges to zero. Therefore, when time t is large, $\tilde{\mathbf{a}}$ is almost constant, and

$$r(t)\tilde{a}_r + y(t)\tilde{a}_y = 0,$$

or

$$\mathbf{v}^\top(t)\tilde{\mathbf{a}} = 0, \tag{3.8}$$

with $\mathbf{v} = [r,y]^\top$ and $\mathbf{a} = [a_r,a_y]^\top$. The issue of parameter convergence is reduced to the question of what conditions the vector $[r(t),y(t)]^\top$ should satisfy the equation to have a unique zero solution.

If $r(t)$ is a constant r_0 $(r_0 \neq 0)$, then for large t, $y(t) = y_m = \alpha r_0$ with α being the d.c. gain of the reference model. Thus, $[r,y] = [1,\alpha]r_0$. Equation (3.8) becomes $\tilde{a}_r + \alpha\tilde{a}_y = 0$, which implies that the estimated parameters, instead of converging to zero, converge to a straight line in parameter space.

However, when $r(t)$ is such that the corresponding signal vector $\mathbf{v}(t)$ satisfies the so-called "persistent excitation" (PE) condition, we can show that (3.7) will guarantee parameter convergence. By persistent excitation of $\mathbf{v}$, we mean that there exist strictly positive constants α_1 and T such that for any t > 0,

$$\int_t^{t+T} \mathbf{v}\mathbf{v}^\top\, dr \geq \alpha_1 \mathbf{I}. \tag{3.9}$$

To show parameter convergence, we note that multiplying (3.8) by $\mathbf{v}(t)$ and integrating the equation for a period of time T, leads to

$$\int_t^{t+T} \mathbf{v}\mathbf{v}^\top\, dr\, \tilde{\mathbf{a}} = 0.$$

(3.9) implies that the only solution of this equation is $\tilde{\mathbf{a}} = 0$, i.e., the parameter error converges to zero.

It is seen from the above analysis that the proposed MRAC method not only ensures that the tracking error converges to zero as time goes to infinity, but also guarantees that all signals in the closed-loop systems are bounded. However, it is also noted that the convergence of parameter estimates depends on the PE condition. As the PE condition is a quite conservative condition and is also difficult (even impossible) to be ensured for the adaptive control of uncertain systems (especially for the uncertain nonlinear systems), then the convergence of parameter estimates is not the focus of adaptive control. Therefore, we will not analyze the convergence of parameter estimates in this book. Interested readers may refer to references [1], [9], and [22].

3.3.2   LYAPUNOV-BASED ADAPTIVE CONTROL

In this section, we will focus on describing the idea of Lyapunov-based adaptive control without requiring the reference model.

To clearly show the differences between the MRAC method and the Lyapunov-based adaptive control, we first consider the following sample scalar plant:

$$\begin{cases} \dot{x} = u + ax, \\ y = x, \end{cases}$$

where $x \in \mathbb{R}$ denotes the system state, $y = x$ is the system output, $a \in \mathbb{R}$ represents the unknown system parameter, and $u \in \mathbb{R}$ is the control input. As a is unknown for control design, we attempt to employ the estimate of a, ($\hat{a}$), instead of a itself in the control design. Here the objective is to develop an adaptive control scheme such that the tracking error $e = y - y_d$ converges to zero as time goes to infinity by using the Lyapunov method directly, where $y_d \in \mathbb{R}$ is the reference signal.

The key point is to seek a parameter update law for the estimate $\hat{a}(t)$ (i.e., $\dot{\hat{a}} = \tau(x, y_d, \hat{a})$) and an adaptive controller $u = \beta(x, y_d, \dot{y}_d, \hat{a})$ such that the Lyapunov function candidate

$$V(x, y_d, \hat{a}) = \frac{1}{2}e^2 + \frac{1}{2}(a - \hat{a})^2$$

is non-increasing w.r.t. time, i.e.,

$$V(e(t), \hat{a}(t)) \leq V(e(t_0), \hat{a}(t_0)), \quad \forall t \geq t_0 \geq 0.$$

To this end, we express $\dot{V}$ as a function of u and $\dot{\hat{a}}$, and seek $\beta(x, y_d, \dot{y}_d, \hat{a})$ and $\tau(x, y_d, \hat{a})$ to guarantee that $\dot{V} \leq -ce^2$ with $c > 0$, namely,

$$\dot{V} = e(u + ax - \dot{y}_d) - (a - \hat{a})\dot{\hat{a}} \leq -ce^2,$$

which can be rewritten as:

$$eu - e\dot{y}_d + \hat{a}\dot{\hat{a}} + a(ex - \dot{\hat{a}}) \leq -ce^2.$$

As the parameter a is not allowed to be included in u and $\dot{\hat{a}}$, then the adaptive law can be chosen as:

$$\dot{\hat{a}} = ex, \tag{3.10}$$

which leads to

$$eu - e\dot{y}_d + \hat{a}\dot{\hat{a}} \leq -ce^2.$$

Therefore, the choice of such form of adaptive law results in the following adaptive controller:

$$u = -ce + \dot{y}_d - \hat{a}x. \tag{3.11}$$

Recall the expressions of the adaptive controller (3.11) and adaptive law (3.10), the closed-loop adaptive system is:

$$\begin{cases} \dot{e} = -ce + (a - \hat{a})x, \\ \dot{\hat{a}} = ex. \end{cases}$$

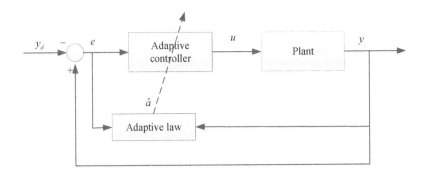

FIGURE 3.6 Lyapunov-based adaptive system.

According to the standard Lyapunov stability method, it is not difficult to conclude that not only all signals in the closed-loop system are bounded, but also the tracking error converges to zero as $t \to \infty$. You may read the detailed analysis in Chapter 4.

The above design procedure that the Lyapunov-based adaptive control system can be schematically represented by Fig. 3.6. Compared with the MRAC method, it is easily to see that the reference model in MRAC is not required, and the controller design and adaptive law design are easy to understand under the framework of Lyapunov theorem.

Furthermore, an adaptive controller is formed by combining an online parameter estimator with a controller. According to different estimates, adaptive control can be divided into two approaches. The first approach is referred to as indirect adaptive control, in which the plant parameters are estimated online and used to calculate the controller parameters. Obviously, the Lyapunov-based adaptive control in this section belongs to the indirect adaptive control. In the second adaptive approach, referred to as direct adaptive control, the controller parameters are estimated directly without intermediate calculations and without involving plant parameter estimates. The detailed discussion and other classification can be found in references [9] and [22].

Note that robust control can also be used to deal with parameter uncertainty. Thus, one may naturally wonder about the differences and relations between the robust approach and the adaptive approach. In principle, adaptive control is superior to robust control in dealing with uncertainties in *constant or slowly-varying parameters*. The basic reason lies in the learning behavior of adaptive control systems: an adaptive controller improves its performance as adaptation goes on, while a robust controller simply attempts to keep consistent performance. Another reason is that an adaptive controller requires little or no a *priori* information about the unknown parameters, while a robust controller usually requires reasonable a *priori* estimates of the parameter bounds. Conversely, robust control has some desirable features which adaptive control does not have, such as its ability to deal with disturbances, quickly varying parameters, and unmodeled dynamics. Such features actually may be combined with adaptive control, leading to robust adaptive controllers in which uncertainties on constant or slowly-varying parameters are reduced by parameter adaptation and other sources of uncertainty are handled by robustification techniques. It is also important to point out that existing adaptive techniques for nonlinear systems generally require a linear parametrization of the plant dynamics, i.e., that the parametric uncertainty is expressed linearly by a set of unknown parameters. In some cases, linear parametrization cannot be achieved, thus adaptive control is not effective any more, but robust control (or adaptive control with robust terms) may be possible, which will be introduced in Chapter 4.

3.4 BACKSTEPPING METHOD

Based on the Lyapunov stability theorem, Kanellakopoulos et al. proposed a backstepping method in the 1990s [38], in which by introducing a coordinate transformation, the control of considered nonlinear systems can be converted into a series of recursive design of first-order systems. Take the strict-feedback system as an example, the specific design process can be summarized as follows.

Step 1. Designing the virtual controller of the first subsystem to ensure the stability of the corresponding subsystem.

Step 2. Choosing the Lyapunov function candidate and utilizing the Lyapunov stability theorem to design the virtual controller in the second subsystem so that the stability of the first and the second subsystem is guaranteed.

Step i $(i = 3, 4, \cdots, n)$. Employing the same processes in Step 1 and Step 2, with the aid of the Lyapunov stability theorem, we are able to design the following virtual controllers from Step 3 to Step $n - 1$ and the actual controller to ensure the stability of the whole closed-loop system.

In order to grasp the design idea of backstepping clearly, we employ the following third-order strict-feedback nonlinear system as an example to introduce the design procedure and to make a detailed stability analysis:

$$\begin{cases} \dot{x}_k = f_k(\bar{x}_k) + x_{k+1}, \quad k = 1, 2, \\ \dot{x}_3 = f_3(\bar{x}_3) + u, \\ y = x_1, \end{cases} \tag{3.12}$$

where $x_1 \in \mathbb{R}$, $x_2 \in \mathbb{R}$, $x_3 \in \mathbb{R}$ are the system states and $\bar{x}_k = [x_1, \cdots, x_k]^\top \in \mathbb{R}^k$ ($k = 1, 2, 3$), $f_k \in \mathbb{R}$ is a known smooth nonlinear function, $u \in \mathbb{R}$ and $y \in \mathbb{R}$ represent the control input and system output, respectively.

For the third-order strict-feedback nonlinear system (3.12), the controller design requires three steps. The virtual controllers are designed in Step 1 and Step 2, and the actual controller u is given in Step 3.

Now we carry out the control design.

Step 1. Define the following coordinate transformation:

$$z_1 = x_1,$$

then the derivation of z_1 is:

$$\dot{z}_1 = \dot{x}_1 = f_1(\bar{x}_1) + x_2. \tag{3.13}$$

To establish a relationship between the first subsystem and the second subsystem, the following coordinate transformation is given:

$$z_2 = x_2 - \alpha_1,$$

in which z_2 is the virtual error and α_1 is the virtual controller.

Upon using the definition of z_2, we have:

$$x_2 = z_2 + \alpha_1. \tag{3.14}$$

Substituting (3.14) into (3.13), it obtains:

$$\dot{z}_1 = f_1(\bar{x}_1) + z_2 + \alpha_1, \tag{3.15}$$

then the derivative of quadratic function $V_1 = \frac{1}{2}z_1^2$ is:

$$\dot{V}_1 = z_1 \dot{z}_1 = z_1 (f_1 + z_2 + \alpha_1). \tag{3.16}$$

The virtual controller α_1 is designed as:

$$\alpha_1 = -c_1 z_1 - f_1, \tag{3.17}$$

with $c_1 > 0$.

Substituting the expression of the virtual controller as given in (3.17) into (3.15) and (3.16), one obtains:

$$\begin{cases} \dot{z}_1 = -c_1 z_1 + z_2, \\ \dot{V}_1 = -c_1 z_1^2 + z_1 z_2, \end{cases} \tag{3.18}$$

where the second term of (3.18) on the right-hand side, $z_1 z_2$, will be handled in the next step.

Step 2. The derivation of the virtual error z_2 is:

$$\dot{z}_2 = \dot{x}_2 - \dot{\alpha}_1 = f_2 + x_3 - \dot{\alpha}_1.$$

Since the virtual controller α_1 is the function of variable x_1, then the derivative of α_1 is:

$$\dot{\alpha}_1 = \frac{\partial \alpha_1}{\partial x_1} \dot{x}_1 = \frac{\partial \alpha_1}{\partial x_1} (f_1 + x_2),$$

then we have:

$$\dot{z}_2 = f_2 + x_3 - \frac{\partial \alpha_1}{\partial x_1} (f_1 + x_2). \tag{3.19}$$

Let

$$z_3 = x_3 - \alpha_2,$$

then (3.19) can be further rewritten as:

$$\dot{z}_2 = f_2 + z_3 + \alpha_2 - \frac{\partial \alpha_1}{\partial x_1} (f_1 + x_2). \tag{3.20}$$

The virtual controller α_2 is designed as:

$$\alpha_2 = -c_2 z_2 - f_2 + \frac{\partial \alpha_1}{\partial x_1}(f_1 + x_2) - z_1, \tag{3.21}$$

with $c_2 > 0$.

Substituting the expression of α_2 as defined in (3.21) into (3.20), one has:

$$\dot{z}_2 = -c_2 z_2 + z_3 - z_1. \tag{3.22}$$

Choosing the quadratic function as $V_2 = \frac{1}{2} \sum\limits_{k=1}^{2} z_k^2$, together with (3.18) and (3.22), the derivative of V_2 is:

$$\dot{V}_2 = -\sum_{k=1}^{2} c_k z_k^2 + z_2 z_3, \tag{3.23}$$

in which the term $z_2 z_3$ will be coped with in the next step.

Step 3. The actual controller u is given in this step. The derivative of the virtual error z_3 is:

$$\dot{z}_3 = \dot{x}_3 - \dot{\alpha}_2 = f_3 + u - \dot{\alpha}_2. \tag{3.24}$$

The virtual controller α_2 is the function of variables x_1, x_2 and virtual errors z_1, z_2. Noting that the virtual error z_2 can be expressed by states x_1, x_2 and α_1, then it is seen that the virtual controller α_2 is the function of states x_1 and x_2, therefore its derivation is:

$$\dot{\alpha}_2 = \frac{\partial \alpha_2}{\partial x_1}\dot{x}_1 + \frac{\partial \alpha_2}{\partial x_2}\dot{x}_2 = \frac{\partial \alpha_2}{\partial x_1}(f_1 + x_2) + \frac{\partial \alpha_2}{\partial x_2}(f_2 + x_3),$$

then (3.24) becomes:

$$\dot{z}_3 = f_3 + u - \sum_{k=1}^{2} \frac{\partial \alpha_2}{\partial x_k}(f_k + x_{k+1}). \tag{3.25}$$

The actual controller u is designed as:

$$u = -c_3 z_3 - f_3 + \sum_{k=1}^{2} \frac{\partial \alpha_2}{\partial x_k}(f_k + x_{k+1}) - z_2, \tag{3.26}$$

with $c_3 > 0$.

Involving (3.26) into (3.25), one has:

$$\dot{z}_3 = -c_3 z_3 - z_2. \tag{3.27}$$

The Lyapunov function candidate is selected as:

$$V = \frac{1}{2} \sum_{k=1}^{3} z_k^2.$$

With the aid of (3.23) and (3.27), its derivative is:

$$\dot{V} = -\sum_{k=1}^{3} c_k z_k^2 \leq 0. \tag{3.28}$$

Now we carry out the stability analysis of the closed-loop systems (3.18), (3.22), and (3.27).

Firstly, we prove the boundedness of all signals in the closed-loop system. According to (3.28), it is not difficult to get that:

$$V(t) - V(0) = -\sum_{k=1}^{3} \int_0^t c_k z_k^2(\tau) d\tau \leq 0, \tag{3.29}$$

or

$$V(t) + \sum_{k=1}^{3} \int_0^t c_k z_k^2(\tau) d\tau = V(0),$$

where $V(0)$ denotes the initial value of the Lyapunov function candidate (the initial time t_0 is set as $t_0 = 0$). According to (3.29), we have $V(t) \in \mathcal{L}_{\infty}$ and $z_k \in \mathcal{L}_2$, then it follows that $z_k \in \mathcal{L}_{\infty}$, $(k = 1, 2, 3)$. As $z_1 = x_1$ and $f_1(x_1)$ is smooth, f_1 is bounded, which further implies that the virtual control α_1 and the partial derivative $\frac{\partial \alpha_1}{\partial x_1}$ are bounded. Upon using the definition of z_2, it is ensured that $x_2 \in \mathcal{L}_{\infty}$. As f_2 is smooth and the states x_1, x_2 are bounded, one has $f_2 \in \mathcal{L}_{\infty}$, it is not difficult to get the boundedness of α_2. By using the similar analysis process, it is able to get the conclusion that $x_3 \in \mathcal{L}_{\infty}$, $f_3 \in \mathcal{L}_{\infty}$, $u \in \mathcal{L}_{\infty}$, and $\dot{z}_k \in \mathcal{L}_{\infty}$.

Secondly, we prove that $\lim_{t \to \infty} x_k(t) = 0$, $k = 1, 2, 3$. As $z_k \in \mathcal{L}_{\infty} \cap \mathcal{L}_2$ and $\dot{z}_k \in \mathcal{L}_{\infty}$, by utilizing Barbalat's Lemma, it is straightforward to show that $\lim_{t \to \infty} z_k(t) = 0$.

3.5   FINITE-/PRESCRIBED-TIME CONTROL

In most of the existing control schemes, the convergence rate is asymptotic (or is at best exponential) with infinite settling time [1, 9, 22]. That is, it needs infinite time to achieve stabilization, regulation, or tracking. The control results corresponding to the nonfinite time convergence include the asymptotic stability [9], exponential stability [39, 40], and UUB [41, 42].

Finite-time control means that the system state (or tracking error) converges to zero within a finite time under a finite-time control algorithm. Compared with nonfinite-time control, finite-time control not only ensures that the closed-loop system has a faster convergence rate and higher convergence accuracy, but also guarantees that the corresponding control has stronger disturbance rejection and better robustness when there are external disturbances [43]. Therefore, the study of finite-time control has obtained more and more attention in the control community.

Weiss and Infante [44] gave the concept of finite-time stability in 1965 and extended the finite-time stability theorem to the non-autonomous systems under the influence of external forces [45], which were the embryo of the finite-time stability theorem. However, these results obtained in references [44] and [45] are only about the analysis of finite-time stability, there is no analysis on how to design the finite-time controller. Generally speaking, according to the continuity of the control input signals, the finite-time control methods in existing literature are classified into three categories:

1. discontinuous finite-time control;
2. continuous but non-smooth finite-time control;
3. prescribed-time control.

We will briefly introduce the above finite-time control schemes one by one.

3.5.1 DISCONTINUOUS FINITE-TIME CONTROL

Generally speaking, the discontinuous finite-time controls mainly include

(1) signum function feedback-based control;
(2) terminal sliding mode (TSM)-based control.

Signum function feedback-based control. The signum function feedback-based control is a class of discontinuous controls based on the signs of system states. Take the simple first-order linear system as an example:

$$\dot{x} = u,$$

where $x \in \mathbb{R}$ and $u \in \mathbb{R}$ denote the system state and control input, respectively.

The signum function feedback-based control is of the basic form:

$$u = -c \cdot \text{sgn}(x),$$

where $c > 0$ is a freely chosen constant parameter. As this finite-time control is based on the signum function, then it is obvious that the control action is discontinuous. Under such a discontinuous control, the system states converge to the origin in a finite time T, satisfying $T \leq \frac{|x(0)|}{c}$. The signum function feedback-based control is able to ensure the finite-time stability of systems and also possess better disturbance rejection, however, the discontinuous terms in the control input might cause the notorious chattering phenomenon during system operation.

TSM-based control. Sliding mode control is a special kind of variable structure control proposed by Emelyanov and Taranm in reference [46]. Sliding mode control is known to drive the system states to the sliding surface in finite time, and further, it is indicated that the sliding mode control has strong robustness and thus provides an effective control method for the nonlinear control systems. Yu and Man introduced the TSM control method in reference [47], in which the TSM was defined by the first-order dynamics:

$$\dot{x} + \beta x^{\frac{q}{p}} = 0, \tag{3.30}$$

where $x \in \mathbb{R}$ is a scalar variable, $\beta > 0$, and $p > q > 0$ are odd integers such that for any real number x, $x^{q/p}$ returns a real number. For (3.30), the trajectory of x converges to an equilibrium state in a finite time T, satisfying:

$$T = \frac{p}{\beta(p-q)}|x(0)|^{\frac{p-q}{p}}.$$

Furthermore, the dynamics (3.30) can be extended to higher dimensional TSM case by introducing the following hierarchical sliding mode structure:

$$\begin{cases} s_1 = \dot{s}_0 + \beta_1 s_0^{q_1/p_1}, \\ s_2 = \dot{s}_1 + \beta_2 s_1^{q_2/p_2}, \\ \quad\vdots \\ s_{n-1} = \dot{s}_{n-2} + \beta_{n-1} s_{n-2}^{q_{n-1}/p_{n-1}}, \end{cases} \tag{3.31}$$

where $s_0 = x$, $\beta_i > 0$, and $p_i > q_i$ are positive odd integers, and the finite convergence time T^s for system (3.31) satisfies:

$$T^s = t_1^s + \sum_{i=1}^{n-1} \frac{p_{n-i}}{\beta_{n-i}(p_{n-i} - q_{n-i})}|s_{n-i-1}(t_i^s)|^{(p_{n-i}-q_{n-i})/p_{n-i}},$$

where t_i^s is the reaching time of the TSM $s_{n-i} = 0$ for $i = 1, \cdots, n-1$. The TSM-based control can ensure the finite-time convergence of systems and is robust to input disturbances [48]. However, this method may cause a singularity problem around the equilibrium.

The above-mentioned two methods are discontinuous, and the discontinuous control terms would lead to chattering in control systems, which is not desired in the actual control. Thus, it is highly desired to develop continuous finite-time control methods for systems to avoid chattering.

3.5.2   CONTINUOUS BUT NON-SMOOTH FINITE-TIME CONTROL

Most of the existing works on continuous finite-time control are based on the theory of finite-time stability proposed in references [31], [43], and [49], which can mainly be divided into the following categories:

(1)   homogeneous finite-time control;
(2)   fractional power state feedback-based control;
(3)   adding power integrator-based control.

Homogeneous finite-time control. The finite-time stability of homogeneous systems was proposed by Bhat in reference [50]. It is stated in reference [50] that a homogeneous system is finite-time stable if and only if it is asymptotically stable and has a negative degree of homogeneity. Hong et al. [51, 52] further deepened this theory and applied it to finite-time control of homogeneous systems. This approach is mainly based on the properties of homogeneous systems. Subsequently, by merging the theories of semi-stability and finite-time stability, Hui et al. [53] developed a rigorous framework for finite-time semi-stability, in which sufficient and necessary conditions for finite-time semi-stability of homogeneous systems were addressed by exploiting the fact that a homogeneous system is finite-time semi-stable if and only if it is semi-stable and has a negative degree of homogeneity. Upon using the finite-time control technique and homogeneous system theory, there are many meaningful results published in recent years [54, 55, 56, 57, 58, 59].

Fractional power state feedback-based control. Fractional power state feedback-based control was proposed by Haimo in reference [31], in which a Lyapunov-based finite-time stability theorem was developed. Compared with the regular state feedback-based controls, the key idea of the fractional power state feedback-based control algorithm is to make the index of the feedback state smaller than 1. It is such a feature that offers numerous benefits including faster convergence rate, better disturbance rejection and robustness against uncertainties [60, 61, 62, 63, 64].

Adding power integrator-based control. Adding a power integrator, as a feedback design tool was first introduced by Caron in reference [65] for the stabilization problem, and then was applied extensively by Lin and Qian [66, 67, 68] to solve the problem of stabilization for uncertain nonlinear systems with unmatched uncertainties. Due to the introduction of the power integrator, the corresponding finite-time control has more advantages compared with some other finite-time controls, for example:

(1) Compared with the TSM-based control, the controller designed by adding a power integrator not only has no singular problems but also has the complete expression of the upper bound of the settling time.
(2) It is more suitable to cope with the problem of finite-time control for high-order nonlinear systems with or without unmatched uncertainties (such as strict-feedback nonlinear systems).
(3) It has more excellent results than homogeneous methods and TSM-based methods, such as the control performance for systems with disturbances, and the calculation on the upper bound of convergence time.

However, since a large number of inequalities are utilized in the stability analysis, the finite-time stability of the closed-loop system can be guaranteed if and only if the design parameters are large enough, which implies that there is an extra constraint on the design parameters, making it less applicable than homogeneous and TSM-based controls in practical systems. Therefore, how to reduce the conservativeness of parameter selection represents a difficult yet challenging work.

3.5.3 PRESCRIBED-TIME CONTROL

Prescribed-time control, a time-varying gain-based regular state feedback control, was introduced by Song et al. in references [18] and [69] for the robust regulation of normal-form nonlinear systems in prescribed time, and then was applied extensively in reference [70] to solve the prescribed time leader-following control problem of high-order linear multi-agent systems under directed graphs. Such a control method is based on the time-varying control gain together with the regular state feedback (rather than the fractional power state feedback or signum function feedback), and thus the control input signal is not only continuous but also $\mathscr{C}^1$ smooth.

Mathematically, the prescribed-time stabilization is attained via control gains that go to infinity as the time t approaches the prescribed terminal time, while however ensuring that the state and the control input remain bounded (and actually converge to zero). Particularly, the convergence time is not only finite but can be pre-specified by the designer and is independent of other design parameters and initial conditions. Furthermore, it is worth noting that the well-known proportional navigation law in tactical and strategic missile guidance provides the original idea for prescribed-time control with early theoretical results appearing in optimal control over finite-time intervals [71, 72, 73, 74].

Now we take the following simple first-order system as an example:

$$\dot{x} = u, \quad x(0) = x_0, \tag{3.32}$$

where $x \in \mathbb{R}$ and $u \in \mathbb{R}$ denote the system state and control input, respectively, x_0 is the initial value of the system state.

The time-varying gain-based regular state feedback control is designed as:

$$\begin{cases} u = -k(t)x, \\ k(t) = \dfrac{1}{t_f - t}, \end{cases} \tag{3.33}$$

where $t_f \in \mathbb{R} > 0$ denotes the settling time and here the initial time is $t_0 = 0$.

Substituting the prescribed-time controller (3.33) into (3.32) and solving such a differential equation, we have:

$$x(t) = x_0 \cdot (t_f - t). \tag{3.34}$$

From (3.34), it is seen that $x(t) = 0$ at the finite time $t = t_f$, which implies that the system state converges to zero at the prescribed time t_f.

Actually, except for finite-time control and prescribed-time control, another control algorithm has the ability to make the system state converge to zero within a finite time, which is called fixed-time control [75]. It is worth mentioning that most of the algorithms for finite-time control introduced above can be employed to design the corresponding fixed-time controllers. But, compared with finite-time control, the most significant advantage of fixed-time control is that the upper bound of settling/convergence time is independent of initial conditions. Nonetheless, prescribed-time control is more flexible as the settling time in fixed-time control still cannot be preassigned arbitrarily within any physically possible range as the upper bound is subject to certain restrictions. We will clarify the main differences among finite-time control, fixed-time control, and prescribed-time control in Chapters 10-13.

3.6 NOTES

The introduction of control design steps can be found in references [1] and [22]. The writing of adaptive control and backstepping is influenced by the clear presentation in reference [9]. Interested readers may also consult the classical books [1, 9, 22], which contain many meaningful results. Furthermore, there is more detailed discussion about finite-time control in reference [76].

Section II

Control of Nonlinear Systems

4 Control of First-order Systems

In the practical engineering systems, many systems have nonlinear characteristics. Although the Lyapunov's linearization method in Chapter 2 is able to handle the weak nonlinearity, it may fail if the system nonlinearity is strong, so we need to design more advanced control algorithms for nonlinear systems. In this chapter, we mainly introduce some classical control design methods for first-order nonlinear systems.

Consider the following first-order nonlinear systems:

$$\dot{x}(t) = f(x, \theta) + g(x) u(t), \tag{4.1}$$

where $x \in \mathbb{R}$ denotes the system state, $u \in \mathbb{R}$ is the control input, $f(x, \theta) \in \mathbb{R}$ is a nonlinear smooth function, $g(x) \in \mathbb{R}$ represents the control coefficient (or control gain). To guarantee the considered system is controllable, the control coefficient cannot be chosen as zero, i.e., $g(x) \neq 0$. Otherwise, the system is not controllable anymore. For convenience, the arguments of the function are omitted if no confusion is caused in the following. For example, system (4.1) is abbreviated as:

$$\dot{x} = f + gu. \tag{4.2}$$

In Section 4.1, we introduce the model-dependent controller design and stability analysis for first-order nonlinear systems. Noting that in most practical engineering systems, the system parameter is normally unknown and is unavailable for control design, the corresponding model-dependent control scheme in Section 4.1 is not applicable anymore; then we present an adaptive control method for parametric nonlinear systems in Section 4.2. However, if the uncertainties in nonlinear systems don't satisfy the parameter decomposition condition, how to design effective control methods to handle the non-parametric uncertainties requires further studies. In Section 4.3, we elaborate on the core function-based method and neural network (NN)-based method to deal with such an uncertainty, respectively, and then we introduce some classical control schemes (such as robust control, robust adaptive control). Furthermore, if the control direction of nonlinear systems is fixed but unknown, the traditional methods will become invalid. Therefore, in Section 4.4, we utilize the Nussbaum function to develop an adaptive control algorithm, which provides a perfect solution for handling the problem of unknown directions in nonlinear systems.

DOI: 10.1201/9781003474364-4

4.1 MODEL-DEPENDENT CONTROL

4.1.1 MODEL-DEPENDENT STABILIZATION CONTROL

In this section, we consider the case that the system model (i.e., (4.1)) is completely available for control design, i.e., the nonlinear function f and the control coefficient g are known. To guarantee the stabilization, we impose the following assumptions.

Assumption 4.1 *The system state x is measurable.*

Assumption 4.2 *The functions f and g are smooth and available for control design. In addition, there exist some constants $\underline{g}$ and $\bar{g}$ such that $0 < \underline{g} \le g(\cdot) \le \bar{g} < \infty$.*

Combining with (4.2), the model-dependent stabilization controller is:

$$u = \frac{1}{g}(-cx - f), \tag{4.3}$$

with $c > 0$.

Theorem 4.1 *Consider the first-order nonlinear system (4.2) under Assumptions 4.1 and 4.2, the proposed stabilization controller (4.3) not only ensures that the system state converges to zero exponentially, but also guarantees the boundedness of all signals in the closed-loop system.*

Proof. Substituting the controller as given in (4.3) into (4.2), we have:

$$\dot{x} = -cx.$$

Solving the above differential equation, one has:

$$x(t) = \exp(-ct)x(0), \tag{4.4}$$

where $\exp(\cdot)$ denotes an exponential function. According to (4.4), it is seen that for any bounded initial state $x(0)$, the system state is not only bounded but also converges to zero exponentially. With the aid of Assumption 4.2, it follows that $f \in \mathcal{L}_\infty$, then it is shown that $u \in \mathcal{L}_\infty$ and $\dot{x} \in \mathcal{L}_\infty$. The proof is completed.

4.1.2 MODEL-DEPENDENT TRACKING CONTROL

In this section, we consider the tracking control of the nonlinear system (4.2). To this end, except for Assumptions 4.1 and 4.2, we impose the following assumption.

Assumption 4.3 *The reference trajectory $y_d(t)$ and its derivative $\dot{y}_d(t)$ are bounded, known, and piecewise continuous.*

Let

$$e = x - y_d$$

be the tracking error, the derivative of e along (4.2) yields:

$$\dot{e} = \dot{x} - \dot{y}_d = f + gu - \dot{y}_d. \qquad (4.5)$$

Designing the model-dependent tracking controller as:

$$u = \frac{1}{g}(-ce - f + \dot{y}_d), \qquad (4.6)$$

with $c > 0$. By utilizing the controller (4.6), we develop the following theorem.

Theorem 4.2 *Consider the first-order nonlinear system (4.2) under Assumptions 4.1-4.3, the proposed controller (4.6) not only ensures that the tracking error converges to zero exponentially, but also guarantees the boundedness of all signals in the closed-loop system.*

Proof. Substituting the controller as given in (4.6) into (4.5), one obtains:

$$\dot{e} = -ce \quad \text{and} \quad e(t) = \exp(-ct)e(0). \qquad (4.7)$$

Then it is shown from (4.7) that the tracking error converges to zero as time goes to infinity. Noting that the reference signal is bounded, then it ensures that the system state x is also bounded. With the aid of Assumption 4.2, the smooth function f is bounded, which indicates that $u \in \mathcal{L}_\infty$ and $\dot{e} \in \mathcal{L}_\infty$, i.e., all signals in the closed-loop system are bounded. The proof is completed.

4.2 ADAPTIVE CONTROL OF PARAMETRIC SYSTEMS

In this section, we consider the tracking control of nonlinear systems with parametric uncertainty. The control objective is to design an adaptive controller for the nonlinear system (4.1) such that:

(*i*)   all signals in the closed-loop system are bounded;
(*ii*) the tracking error converges to zero asymptotically.

To this end, except for Assumptions 4.1 and 4.3, we impose the following assumptions.

Assumption 4.4 *The uncertain function $f(x, \theta)$ satisfies the following parameter decomposition condition, i.e., $f(x, \theta) = \theta^\top \varphi(x)$, with $\theta \in \mathbb{R}^n$ being an unknown constant and $\varphi(x) \in \mathbb{R}^n$ being an available smooth function vector.*

Assumption 4.5 *The control coefficient $g(x)$ is known and smooth, and there exist some constants $\underline{g}$ and $\overline{g}$ such that $0 < \underline{g} \le g(x) \le \overline{g} < \infty$.*

4.2.1 CONTROL DESIGN

Upon using Assumption 4.4, (4.5) can be rewritten as:

$$\dot{e} = \dot{x} - \dot{y}_d = \theta^\top \varphi + gu - \dot{y}_d. \tag{4.8}$$

If the system model and the system parameter are available for control design, the controller (similar to that in Section 4.1.2) is designed as:

$$u = \frac{1}{g}\left(-ce - \theta^\top \varphi + \dot{y}_d\right). \tag{4.9}$$

However, as the system parameter θ is unknown, the above controller (4.9) is no longer applicable for nonlinear system (4.2). Noting that adaptive control has the ability to estimate unknown system parameters, i.e., we are able to use the estimate $\hat{\theta}$ to replace the parameter θ in (4.9), the adaptive controller with the adaptive law is designed as:

$$\begin{cases} u = \frac{1}{g}\left(-ce - \hat{\theta}^\top \varphi + \dot{y}_d\right), \\ \dot{\hat{\theta}} = \Gamma \varphi e, \end{cases} \tag{4.10}$$

where $c > 0$ is a design parameter, $\Gamma = \Gamma^\top \in \mathbb{R}^{n \times n}$ is a positive definite matrix, and $\hat{\theta}$ is the estimate of the unknown parameter θ.

4.2.2 THEOREM AND STABILITY ANALYSIS

According to the adaptive controller (4.10), we state the following theorem.

Theorem 4.3 *Consider the first-order nonlinear system (4.2) with Assumptions 4.1, 4.3-4.5, if the adaptive controller and the parameter update law as shown in (4.10) are applied, then the objectives (i) and (ii) can be ensured.*

Proof. Substituting the adaptive controller as presented in (4.10) into (4.8), it follows that:

$$\dot{e} = -ce + \left(\theta - \hat{\theta}\right)^\top \varphi. \tag{4.11}$$

Constructing the Lyapunov function candidate as $V = \frac{1}{2}e^2 + \frac{1}{2}\tilde{\theta}^\top \Gamma^{-1}\tilde{\theta}$, where $\tilde{\theta} = \theta - \hat{\theta}$ denotes the parameter estimation error. Then its derivative along (4.11) yields:

$$\dot{V} = e\dot{e} + \tilde{\theta}^\top \Gamma^{-1}\dot{\tilde{\theta}}. \tag{4.12}$$

As θ is an unknown parameter, then $\dot{\tilde{\theta}} = -\dot{\hat{\theta}}$, (4.12) can be rewritten as:

$$\dot{V} = e\dot{e} - \tilde{\theta}^{\top}\Gamma^{-1}\dot{\hat{\theta}}. \tag{4.13}$$

Substituting the error dynamics (4.11) into (4.13), we derive that:

$$\dot{V} = -ce^2 + \tilde{\theta}^{\top}\varphi e - \tilde{\theta}^{\top}\Gamma^{-1}\dot{\hat{\theta}}. \tag{4.14}$$

Inserting the adaptive law $\dot{\hat{\theta}}$ as given in (4.10) into (4.14), one has: $\dot{V} = -ce^2 \leq 0$, then it follows that:

$$V(t) - V(0) = -c\int_0^t e^2(\tau)\,d\tau. \tag{4.15}$$

Therefore, it is seen from (4.15) that $\int_0^t e^2(\tau)\,d\tau$ and $V(t)$ are bounded, i.e., $e \in \mathcal{L}_2$, $e \in \mathcal{L}_\infty$, and $\tilde{\theta} \in \mathcal{L}_\infty$. As θ is an unknown yet bounded constant, according to the definition of $\tilde{\theta} = \theta - \hat{\theta}$, one has $\hat{\theta} \in \mathcal{L}_\infty$. Noting that y_d and its derivative $\dot{y}_d$ are bounded, it follows that $x \in \mathcal{L}_\infty$. Furthermore, due to the smoothness of φ and g, we have $\varphi \in \mathcal{L}_\infty$, $g \in \mathcal{L}_\infty$, and $f \in \mathcal{L}_\infty$, it is further shown from the definitions of the controller and adaptive law as given in (4.10) that $\dot{e} \in \mathcal{L}_\infty$, $u \in \mathcal{L}_\infty$, $\dot{\hat{\theta}} \in \mathcal{L}_\infty$, therefore, all signals in the closed-loop systems are bounded. As $e \in \mathcal{L}_2 \cap \mathcal{L}_\infty$ and $\dot{e} \in \mathcal{L}_\infty$, by utilizing Barbalat's Lemma, we have $\lim_{t\to\infty} e(t) = 0$. The proof is completed.

Remark 4.1 *It should be emphasized that the adaptive law in adaptive control cannot be designed arbitrarily, actually, it is derived from the Lyapunov stability theory. For example, the selection of the adaptive law as shown in (4.14) is to ensure $\dot{V} \leq 0$ (i.e., $\tilde{\theta}^{\top}\varphi e - \tilde{\theta}^{\top}\Gamma^{-1}\dot{\hat{\theta}} = 0$), therefore, it is not difficult to choose the adaptive law as $\Gamma\varphi e$ such that $\dot{V} \leq 0$.*

4.2.3 NUMERICAL SIMULATION

To valid the effectiveness of the developed adaptive control (4.2), the system model is chosen as: $f(\theta,x) = \theta_1 \sin(x) + \theta_2 x$ and $g(x) = 1$. From the expression of f, it is seen that the uncertain function f satisfies parameter decomposition condition, i.e., $f(\theta,x) = \theta^{\top}\varphi(x)$ with $\theta = [\theta_1, \theta_2]^{\top}$ and $\varphi(x) = [\sin(x), x]^{\top}$.

In the simulation, the system parameter is selected as: $\theta = [\theta_1, \theta_2]^{\top} = [1,1]^{\top}$. The desired trajectory is chosen as: $y_d(t) = \sin(t)$. The initial conditions are given as: $x(0) = 1$, $\hat{\theta}(0) = [0,0]^{\top}$, the design parameters are chosen as: $c = 2$ and $\Gamma = \text{diag}\{2\}$. The simulation results are shown in Fig. 4.1-Fig. 4.4. The evolutions of the system state and reference signal are shown in Fig. 4.1, the corresponding tracking error is given in Fig. 4.2, from which it is seen that the tracking error converges to zero asymptotically, confirming the theoretical analysis. The evolutions of the control input and the norm of the parameter estimate are plotted in Fig. 4.3 and Fig. 4.4, respectively, which are bounded for $\forall t \geq 0$.

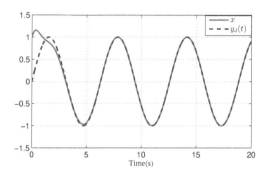

FIGURE 4.1   The evolutions of x and y_d.

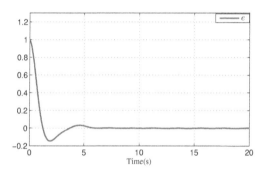

FIGURE 4.2   The evolution of the tracking error e.

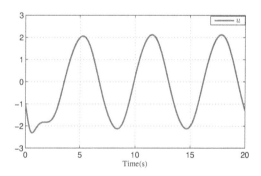

FIGURE 4.3   The evolution of the control input u.

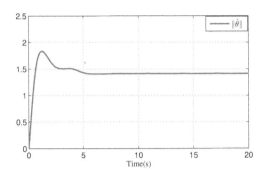

FIGURE 4.4 The evolution of the norm of parameter estimate $\|\hat{\theta}\|$.

4.3 ROBUST ADAPTIVE CONTROL OF NON-PARAMETRIC SYSTEMS

If the nonlinear function $f(x,\theta)$ does not meet the parameter decomposition condition, the corresponding controller developed in Section 4.2 is not applicable at all. Therefore, the control objective of this section is to develop an effective control scheme such that:

(i)   the nonlinear term $f(\theta,x)$ can be handled;
(ii)   the tracking error converges to a small neighborhood of zero;
(iii) all signals in the closed-loop systems are bounded.

Before control design, we impose the following two methods dealing with the non-parametric uncertainties, i.e., the core function-based method and NN-based method.

4.3.1 CORE FUNCTION

For the nonlinear function $f(\theta,x)$, if there exist a smooth and known function $\phi(x) \geq 0$ and an unknown constant $a \geq 0$ such that:

$$|f(\theta,x)| \leq a\phi(x),$$

where $\phi(x) \geq 0$ is called the core function and the corresponding control is the core function-based method [42]. Furthermore, it should be mentioned that compared with the method in Section 4.2, the parametric uncertainty $f(\theta,x) = \theta^\top \varphi(x)$ can also be handled by using the core function-based method, i.e.,

$$|f| \leq \|\theta\| \|\varphi(x)\| \leq a\phi(x),$$

in which $a = \|\theta\|$ and $\phi(x) = \|\varphi(x)\|^2 + 1$.

Now we give the following example to show the advantage of the core function-based method. For example, the nonlinear function $L(\rho,x)$ is selected as:

$$L(\rho,x) = \rho_1 \cos(\rho_2 x) + x\exp(-|\rho_3 x|),$$

where $\rho = [\rho_1, \rho_2, \rho_3] \in \mathbb{R}^3$ is an unknown parameter vector. It is obvious that the system parameter ρ can't be directly exacted from $L(\cdot)$. However, according to the definition of the core function, it is easy to find that there exists a function $\phi(x) = 2 + x^2$ such that $|L(\rho, x)| \leq a\phi(x)$, where $a = \max\{|\rho_1|, 1\}$.

4.3.2 RADIAL BASIS FUNCTION NEURAL NETWORK

In this subsection, we focus on illustrating the Radial Basis Function Neural Network (RBFNN), which was proposed in the 1980s and was a kind of feed-forward neural networks with a single hidden layer. Due to its simple structure and satisfactory approximation ability over some compact sets, it has been widely used by and received attention from researchers [16, 77, 78]. The schematic diagram of RBFNN is shown in Fig. 4.5, which can be expressed as:

$$f_{NN}(Z) = W^\top S(Z),$$

where $W = [w_1, w_2, \cdots, w_p]^\top \in \mathbb{R}^p$ is the ideal weight vector with the number of p nodes, $Z \in \Omega \subset \mathbb{R}^q$ denotes the input of NN, $S(Z) = [S_1(Z), \cdots, S_p(Z)]^\top \in \mathbb{R}^p$ is the basis function vector and the basis function $S_i(Z)$ is the Gaussian function with the form:

$$S_i(Z) = \exp\left(-\frac{(Z - \tau_i)^\top (Z - \tau_i)}{\psi^2}\right), \quad i = 1, 2, \cdots, p,$$

in which $\tau_i \in \mathbb{R}^q$ and $\psi \in \mathbb{R}$ are the center and width of the basis function, respectively.

According to the approximation ability of NN, for any continuous function $\Pi(Z)$ in a compact set Ω, it can be approximated by NN in the following form:

$$\Pi(Z) = f_{NN}(Z) + \varepsilon(Z), \quad |\varepsilon(Z)| \leq \bar{\varepsilon}, \quad \forall Z \in \Omega,$$

where $\varepsilon(Z)$ denotes the approximation error, and $\bar{\varepsilon}$ is the upper bound of the approximation error. Widespread practical applications of NN show that if the NN node number is chosen large enough, $\varepsilon(Z)$ can be reduced to an arbitrarily small value in a compact set.

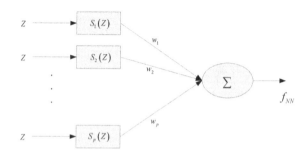

FIGURE 4.5 The schematic diagram of RBFNN.

By utilizing the description of RBFNN, if the variable of the uncertain continuous function $f(\theta, x)$ in (4.2) is within a compact set, namely, $x \in \Omega_x \subset \mathbb{R}$, then it can be approximated by NN over a compact set and can be mathematically stated as:

$$f(\theta, x) = W^{\top} S(x) + \varepsilon(x), \quad |\varepsilon(x)| \leq \bar{\varepsilon}, \quad \forall x \in \Omega_x.$$

Remark 4.2 *It should be mentioned that the NN-based method is also a special case of the core function-based method. That is, $f(\theta, x) = W^{\top} S(x) + \varepsilon(x)$ can be written as the following form:*

$$|f(\theta, x)| \leq \|W\| \|S(x)\| + |\varepsilon(x)| \leq a\phi(x),$$

where $a = \max\{\|W\|, \bar{\varepsilon}\}$ is an unknown parameter and $\phi = \|S(x)\|^2 + 2$ is the core function. In addition, the core function method is also able to reduce the number of parameter estimates (from p-dimensional vector W to 1-dimensional vector a).

In the following, we introduce some typical control schemes to handle the non-parametric uncertainties.

4.3.3   ROBUST CONTROL

To develop the robust control, except for Assumptions 4.1 and 4.3, we impose the following assumption.

Assumption 4.6 *The control gain $g(x)$ is unknown and time-varying, and there exist unknown positive constants $\underline{g}$ and $\bar{g}$ such that $0 < \underline{g} \leq g(x) \leq \bar{g} < \infty$.*

According to the error dynamics (4.5), the derivative of the Lyapunov function candidate $V = \frac{1}{2}e^2$ is:

$$\dot{V} = e\dot{e} = e(f + gu - \dot{y}_d). \tag{4.16}$$

Upon using Young's inequality and employing the core function technique, one has:

$$ef \leq |e| a\phi \leq \underline{g}e^2\phi^2 + \frac{a^2}{4\underline{g}}, \quad -e\dot{y}_d \leq |e||\dot{y}_d| \leq \underline{g}e^2\dot{y}_d^2 + \frac{1}{4\underline{g}}.$$

Therefore,

$$e(f - \dot{y}_d) \leq \underline{g}e^2\phi^2 + \frac{a^2}{4\underline{g}} + \underline{g}e^2\dot{y}_d^2 + \frac{1}{4\underline{g}} = \underline{g}e^2\Phi_1 + \Delta_1,$$

where $\Delta_1 = \frac{a^2}{4\underline{g}} + \frac{1}{4\underline{g}} > 0$ is an unknown constant, and $\Phi_1 = \phi^2 + \dot{y}_d^2 \geq 0$ is a computable function for control design.

The robust controller is designed as:

$$u = -ce - e\Phi_1, \tag{4.17}$$

with $c > 0$.

Now we give the following theorem.

Theorem 4.4 *For the nonlinear system (4.2) with Assumptions 4.1, 4.3, and 4.6, the proposed robust control (4.17) guarantees that the objectives (i)-(iii) are achieved.*

Proof. According to the robust controller, as given in (4.17), we have:

$$egu = g\left(-ce^2 - \Phi_1 e^2\right).$$

As $-ce^2 - \Phi_1 e^2 \leq 0$ and $g \geq \underline{g} > 0$, then it is seen that:

$$egu \leq -\underline{g}\left(ce^2 + \Phi_1 e^2\right). \tag{4.18}$$

Substituting (4.18) into (4.16), one has:

$$\dot{V} \leq -\underline{g}\left(ce^2 + \Phi_1 e^2\right) + ge^2\Phi_1 + \Delta_1 = -\underline{g}ce^2 + \Delta_1 \leq -l_1 V + l_2, \tag{4.19}$$

where $l_1 = 2\underline{g}c > 0$ and $l_2 = \Delta_1 > 0$.

We now prove the boundedness of all signals in the closed-loop system. In the view of (4.19), we have:

$$\dot{V} + l_1 V \leq l_2. \tag{4.20}$$

Multiplying $\exp(l_1 t)$ in both sides of (4.20), we have:

$$\dot{V} \exp(l_1 t) + l_1 V \exp(l_1 t) \leq l_2 \exp(l_1 t),$$

or

$$\frac{d}{dt}(V \exp(l_1 t)) \leq l_2 \exp(l_1 t). \tag{4.21}$$

Integrating (4.21) on the interval $[0, t]$, it follows that:

$$(V(\tau) \exp(l_1 \tau))\big|_0^t \leq \frac{l_2}{l_1} \exp(l_1 \tau)\big|_0^t,$$

which implies that:

$$V(t) \exp(l_1 t) \leq \frac{l_2}{l_1}(\exp(l_1 t) - 1) + V(0). \tag{4.22}$$

Multiplied by $\exp(-l_1 t)$ in both sides of (4.22), we have:

$$V(t) \leq \frac{l_2}{l_1}(1 - \exp(-l_1 t)) + V(0) \exp(-l_1 t)$$

$$= \frac{l_2}{l_1} + \left(V(0) - \frac{l_2}{l_1}\right) \exp(-l_1 t), \tag{4.23}$$

which means that the Lyapunov function candidate V is bounded, it follows that $e \in \mathscr{L}_\infty$. With Assumption 4.3, it is seen that the system state $x = e + y_d$ is bounded, which implies that $\phi(x) \in \mathscr{L}_\infty$, $f \in \mathscr{L}_\infty$, and $\Phi_1 \in \mathscr{L}_\infty$, according to the expression of the controller as given in (4.17), one has $u \in \mathscr{L}_\infty$, it further follows that $\dot{e} \in \mathscr{L}_\infty$. Therefore, all signals in the closed-loop system are bounded.

Next, we prove that the tracking error converges to a small neighborhood around zero. (4.19) can be rewritten as:

$$\dot{V} \le -\underline{g}ce^2 + l_2.$$

When $|e| > \sqrt{\frac{l_2+v}{\underline{g}c}}$, we have $\dot{V} < 0$, in which v is an unknown small constant, then it is further ensured that the tracking error may enter and keep in the compact set $\Omega_e = \left\{ e \in \mathbb{R} \,\middle|\, |e| \le \sqrt{\frac{\eta_2+v}{\underline{g}c}} \right\}$. It is not difficult to see that the size of Ω_e can be reduced by increasing the design parameter c. The proof is completed.

4.3.4 ROBUST ADAPTIVE CONTROL

In this section, we present the robust adaptive control design and stability analysis. Upon using the error dynamics (4.5), the derivative of quadratic function $\frac{1}{2}e^2$ is:

$$e\dot{e} = e\left(f + gu - \dot{y}_d\right).$$

By using Young's inequality, one has:

$$ef \le |e|a\phi \le \underline{g}a^2e^2\phi^2 + \frac{1}{4\underline{g}}, \quad -e\dot{y}_d \le |e||\dot{y}_d| \le \underline{g}e^2\dot{y}_d^2 + \frac{1}{4\underline{g}}.$$

Therefore, the term $ef - e\dot{y}_d$ can be upper bounded by:

$$e\left(f - \dot{y}_d\right) \le \underline{g}a^2e^2\phi^2 + \frac{1}{4\underline{g}} + \underline{g}e^2\dot{y}_d^2 + \frac{1}{4\underline{g}} \le \underline{g}be^2\Phi_2 + \Delta_2,$$

where $\Delta_2 = \frac{1}{2\underline{g}} > 0$ is an unknown constant,

$$b = \max\left\{a^2, 1\right\} > 0$$

is an unknown parameter, and

$$\Phi_2 = \phi^2 + \dot{y}_d^2 \ge 0$$

denotes a computational function. As the parameter b has no actual meaning in the nonlinear system, so we call it the virtual parameter.

Designing the robust adaptive controller as:

$$u = -ce - e\hat{b}\Phi_2, \tag{4.24}$$

where $c > 0$ is a design parameter, and $\hat{b}$ is the estimate of the unknown parameter b, which is updated by:

$$\dot{\hat{b}} = \gamma e^2\Phi_2 - \sigma\hat{b}, \quad \hat{b}(0) \ge 0, \tag{4.25}$$

with $\gamma > 0$ and $\sigma > 0$ being design parameters and $\hat{b}(0)$ being the initial value of the parameter estimate.

Remark 4.3 *Solving the equation as defined in (4.25), one has:*

$$\hat{b}(t) = \exp(-\sigma t)\,\hat{b}(0) + \exp(-\sigma t)\int_0^t \exp(\sigma\tau)\,\gamma e^2(\tau)\,\Phi_2(\tau)\,d\tau.$$

As $\hat{b}(0) \geq 0$ and $\gamma e^2 \Phi_2 \geq 0$, then $\hat{b}(t) \geq 0$ holds for $t \geq 0$.

According to the developed robust adaptive control, now we state the following theorem.

Theorem 4.5 *For the nonlinear system (4.2) with Assumptions 4.1, 4.3, and 4.6, the developed control (4.24)-(4.25) guarantees that the objectives (i)-(iii) are achieved.*

Proof. According to the robust adaptive controller (4.24), it is checked that:

$$egu = g\left(-ce^2 - \hat{b}\Phi_2 e^2\right).$$

As $-ce^2 - \hat{b}\Phi_2 e^2 \leq 0$ and $g \geq \underline{g} > 0$, then it is shown that:

$$egu \leq -\underline{g}\left(ce^2 + \hat{b}\Phi_2 e^2\right). \tag{4.26}$$

Constructing the Lyapunov function candidate as $V = \frac{1}{2}e^2 + \frac{g}{2\gamma}\tilde{b}^2$, where $\tilde{b} = b - \hat{b}$ is the estimate error. With the aid of (4.26), the derivative of V is:

$$\dot{V} = e\dot{e} + \frac{g}{\gamma}\tilde{b}\dot{\tilde{b}} \leq -\underline{g}\left(ce^2 + \hat{b}\Phi_2 e^2\right) + gbe^2\Phi_2 + \Delta_2 + \frac{g}{\gamma}\tilde{b}\dot{\tilde{b}}. \tag{4.27}$$

Since b is an unknown parameter, then $\dot{b} = 0$, which implies that $\dot{\tilde{b}} = \dot{b} - \dot{\hat{b}} = -\dot{\hat{b}}$, then we have:

$$\dot{V} \leq -\underline{g}ce^2 + g\underline{b}e^2\Phi_2 + \Delta_2 - \frac{g}{\gamma}\tilde{b}\dot{\hat{b}}. \tag{4.28}$$

Substituting the adaptive law as given in (4.25) into (4.28), one has:

$$\dot{V} \leq -\underline{g}ce^2 + \Delta_2 + \frac{g\sigma}{\gamma}\tilde{b}\hat{b}. \tag{4.29}$$

As

$$\tilde{b}\hat{b} = \tilde{b}(b - \tilde{b}) = \tilde{b}b - \tilde{b}^2 \leq \frac{1}{2}\tilde{b}^2 + \frac{1}{2}b^2 - \tilde{b}^2 = -\frac{1}{2}\tilde{b}^2 + \frac{1}{2}b^2,$$

then (4.29) can be rewritten as:

$$\dot{V} \leq -\underline{g}ce^2 - \frac{g\sigma}{2\gamma}\tilde{b}^2 + \frac{g\sigma}{2\gamma}b^2 + \Delta_2 \leq -l_1 V + l_2,$$

where $l_1 = \min\{2\underline{g}c, \sigma\} > 0$, $l_2 = \frac{g\sigma}{2\gamma}b^2 + \Delta_2 > 0$. The analysis of stability and tracking performance is similar to that in Section 4.3.3, so it is omitted.

Although the developed control schemes in Section 4.3.3 and Section 4.3.4 are able to deal with the problem of non-parametric uncertainties, due to the constant term Δ_i $(i = 1,2)$ appears in the stability analysis, it is difficult or even impossible to achieve the zero-error tracking. In the following subsection, we introduce a new robust adaptive control scheme for the nonlinear system (4.2) such that not only the boundedness of all signals in the closed-loop system is ensured, but also zero-error tracking can be guaranteed.

4.3.5   ZERO-ERROR ADAPTIVE CONTROL

In this section, we develop an adaptive zero-error tracking control scheme for the first-order nonlinear system (4.2).

Noting that $e\dot{e} = e(f + gu - \dot{y}_d)$, with the aid of the core function technique, we have:

$$ef \le |e|\,a\phi, \quad -e\dot{y}_d \le |e|\,|\dot{y}_d|,$$

then it follows that:

$$e(f - \dot{y}_d) \le |e|\,a\phi + |e|\,|\dot{y}_d| \le p|e|\,\Phi_3, \tag{4.30}$$

where $p = \max\{a, 1\}$ is an unknown virtual parameter, and $\Phi_3 = \phi + \dot{y}_d^2 + \frac{1}{4}$ is a computable function.

Upon using Theorem 2.11, (4.30) can be upper bounded by:

$$e(f - \dot{y}_d) \le p|e|\,\Phi_3 \le p\varepsilon(t) + \frac{pe^2\Phi_3^2}{\sqrt{e^2\Phi_3^2 + \varepsilon^2(t)}},$$

where $\varepsilon(t) > 0$ is smooth and satisfies $\int_0^t \varepsilon(\tau)d\tau \le \bar{\varepsilon} < \infty$ with $\bar{\varepsilon}$ being a positive constant, then one further has:

$$e\dot{e} \le egu + p\varepsilon(t) + \frac{pe^2\Phi_3^2}{\sqrt{e^2\Phi_3^2 + \varepsilon^2(t)}}. \tag{4.31}$$

The robust adaptive control is designed as:

$$u = -ce - \frac{\hat{p}e\Phi_3^2}{\sqrt{e^2\Phi_3^2 + \varepsilon^2(t)}}, \tag{4.32}$$

where $c > 0$ is a design parameter, $\hat{p}$ is the estimate of the unknown parameter p and is updated by:

$$\dot{\hat{p}} = \frac{\gamma e^2 \Phi_3^2}{\sqrt{e^2\Phi_3^2 + \varepsilon^2(t)}}, \quad \hat{p}(0) \ge 0, \tag{4.33}$$

with $\gamma > 0$ being the design parameter and $\hat{p}(0)$ being the initial value of the parameter estimate $\hat{p}$. As $\frac{\gamma e^2 \Phi_3^2}{\sqrt{e^2\Phi_3^2 + \varepsilon^2(t)}} \ge 0$, then according to (4.33), $\hat{p}(t) \ge 0$ holds for $t \ge 0$.

Upon using (4.32), we present the following theorem.

Theorem 4.6 *For the nonlinear system (4.2) with Assumptions 4.1, 4.3, and 4.6, the developed control (4.32) and adaptive law (4.33) guarantee that not only all signals in the closed-loop system are bounded, but also the asymptotic tracking can be achieved.*

Proof. Substituting the control law as given in (4.32) into the term egu, we have:

$$egu = -\underline{g}ce^2 - \frac{g\hat{p}e^2\Phi_3^2}{\sqrt{e^2\Phi_3^2 + \varepsilon^2(t)}}.$$

As $g \geq \underline{g} > 0$ and $\hat{p} \geq 0$, then one further has:

$$egu \leq -\underline{g}ce^2 - \frac{\underline{g}\hat{p}e^2\Phi_3^2}{\sqrt{e^2\Phi_3^2 + \varepsilon^2(t)}},$$

then it follows from (4.31) that:

$$e\dot{e} \leq -\underline{g}ce^2 - \frac{\underline{g}\hat{p}e^2\Phi_3^2}{\sqrt{e^2\Phi_3^2 + \varepsilon^2(t)}} + p\varepsilon(t) + \frac{pe^2\Phi_3^2}{\sqrt{e^2\Phi_3^2 + \varepsilon^2(t)}}$$

$$= -\underline{g}ce^2 + (p - \underline{g}\hat{p})\frac{e^2\Phi_3^2}{\sqrt{e^2\Phi_3^2 + \varepsilon^2(t)}} + p\varepsilon(t).$$

Constructing the Lyapunov function candidate as $V = \frac{1}{2}e^2 + \frac{1}{2g\gamma}(p - \underline{g}\hat{p})^2$, where $\gamma > 0$ is a design parameter, then the derivative of V is:

$$\dot{V} \leq -\underline{g}ce^2 + (p - \underline{g}\hat{p})\frac{e^2\Phi_3^2}{\sqrt{e^2\Phi_3^2 + \varepsilon^2(t)}} + p\varepsilon(t) + \frac{1}{\gamma}(p - \underline{g}\hat{p})(-\dot{\hat{p}}). \qquad (4.34)$$

Inserting the adaptive law as shown in (4.33) into (4.34), one has:

$$\dot{V} \leq -\underline{g}ce^2 + p\varepsilon(t). \qquad (4.35)$$

Integrating (4.35) on $[0,t]$, we have:

$$V(t) - V(0) \leq -\underline{g}c\int_0^t e^2(\tau)d\tau + p\int_0^t \varepsilon(t)d\tau \leq -\underline{g}c\int_0^t e^2(\tau)d\tau + p\bar{\varepsilon},$$

which further implies that:

$$V(t) + \underline{g}c\int_0^t e^2(\tau)d\tau \leq V(0) + p\bar{\varepsilon}. \qquad (4.36)$$

Now we first prove the boundedness of all signals in the closed-loop system. According to (4.36), one has $V \in \mathscr{L}_\infty$ and $e \in \mathscr{L}_2$. Upon using the definition of the Lyapunov function candidate, it is shown that the tracking error e is bounded and the estimate error $p - \underline{g}\hat{p}$ is also bounded. As $e = x - y_d$ and the reference signal y_d is bounded, it follows that the system state x is bounded, which further implies that $\phi \in \mathscr{L}_\infty$. Noting that p and $\underline{g} > 0$ are bounded, one has $\hat{p} \in \mathscr{L}_\infty$, then it is seen from the expressions in (4.32) and (4.33) that $u \in \mathscr{L}_\infty$ and $\dot{\hat{p}} \in \mathscr{L}_\infty$. As $|f| \leq a\phi$, then the nonlinear function f is bounded, it further follows that $\dot{e} \in \mathscr{L}_\infty$.

Next, we prove that the tracking error converges to zero asymptotically. Note that $e \in \mathscr{L}_\infty \cap e \in \mathscr{L}_2$ and $\dot{e} \in \mathscr{L}_\infty$, according to Barbalat's Lemma, it is seen that $\lim_{t\to\infty} e(t) = 0$. The proof is completed.

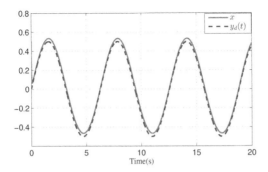

FIGURE 4.6 The evolutions of x and y_d.

4.3.6 NUMERICAL SIMULATION

In this section, we verify the effectiveness of the proposed control methods in Section 4.3.3 and Section 4.3.5, respectively.

Test 1. We first verify the effectiveness of the developed control in Section 4.3.3. The nonlinear function is selected as $f(\theta, x) = 3x \sin(\theta_1 x) + \exp(-\theta_2 x^2)$, where $\theta = [\theta_1, \theta_2]^\top = [1, 1]^\top$. It is seen from the expression of f that it does not satisfy the parameter decomposition condition, however, by using the core function technique, we have $\phi(x) = x^2 + \frac{5}{4}$. In the simulation, the initial condition is given as $x(0) = 0.5$, the reference signal is $y_d(t) = 0.5 \sin(t)$, the design parameter is chosen as $c = 20$, the simulation results are shown in Fig. 4.6–Fig. 4.8, the evolutions of the system state and the reference signal are plotted in Fig. 4.6, the evolution of the tracking error is given in Fig. 4.7, and the evolution of the control input is plotted in Fig. 4.8, from which it is seen that all signals are bounded for all time.

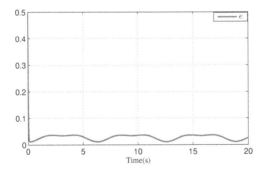

FIGURE 4.7 The evolution of the tracking error e.

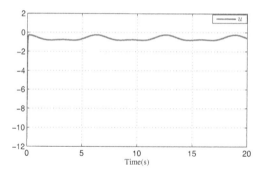

FIGURE 4.8 The evolution of the control input u.

Test 2. In this part we verify the effectiveness of the proposed control in Section 4.3.5. In the simulation, the expression of f is identical to that in Test 1, the initial conditions are given as: $x(0) = 1$ and $\hat{p}(0) = 0$, the reference signal is $y_d(t) = 0.5\sin(t)$, the design parameters are chosen as: $c = 10$, $\gamma = 3$, and the integral function is $\varepsilon(t) = \exp(-0.4t)$, the simulation results are shown in Fig. 4.9-Fig. 4.12. The evolutions of the system state and the reference signal are shown in Fig. 4.9, the evolution of the tracking error is plotted in Fig. 4.10, from which it is seen that the tracking error converges to zero asymptotically, confirming the effectiveness of the proposed control. The evolutions of the control input and parameter estimate are plotted in Fig. 4.11 and Fig. 4.12, respectively, which are bounded for all time.

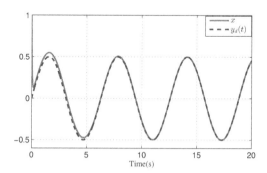

FIGURE 4.9 The evolutions of x and y_d.

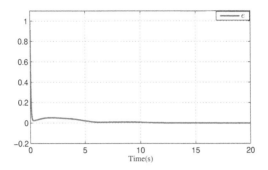

FIGURE 4.10 The evolution of the tracking error e.

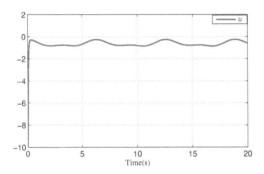

FIGURE 4.11 The evolution of the control input u.

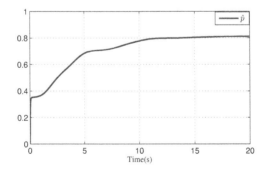

FIGURE 4.12 The evolution of the parameter estimate $\hat{p}$.

4.4   ADAPTIVE CONTROL WITH UNKNOWN CONTROL DIRECTIONS

In Section 4.3 we have illustrated some typical control design methods for nonlinear systems, however, the precondition is that the control direction is known. If the control direction is unavailable for control design, the corresponding control algorithms in Section 4.3 are not effective any more. Therefore, in this section, we develop a Nussbaum gain-based adaptive control for nonlinear systems in the presence of unknown control directions.

4.4.1   NUSSBAUM FUNCTION

Definition 4.1 *[79, 80, 81] If a function $N(\chi)$ satisfies the following properties:*

$$\limsup_{s \to \infty} \frac{1}{s} \int_0^s N(\chi) d\chi = +\infty, \quad \liminf_{s \to \infty} \frac{1}{s} \int_0^s N(\chi) d\chi = -\infty,$$

then $N(\chi)$ is called a Nussbaum function.

The normally used Nussbaum functions are $\exp(\chi^2) \cos((\pi/2)\chi)$, $\chi^2 \sin(\chi)$, and $\chi^2 \cos(\chi)$. In this section the Nussbaum even function $N(\chi) = \exp(\chi^2) \cos((\pi/2)\chi)$ is used here.

Lemma 4.1 *[81] Let $V(\cdot)$ and $\chi(\cdot)$ be continuously differentiable functions defined on $[0, t_s)$ with $V(t) \geq 0, \forall t \in [0, t_s)$, while $N(\cdot)$ is a Nussbaum even function. If the following inequality*

$$V(t) \leq cons + \frac{1}{\gamma_x} \int_0^t (g(\tau)N(\chi) + 1)\dot{\chi}(\tau)d\tau, \quad \forall t \in [0, t_s)$$

holds, where $g(\cdot)$ is a time-varying unknown function which takes a value in the unknown closed interval $I := [l^-, l^+]$, $0 \notin I$, "cons" denotes a positive constant, and $\gamma_x > 0$ represents a constant. Then $V(t)$, χ, and $\int_0^t (g(\tau)N(\chi) + 1)\dot{\chi}(\tau)d\tau$ must be bounded on $[0, t_s)$.

Remark 4.4 *It is worth pointing out that most of the existing results dealing with unknown control directions are based on the following lemma.*

Lemma 4.2 *[79] Let $V(\cdot)$ and $\chi(\cdot)$ be continuously differentiable functions defined on $[0, t_s)$ with $V(t) \geq 0, \forall t \in [0, t_s)$, while $N(\cdot)$ is a Nussbaum even function. If the following inequality*

$$V(t) \leq cons + \frac{1}{\gamma_x} \int_0^t [g(\tau)N(\chi) + 1]\dot{\chi}(\tau) \exp(-\mu(t - \tau)) d\tau$$

holds, where "cons", γ_x, and μ are some positive constants, while $g(\cdot)$ is a time-varying function or constant which takes a value in the unknown closed interval $I := [l^-, l^+]$, $0 \notin I$, then $V(t)$, $\chi(t)$, and $\int_0^t [g(\tau)N(\chi) + 1]\dot{\chi}(\tau)d\tau$ are bounded on $[0, t_s)$.

However, by carefully checking the proof processes of Lemma 4.1 and Lemma 4.2, it is shown that the boundedness of $V(t)$ and $\chi(t)$ in Lemma 4.1 is independent of t_s, whereas the boundedness of signals in Lemma 4.2 relies on the finite time t_s, that is, whether the finite time t_s in Lemma 4.2 can be extended to infinity is still to be discussed in reference [82].

4.4.2 CONTROL DESIGN

We consider the first-order nonlinear system (4.1) with constant control gain (but the control direction is unknown):

$$\dot{x} = f(x, \theta) + gu, \qquad (4.37)$$

where $x \in \mathbb{R}$ represents the system state, $f \in \mathbb{R}$ denotes a smooth uncertain function, $u \in \mathbb{R}$ is the control input, $\theta \in \mathbb{R}^n$ is an unknown constant parameter, $g \neq 0$ is an unknown constant and its sign is also unknown, i.e., $\mathrm{sgn}(g) = 1$ or -1.

The control objective of this section is to design a robust adaptive control scheme by employing the Nussbaum function such that:

(i)   the tracking error $e = x_1 - y_d$ converges to zero asymptotically;
(ii) all signals in the closed-loop systems are bounded.

The derivative of the Lyapunov function candidate $V_1 = \frac{1}{2}e^2$ along $\dot{e} = f + gu - \dot{y}_d$ yields:

$$\dot{V}_1 = e\dot{e} = e(f + gu - \dot{y}_d). \qquad (4.38)$$

With the aid of Theorem 2.11 and the core function technique, one has:

$$ef \leq |e|\,|f| \leq |e|\,a\phi(x) \leq a\varepsilon(t) + \frac{ae^2\phi^2}{\sqrt{e^2\phi^2 + \varepsilon^2(t)}},$$

$$-e\dot{y}_d \leq |e|\,|\dot{y}_d| \leq \varepsilon(t) + \frac{e^2\dot{y}_d^2}{\sqrt{e^2\dot{y}_d^2 + \varepsilon^2(t)}},$$

where $\varepsilon(t)$ denotes a positive function satisfying $\int_0^t \varepsilon(\tau)d\tau \leq \bar{\varepsilon} < \infty$, where $\bar{\varepsilon}$ is an unknown positive constant. Therefore, one further has:

$$ef - e\dot{y}_d \leq b\Phi e^2 + \hbar\varepsilon(t),$$

where $b = \max\{a, 1\}$ is an unknown virtual parameter,

$$\Phi = \frac{\phi^2}{\sqrt{e^2\phi^2 + \varepsilon^2(t)}} + \frac{\dot{y}_d^2}{\sqrt{e^2\dot{y}_d^2 + \varepsilon^2(t)}} \geq 0$$

is a computable scalar function, and $\hbar = a + 1$ is an unknown constant.

Then, (4.38) can be rewritten as:

$$\dot{V}_1 \leq egu + b\Phi e^2 + \hbar\varepsilon(t). \qquad (4.39)$$

We design the Nussbaum gain-based robust adaptive control as:

$$\begin{cases} u = N(\chi)w, \\ \dot{\chi} = \gamma_x ew, \\ w = (c + \hat{b}\Phi)e, \\ \dot{\hat{b}} = \sigma\Phi e^2, \end{cases} \tag{4.40}$$

where $\gamma_x > 0$, $c > 0$, and $\sigma > 0$ are design parameters, $\hat{b}$ represents the estimate of the virtual parameter b.

According to the controller, as given in (4.40), we have:

$$egu = egN(\chi)w. \tag{4.41}$$

Adding and subtracting the term ew in the right-hand side of (4.41), one has:

$$egu = [gN(\chi) + 1]ew - ew. \tag{4.42}$$

Inserting $\dot{\chi}$ and w as shown in (4.40) into (4.42), it follows that:

$$egu = \frac{1}{\gamma_x}[gN(\chi) + 1]\dot{\chi} - ce^2 - \hat{b}\Phi e^2,$$

which leads to

$$\begin{aligned} \dot{V}_1 &\leq \frac{1}{\gamma_x}[gN(\chi) + 1]\dot{\chi} - ce^2 - \hat{b}\Phi e^2 + b\Phi e^2 + \hbar\varepsilon(t) \\ &= \frac{1}{\gamma_x}[gN(\chi) + 1]\dot{\chi} - ce^2 + \tilde{b}\Phi e^2 + \hbar\varepsilon(t), \end{aligned} \tag{4.43}$$

where $\tilde{b} = b - \hat{b}$ represents the estimation error.

Upon using the proposed robust adaptive control, we get the following theorem.

4.4.3 THEOREM AND STABILITY ANALYSIS

Theorem 4.7 *Consider the nonlinear system (4.37) with Assumptions 4.1 and 4.3, the proposed control as given in (4.40) ensures that the objectives (i)-(ii) are achieved.*

Proof. Choosing the Lyapunov function candidate as $V = V_1 + \frac{1}{2\sigma}\tilde{b}^2$, then it is seen from (4.43) that:

$$\dot{V} \leq \frac{1}{\gamma_x}[gN(\chi) + 1]\dot{\chi} - ce^2 + \tilde{b}\Phi e^2 - \frac{1}{\sigma}\tilde{b}\dot{\hat{b}} + \hbar\varepsilon(t). \tag{4.44}$$

Substituting the adaptive law for $\hat{b}$ as shown in (4.40) into (4.44), it is checked that:

$$\dot{V} \leq \frac{1}{\gamma_x}[gN(\chi) + 1]\dot{\chi} - ce^2 + \hbar\varepsilon(t). \tag{4.45}$$

Integrating (4.45) on $[0,t]$, we obtain:

$$V(t) - V(0) \leq \frac{1}{\gamma_x} \int_0^t [gN(\chi(\tau)) + 1]\dot{\chi}(\tau)d\tau - c\int_0^t e^2(\tau)d\tau + \hbar\int_0^t \varepsilon(\tau)d\tau$$

$$\leq \frac{1}{\gamma_x} \int_0^t [gN(\chi(\tau)) + 1]\dot{\chi}(\tau)d\tau - c\int_0^t e^2(\tau)d\tau + \hbar\bar{\varepsilon},$$

which can be rewritten as:

$$V(t) + c\int_0^t e^2(\tau)d\tau \leq c_x + \frac{1}{\gamma_x} \int_0^t [gN(\chi(\tau)) + 1]\dot{\chi}(\tau)d\tau,$$

where $c_x = \hbar\bar{\varepsilon} + V(0) < \infty$.

We first show that all signals in the closed-loop system are bounded. According to Lemma 4.1, $V(t)$, $\int_0^t e^2(\tau)d\tau$, $\chi(t)$, and $\int_0^t [gN(\chi(\tau)) + 1]\dot{\chi}(\tau)d\tau$ are bounded on $[0,t_s)$ and no finite-time escape phenomenon may occur, so $t_s = \infty$ [83]. Due to the boundedness of V, it is shown that $e \in \mathscr{L}_\infty$ and $\hat{b} \in \mathscr{L}_\infty$. Since y_d is bounded, so is x, then we further have $f \in \mathscr{L}_\infty$, $\phi \in \mathscr{L}_\infty$, $\Phi \in \mathscr{L}_\infty$, $w \in \mathscr{L}_\infty$, $\dot{\chi} \in \mathscr{L}_\infty$, and $\dot{\hat{b}} \in \mathscr{L}_\infty$, which indicates that $N(\chi) \in \mathscr{L}_\infty$ and $u \in \mathscr{L}_\infty$. Thus, all signals in the closed-loop system are bounded.

Next, we prove that the tracking error converges to zero as $t \to \infty$. From the boundedness of x, u, and $\dot{y}_d$, one has $\dot{e} \in \mathscr{L}_\infty$. Since e and $\int_0^t e^2 d\tau$ are bounded, then it follows from Barbalat's Lemma that:

$$\lim_{t \to \infty} e(t) = 0 \text{ or } \lim_{t \to \infty} x(t) = y_d(t).$$

The proof is completed.

4.4.4 NUMERICAL SIMULATION

In the simulation, the nonlinear function is chosen as: $f = x + x^2 \sin(\theta x)$ with $\theta = 1$, the control gain is given as $g = 5$. The reference trajectory is selected as $y_d(t) = \sin(t)$, the initial values are selected as $x(0) = 1$, $\hat{b}(0) = 0$, and $\chi(0) = 0$, the design parameters are selected as $c = 1$, $\gamma_x = 1$, $\sigma = 0.1$, and the integrable function is $\varepsilon(t) = \exp(-0.4t)$.

The simulation results are shown in Fig. 4.13-Fig. 4.17. Fig. 4.13 and Fig. 4.14 show the evolutions of the system state and reference signal as well as the tracking error, respectively, from which it is seen that the system state x follows the reference trajectory $y_d(t)$ asymptotically, confirming the theoretical analysis of Section 4.4. The evolution of the control input signal is shown in Fig. 4.15, the evolutions of the parameter estimate $\hat{b}$ and the argument of Nussbaum function χ are plotted in Fig. 4.16 and Fig. 4.17, respectively, which are bounded for all time.

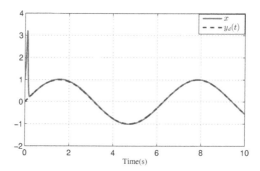

FIGURE 4.13 The evolutions of x and y_d.

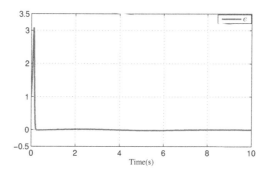

FIGURE 4.14 The evolution of the tracking error e.

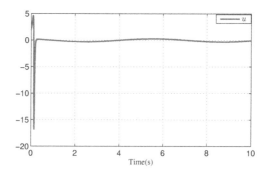

FIGURE 4.15 The evolution of the control input u.

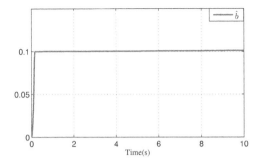

FIGURE 4.16   The evolution of the parameter estimate $\hat{b}$.

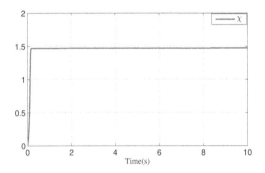

FIGURE 4.17   The evolution of the parameter χ.

4.5   NOTES

In this chapter, we present some typical control methods for first-order nonlinear systems. We first introduce the model-dependent controller design and stability analysis in Section 4.1. In order to handle the parametric or non-parametric uncertainties, we develop some adaptive control schemes and core function-based controls in Section 4.2 and Section 4.3, respectively. Furthermore, we present a Nussbaum gain-based adaptive control algorithm for nonlinear systems with unknown control directions in Section 4.4. The effectiveness of the proposed controls is verified by some numerical simulations.

5 Control of Normal-form Systems

In Chapter 4, we dealt with the stabilization and tracking control of first-order systems. However, such a simple system usually cannot describe the nonlinear dynamics of many practical applications because they have high-order nonlinear properties. Therefore, the control methods in Chapter 4 are no longer applicable. In this chapter, we will focus on coping with the control problem of normal-form nonlinear systems.

5.1 MODEL-DEPENDENT CONTROL OF SECOND-ORDER NORMAL-FORM SYSTEMS

Consider the following second-order normal-form nonlinear systems:

$$\ddot{x} = f(X) + g(X)u, \tag{5.1}$$

where $x \in \mathbb{R}$ and $X = [x, \dot{x}]^{\top} \in \mathbb{R}^2$ represent the system states, $u \in \mathbb{R}$ is the control input, $f(X) \in \mathbb{R}$ is a known and smooth nonlinear function, and $g(X) \neq 0$ is a time-varying but known control coefficient.

The control objective in this section is to design a feasible controller u such that the system states $x(t)$ and $\dot{x}(t)$ asymptotically track the reference trajectories $y_d(t)$ and $\dot{y}_d(t)$, respectively, i.e., $\lim_{t \to \infty}[x(t) - y_d(t)] = 0$ and $\lim_{t \to \infty}[\dot{x}(t) - \dot{y}_d(t)] = 0$.

To this end, the following assumptions are imposed.

Assumption 5.1 *The system state* $X = [x, \dot{x}]^{\top}$ *is measurable.*

Assumption 5.2 *The reference trajectory* $y_d(t)$ *and its derivatives* $y_d^{(k)}(t)$, $k = 1, 2$, *are bounded, known, and piecewise continuous.*

Assumption 5.3 *For the available control coefficient* $g(X)$, *there exist some positive constants* $\underline{g}$ *and* $\bar{g}$ *such that* $0 < \underline{g} \leq g(\cdot) \leq \bar{g} < \infty$.

Let

$$e = x - y_d \quad \text{and} \quad \dot{e} = \dot{x} - \dot{y}_d \tag{5.2}$$

be the tracking error. In order to facilitate the control design, we introduce the following filtered error:

$$s = \lambda e + \dot{e}, \tag{5.3}$$

with $\lambda > 0$.

The derivative of s along (5.2) is:

$$\dot{s} = \ddot{e} + \lambda \dot{e} = \ddot{x} - \ddot{y}_d + \lambda \dot{e} = f + gu - \ddot{y}_d + \lambda \dot{e}. \tag{5.4}$$

Now we state the following theorem.

DOI: 10.1201/9781003474364-5

Theorem 5.1 *Consider the second-order normal-form nonlinear system (5.1). Under Assumptions 5.1-5.3, if the model-dependent control u is designed as:*

$$u = -\frac{1}{g}(cs + f - \ddot{y}_d + \lambda\dot{e}), \tag{5.5}$$

with c > 0, then it is ensured that all signals in the closed-loop system are bounded, and the tracking errors e and $\dot{e}$ asymptotically converge to 0, that is, $\lim_{t\to\infty} e(t) = 0$ *and* $\lim_{t\to\infty} \dot{e}(t) = 0.$

Proof. Substituting (5.5) into (5.4), we obtain $\dot{s} = -cs$, then it further follows that:

$$s(t) = \exp(-ct)s(0). \tag{5.6}$$

Bearing in mind (5.6), it is obvious that $|s(t)| \le |s(0)|$ and $\lim_{t\to\infty} s(t) = 0$. Since $e(0)$ and $\dot{e}(0)$ represent the initial values of the tracking error and its derivative, respectively, it is obvious that they are bounded, then $s(0) = \lambda e(0) + \dot{e}(0)$ is bounded. As $|s(t)| \le |s(0)|$, then $s(t)$ is also bounded, that is, there exists an unknown positive number $\bar{s}$ such that $|s(t)| \le \bar{s} < \infty$.

Multiplying $\exp(\lambda t)$ in both sides of $s = \lambda e + \dot{e}$, one has:

$$s\exp(\lambda t) = \lambda e\exp(\lambda t) + \dot{e}\exp(\lambda t) = (e\exp(\lambda t))'. \tag{5.7}$$

Integrating both sides of (5.7) on $[0, t]$, one has:

$$\int_0^t s(\tau)\exp(\lambda\tau)d\tau = e(t)\exp(\lambda t) - e(0),$$

namely,

$$e(t)\exp(\lambda t) = \int_0^t s(\tau)\exp(\lambda\tau)d\tau + e(0). \tag{5.8}$$

Multiplying $\exp(-\lambda t)$ in both sides of (5.8), it is checked that:

$$e(t) = \exp(-\lambda t)e(0) + \exp(-\lambda t)\int_0^t s(\tau)\exp(\lambda\tau)d\tau. \tag{5.9}$$

Then, it follows from (5.9) that:

$$|e(t)| \le \exp(-\lambda t)|e(0)| + \exp(-\lambda t)\left|\int_0^t s(\tau)\exp(\lambda\tau)d\tau\right|$$

$$\le \exp(-\lambda t)|e(0)| + \exp(-\lambda t)\int_0^t |s(\tau)\exp(\lambda\tau)|d\tau$$

$$\le \exp(-\lambda t)|e(0)| + \exp(-\lambda t)\int_0^t \exp(\lambda\tau)\bar{s}d\tau$$

$$= \exp(-\lambda t)|e(0)| + \frac{\bar{s}}{\lambda}(1 - \exp(-\lambda t)).$$

Note that $0 < \exp(-\lambda t) \leq 1$, then the tracking error $e(t)$ is bounded, from the definition of $s(t)$ as shown in (5.3), one has $\dot{e}(t) \in \mathscr{L}_\infty$. As y_d, $\dot{y}_d$ and $\ddot{y}_d$ are all bounded, then it is readily derived that $x \in \mathscr{L}_\infty$, $\dot{x} \in \mathscr{L}_\infty$, and $f \in \mathscr{L}_\infty$, it further indicates that $u \in \mathscr{L}_\infty$. Furthermore, as $\dot{s} = f + gu - \ddot{y}_d + \lambda\dot{e}$, then $\dot{s}$ is also bounded. Therefore, all signals in the closed-loop system are bounded.

To prove that $\lim_{t\to\infty} e(t) = 0$ and $\lim_{t\to\infty} \dot{e}(t) = 0$, the performance analysis of (5.9) can be divided into the following two situations:

(i)   If $\int_0^t s(\tau)\exp(\lambda\tau)d\tau$ is bounded, $\lim_{t\to\infty} e(t) = 0$ is naturally established;

(ii) If $\int_0^t s(\tau)\exp(\lambda\tau)d\tau$ is unbounded, by utilizing L'Hospital's rule, one has:

$$\lim_{t\to\infty} e(t) = 0 + \lim_{t\to\infty} \frac{\int_0^t s(\tau)\exp(\lambda\tau)d\tau}{\exp(\lambda t)}$$
$$= \lim_{t\to\infty} \frac{s(t)\exp(\lambda t)}{\lambda\exp(\lambda t)} = \lim_{t\to\infty} \frac{s(t)}{\lambda}. \tag{5.10}$$

As $\lim_{t\to\infty} s(t) = 0$ and $\lambda > 0$, then from (5.10) we have $\lim_{t\to\infty} e(t) = 0$, then it further follows from (5.3) that $\lim_{t\to\infty} \dot{e}(t) = 0$. The proof is completed.

5.2 ROBUST ADAPTIVE CONTROL OF SECOND-ORDER NORMAL-FORM SYSTEMS

In Section 5.1, we presented a model-dependent control algorithm for second-order normal-form systems. Although the control design and stability analysis are simple, the system model must be required to be accurately known, which is obviously unrealistic in practical applications. Therefore, in this section, we consider the control problem of second-order normal-form nonlinear systems in the presence of uncertainty.

5.2.1 SYSTEM DESCRIPTION

Consider the following second-order normal-form uncertain nonlinear systems:

$$\ddot{x} = f(X,p) + g(X)u, \tag{5.11}$$

where $x \in \mathbb{R}$ and $X = [x,\dot{x}]^\top \in \mathbb{R}^2$ represent the system states, $u \in \mathbb{R}$ is the control input, $f(X,p) \in \mathbb{R}$ is an uncertain smooth nonlinear function, $g(X) \in \mathbb{R} \neq 0$ denotes the time-varying and unknown control coefficient, and $p \in \mathbb{R}^r$ represents an unknown constant parameter vector.

The control objective is to design a robust adaptive control law u such that:

(i)   the system states $x(t)$ and $\dot{x}(t)$ asymptotically track the reference trajectories $y_d(t)$ and $\dot{y}_d(t)$, respectively;

(ii) all signals in the closed-loop systems are bounded.

Except for Assumptions 5.1 and 5.2, the following assumptions are required.

Assumption 5.4 *The control gain $g(X)$ is time-varying and unknown but bounded away from zero, i.e., there exist some unknown positive constants $\underline{g}$ and $\bar{g}$ such that $0 < \underline{g} \le g(\cdot) \le \bar{g} < \infty$.*

Assumption 5.5 *For the uncertain nonlinear function $f(X,p)$, there exist an unknown constant $a \ge 0$ and a known smooth function $\phi(X) \ge 0$ such that $|f(X,p)| \le a\phi(X)$. In addition, if X is bounded, so are f and ϕ.*

5.2.2 CONTROLLER DESIGN

According to the filtered error (5.3), the Lyapunov function candidate is selected as $V_1 = \frac{1}{2}s^2$, then the derivative of V_1 w.r.t. time is:

$$\dot{V}_1 = s\dot{s} = s(f + gu - \ddot{y}_d + \lambda\dot{e}). \tag{5.12}$$

For the nonlinear term $s(f - \ddot{y}_d + \lambda\dot{e})$, one has $sf \le |s|a\phi$ and $s(\lambda\dot{e} - \ddot{y}_d) \le |s||\lambda\dot{e} - \ddot{y}_d|$, therefore, it follows that:

$$s(f - \ddot{y}_d + \lambda\dot{e}) \le |s|a\phi + |s||\lambda\dot{e} - \ddot{y}_d| \le \underline{g}\theta|s|\Phi,$$

where

$$\theta = \max\left\{\frac{a}{\underline{g}}, \frac{1}{\underline{g}}\right\}$$

is an unknown virtual parameter, and

$$\Phi(\cdot) = \phi^2 + (\lambda\dot{e} - \ddot{y}_d)^2 + \frac{1}{2} > 0$$

is a computable scalar function.

According to Theorem 2.11, we have:

$$\underline{g}\theta|s|\Phi \le \underline{g}\theta\varepsilon(t) + \frac{\underline{g}\theta s^2\Phi^2}{\sqrt{s^2\Phi^2 + \varepsilon^2(t)}}, \tag{5.13}$$

where $\varepsilon(t)$ is a positive function satisfying $\int_0^t \varepsilon(\tau)d\tau \le \bar{\varepsilon} < \infty$ with $\bar{\varepsilon}$ being an unknown positive constant.

Substituting (5.13) into (5.12), we obtain:

$$\dot{V}_1 \le gsu + \underline{g}\theta\varepsilon(t) + \frac{\underline{g}\theta s^2\Phi^2}{\sqrt{s^2\Phi^2 + \varepsilon^2(t)}}. \tag{5.14}$$

Now we get the following theorem.

5.2.3   THEOREM AND STABILITY ANALYSIS

Theorem 5.2 *Consider the second-order normal-form nonlinear system (5.11). Let Assumptions 5.1, 5.2, 5.4, and 5.5 hold. If the adaptive controller with the adaptive law is designed as:*

$$
\begin{cases}
u = -cs - \dfrac{\hat{\theta}s\Phi^2}{\sqrt{s^2\Phi^2+\varepsilon^2(t)}}, \\[2ex]
\dot{\hat{\theta}} = \gamma\dfrac{s^2\Phi^2}{\sqrt{s^2\Phi^2+\varepsilon^2(t)}}, \quad \hat{\theta}(0) \geq 0,
\end{cases}
\tag{5.15}
$$

with $c > 0$ and $\gamma > 0$ being design parameters, and $\hat{\theta}$ being the parameter estimate of the virtual parameter θ, then the objectives (i)-(ii) are ensured.

Proof. Constructing the Lyapunov function candidate as $V = V_1 + \frac{g}{2\gamma}\tilde{\theta}^2$, where $\tilde{\theta} = \theta - \hat{\theta}$ represents the parameter estimation error. Then the derivative of V along (5.14) is:

$$
\dot{V} \leq gsu + \underline{g}\theta\varepsilon(t) + \frac{g\theta s^2\Phi^2}{\sqrt{s^2\Phi^2+\varepsilon^2(t)}} - \frac{g}{\gamma}\tilde{\theta}\dot{\hat{\theta}}.
\tag{5.16}
$$

According to the adaptive law (5.15), it is not difficult to prove that $\hat{\theta}(t) \geq 0$ for any initial value $\hat{\theta}(0) \geq 0$. Substituting the adaptive controller (5.15) into gsu, we have:

$$
gsu = -gcs^2 - \frac{g\hat{\theta}s^2\Phi^2}{\sqrt{s^2\Phi^2+\varepsilon^2(t)}} \leq -\underline{g}cs^2 - \frac{g\hat{\theta}s^2\Phi^2}{\sqrt{s^2\Phi^2+\varepsilon^2(t)}},
$$

where the fact that $0 < \underline{g} < g$ is used.

Then (5.16) can be rewritten as:

$$
\begin{aligned}
\dot{V} &\leq -\underline{g}cs^2 - \frac{g\hat{\theta}s^2\Phi^2}{\sqrt{s^2\Phi^2+\varepsilon^2(t)}} + \frac{g\theta s^2\Phi^2}{\sqrt{s^2\Phi^2+\varepsilon^2(t)}} - \frac{g}{\gamma}\tilde{\theta}\dot{\hat{\theta}} + \underline{g}\theta\varepsilon(t) \\[1ex]
&= -\underline{g}cs^2 + \frac{g\tilde{\theta}s^2\Phi^2}{\sqrt{s^2\Phi^2+\varepsilon^2(t)}} - \frac{g}{\gamma}\tilde{\theta}\dot{\hat{\theta}} + \underline{g}\theta\varepsilon(t).
\end{aligned}
\tag{5.17}
$$

Substituting the adaptive law (5.15) into (5.17), one has:

$$
\dot{V} \leq -\underline{g}cs^2 + \underline{g}\theta\varepsilon(t),
$$

then it leads to $V(t) + \underline{g}c\int_0^t s^2(\tau)d\tau \leq V(0) + \underline{g}\theta\bar{\varepsilon}$.

Since $V(0)$ and $\underline{g}\theta\bar{\varepsilon}$ are finite positive constants, with the fact that $V(t) \geq 0$ and $\underline{g}c\int_0^t s^2(\tau)d\tau \geq 0$, it follows that $V \in \mathscr{L}_\infty$ and $\int_0^t s^2(\tau)d\tau \in \mathscr{L}_\infty$. According to the expression of the Lyapunov function V, one has $\tilde{\theta} \in \mathscr{L}_\infty$ and $s \in \mathscr{L}_\infty$, i.e., there exists a positive constant $\bar{s}$ such that $|s(t)| \leq \bar{s} < \infty$. Note that $\tilde{\theta} = \theta - \hat{\theta}$ is bounded and θ is an unknown bounded constant, then $\hat{\theta} \in \mathscr{L}_\infty$. As $s = \lambda e + \dot{e}$ and $s \in \mathscr{L}_\infty$, similar to the analysis in (5.7)-(5.10), it is readily derived that e and $\dot{e}$ are bounded. In addition, from the boundedness of y_d, $\dot{y}_d$, and $\ddot{y}_d$, we get that $x \in \mathscr{L}_\infty$, $\dot{x} \in \mathscr{L}_\infty$, $f \in \mathscr{L}_\infty$, and $\phi \in \mathscr{L}_\infty$, then it follows that $\Phi(\cdot) \in \mathscr{L}_\infty$. According to the adaptive controller and adaptive law as shown in (5.15), it is ensured that $u \in \mathscr{L}_\infty$ and $\dot{\hat{\theta}} \in \mathscr{L}_\infty$. Furthermore, from the expression of $\dot{s} = f + gu - \ddot{y}_d + \lambda \dot{e}$, the boundedness of $\dot{s}$ is also guaranteed. Therefore, all signals in the closed-loop system are bounded.

As $s \in \mathscr{L}_\infty$, $\dot{s} \in \mathscr{L}_\infty$, and $\int_0^t s^2(\tau)\,d\tau \in \mathscr{L}_\infty$, with the aid of Barbalat's Lemma, we obtain $\lim_{t\to\infty} s(t) = 0$. By utilizing the analysis process similar to that in Section 5.1, it is readily obtained that $\lim_{t\to\infty} e(t) = 0$ and $\lim_{t\to\infty} \dot{e}(t) = 0$. The proof is completed.

5.2.4   NUMERICAL SIMULATION

To verify the effectiveness of the proposed control, we consider the following second-order normal-form system (5.11) with $f(X,p) = p_1 x^2 + \sin(p_2 \dot{x})$, $p = [p_1, p_2]^\top$, and $g(X) = 1 + 0.05 \sin(x\dot{x})$.

In the simulation, the system parameters are given as $p = [p_1, p_2]^\top = [2, 0.3]^\top$, the reference trajectory is selected as $y_d = \sin(t)$, the initial values of the system states and the parameter estimate are selected as $x(0) = 1$, $\dot{x}(0) = 0.5$, and $\hat{\theta}(0) = 0$, the design parameters are selected as $c = 1$, $\lambda = 1$, $\gamma = 0.1$, and $\varepsilon(t) = 2\exp(-t)$. In addition, according to the expression of $f(X,p)$, one has:

$$|f(X,p)| \le |p_1| x^2 + 1 \le a\phi(X),$$

in which $a = \max\{|p_1|, 1\}$ and $\phi(X) = x^2 + 1$. The simulation results are shown in Fig. 5.1-Fig. 5.6. The evolutions of x and $\dot{x}$ are plotted in Fig. 5.1 and Fig. 5.2. The evolutions of the tracking error and its derivative are plotted in Fig. 5.3 and Fig. 5.4, respectively, implying that the tracking errors asymptotically converge to zero, which is consistent with the theoretical analysis. In addition, the evolutions of the control input and parameter estimate are plotted in Fig. 5.5 and Fig. 5.6, respectively.

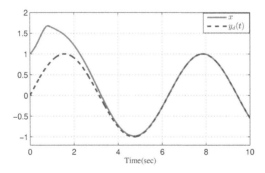

FIGURE 5.1   The evolutions of x and y_d.

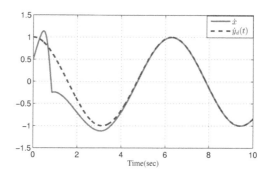

FIGURE 5.2   The evolutions of $\dot{x}$ and $\dot{y}_d$.

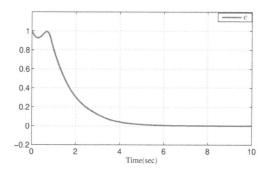

FIGURE 5.3   The evolution of the tracking error e.

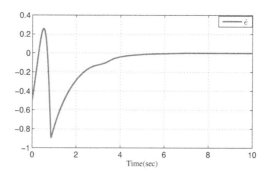

FIGURE 5.4   The evolution of the tracking error $\dot{e}$.

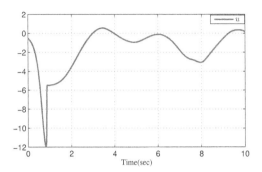

FIGURE 5.5 The evolution of the control input u.

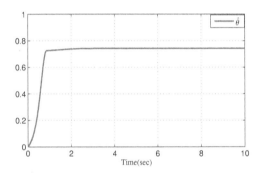

FIGURE 5.6 The evolution of the parameter estimate $\hat{\theta}$.

5.3 ROBUST ADAPTIVE CONTROL OF N-ORDER NORMAL-FORM SYSTEMS

In this section, we extend the filtered error-based robust adaptive control algorithm to n-order normal-form nonlinear systems.

5.3.1 SYSTEM DESCRIPTION

Consider the following n-order normal-form nonlinear systems [42, 84, 85]:

$$x^{(n)} = f(X, p) + g(X)u, \tag{5.18}$$

where $x \in \mathbb{R}$ and $X = \left[x, \dot{x}, \cdots, x^{(n-1)}\right]^{\top} \in \mathbb{R}^n$ denote the system states, $u \in \mathbb{R}$ is the control input, $f(X, p)$ is an uncertain smooth nonlinear function, $g(X) \neq 0$ represents the time-varying and unknown control coefficient, and $p \in \mathbb{R}^r$ denotes an unknown constant parameter vector.

The control objective is to design a robust adaptive control law u such that $x(t)$ and its derivatives $x^{(k)}(t)$, $(k = 1, 2, \cdots, n-1)$, asymptotically track the desired trajectory $y_d(t)$ and its derivatives $y_d^{(k)}(t)$, respectively, i.e., $\lim\limits_{t \to \infty} \left[x^{(i)}(t) - y_d^{(i)}(t) \right] = 0$, $i = 0, 1, \cdots, n-1$.

To this end, the following assumptions are imposed.

Assumption 5.6 *The system state $X = \left[x, \dot{x}, \cdots, x^{(n-1)} \right]^{\top} \in \mathbb{R}^n$ is measurable.*

Assumption 5.7 *The desired trajectory $y_d(t)$ and its derivatives up to n-th are known, bounded, and piecewise continuous.*

Assumption 5.8 *The control gain $g(X)$ is unknown and time-varying but bounded away from zero, i.e., there exist some unknown positive constants $\underline{g}$ and $\bar{g}$ such that $0 < \underline{g} \le g(\cdot) \le \bar{g} < \infty$.*

Assumption 5.9 *For the uncertain nonlinear function $f(X, p)$, there exist an unknown constant $a \ge 0$ and a known and smooth function $\phi(X) \ge 0$ such that $|f(X, p)| \le a\phi(X)$. If X is bounded, so are f and ϕ.*

Before designing the control algorithm, we first introduce the following important Lemma.

5.3.2   FILTERED ERROR-BASED LEMMA

Let

$$e = x - y_d \tag{5.19}$$

be the tracking error, then the filtered error is defined as:

$$s = \lambda_1 e + \lambda_2 \dot{e} + \cdots + \lambda_{n-1} e^{(n-2)} + e^{(n-1)}, \tag{5.20}$$

where λ_k, $k = 1, 2, \cdots, n-1$, denotes a positive constant selected by the designer such that the polynomial $l^{n-1} + \lambda_{n-1} l^{n-2} + \cdots + \lambda_1$ is Hurwitz.

Based on the filtered error (5.20), we present the following lemma.

Lemma 5.1 *For the filter variable s as defined in (5.20), it has the following properties:*

(i)   *If $\lim\limits_{t \to \infty} s(t) = 0$, then $e^{(k)}$, $k = 0, 1, \cdots, n-1$, converges to zero as $t \to \infty$; and*

(ii)   *If $s(t)$ is bounded, so is $e^{(k)}$.*

Proof. To facilitate the proof and analysis in the sequel, (5.20) can be rewritten as the following form:

$$\begin{cases} s = \dot{s}_{n-1} + \alpha_{n-1} s_{n-1}, \\ s_{n-1} = \dot{s}_{n-2} + \alpha_{n-2} s_{n-2}, \\ \vdots \\ s_2 = \dot{s}_1 + \alpha_1 s_1 = \dot{e} + \alpha_1 e, \\ s_1 = e, \end{cases} \tag{5.21}$$

in which the coefficient $\alpha_i > 0$, $i = 1, 2, \cdots, n-1$, is determined by the constant λ_k. The proof of this Lemma includes two steps.

Step 1. We first prove the property (i). By solving the first equation in (5.21), one has:

$$s_{n-1}(t) = \exp(-\alpha_{n-1}t)\, s_{n-1}(0) + \exp(-\alpha_{n-1}t) \int_0^t s(\tau)\exp(\alpha_{n-1}\tau)\,d\tau, \quad (5.22)$$

the boundedness of $s_{n-1}(t)$ can be divided into the following two cases:

If $\int_0^t s(\tau)\exp(\alpha_{n-1}\tau)\,d\tau$ is bounded, it is easy to get that $\lim\limits_{t\to\infty} s_{n-1}(t) = 0$.

If $\int_0^t s(\tau)\exp(\alpha_{n-1}\tau)\,d\tau$ is unbounded, by applying L'Hopital's rule for the term $\exp(-\alpha_{n-1}t)\int_0^t s(\tau)\exp(\alpha_{n-1}\tau)\,d\tau$, one has:

$$\lim_{t\to\infty} s_{n-1}(t) = 0 + \lim_{t\to\infty} \frac{s(t)\exp(\alpha_{n-1}t)}{\alpha_{n-1}\exp(\alpha_{n-1}t)} = \lim_{t\to\infty} \frac{s(t)}{\alpha_{n-1}}. \quad (5.23)$$

As $\lim\limits_{t\to\infty} s(t) = 0$, then it is readily derived from (5.23) that $\lim\limits_{t\to\infty} s_{n-1}(t) = 0$.

From the above analysis, one has $\lim\limits_{t\to\infty} s_{n-1}(t) = 0$ as $\lim\limits_{t\to\infty} s(t) = 0$. Furthermore, according to the first equation in (5.21), $\lim\limits_{t\to\infty} \dot{s}_{n-1}(t) = 0$ is also ensured. Similarly, by solving the second equation in (5.21), it is derived that:

$$s_{n-2}(t) = \exp(-\alpha_{n-2}t)\, s_{n-2}(0) + \exp(-\alpha_{n-2}t) \int_0^t s_{n-1}(\tau)\exp(\alpha_{n-2}\tau)\,d\tau. \quad (5.24)$$

Using the same analysis procedure, $\lim\limits_{t\to\infty} s(t) = 0$ implies that $\lim\limits_{t\to\infty} s_{n-2}(t) = 0$ and $\lim\limits_{t\to\infty} \dot{s}_{n-2}(t) = 0$. Then, using an induction argument for the rest of the equations in (5.21), it follows that $s_k(t)$ and $\dot{s}_k(t)$, $k = 1, 2, \cdots, n-1$, converge to zero as $t \to \infty$.

Upon using the definition of $s_k(t)$, (5.21) can be rewritten as:

$$\begin{cases} s_1 = e, \\ s_2 = \dot{e} + \alpha_1 e, \\ s_3 = \ddot{e} + (\alpha_1 + \alpha_2)\dot{e} + \alpha_1\alpha_2 e, \\ \vdots \\ s_{n-1} = e^{(n-2)} + \cdots, \\ s = e^{(n-1)} + \cdots, \end{cases}$$

then it is readily derived that $e^{(k)}$, $k = 0, 1, \cdots, n-1$, asymptotically converges to zero with $\lim\limits_{t\to\infty} s(t) = 0$.

Step 2. We prove property (ii). Since $s(t)$ is bounded, then there exists an unknown positive constant $\bar{s}$ such that: $|s(t)| \le \bar{s} < \infty$. Then, it is shown from (5.22) that

$$|s_{n-1}(t)| \le \exp(-\alpha_{n-1}t)|s_{n-1}(0)| + \exp(-\alpha_{n-1}t)\left| \int_0^t s(\tau)\exp(\alpha_{n-1}\tau)\,d\tau \right|$$

$$\le \exp(-\alpha_{n-1}t)|s_{n-1}(0)| + \exp(-\alpha_{n-1}t)\int_0^t \bar{s}\exp(\alpha_{n-1}\tau)\,d\tau$$

$$\le \exp(-\alpha_{n-1}t)|s_{n-1}(0)| + \frac{\bar{s}}{\alpha_{n-1}}(1 - \exp(-\alpha_{n-1}t)).$$

Obviously, $s_{n-1}(t)$ is bounded, i.e., there exists a positive constant $\bar{s}_{n-1}$ such that $|s_{n-1}(t)| \leq \bar{s}_{n-1} < \infty$. According to the first equation in (5.21), $\dot{s}_{n-1}$ is also bounded. Similarly, it is readily derived from (5.24) that:

$$|s_{n-2}(t)| \leq \exp(-\alpha_{n-2}t)|s_{n-2}(0)| + \exp(-\alpha_{n-2}t)\int_0^t \bar{s}_{n-1}\exp(\alpha_{n-2}\tau)d\tau$$

$$= \exp(-\alpha_{n-2}t)|s_{n-2}(0)| + \frac{\bar{s}_{n-1}}{\alpha_{n-2}}(1 - \exp(-\alpha_{n-2}t)),$$

which indicates that $s_{n-2}(t)$ and $\dot{s}_{n-2}(t)$ are also bounded. By following the same procedure, $s_k(t)$ and $\dot{s}_k(t)$, $k = 1, 2, \cdots, n-3$, are bounded, which further implies that $e^{(k)}$, $k = 0, \cdots, n-1$, are also bounded. The proof is completed.

5.3.3 CONTROLLER DESIGN

The derivative of the filtered error $s(t)$ as defined in (5.20) w.r.t. time is:

$$\dot{s} = f + gu - y_d^{(n)} + \lambda_1 \dot{e} + \lambda_2 \ddot{e} + \cdots + \lambda_{n-1}e^{(n-1)} = f + gu + L(\cdot), \qquad (5.25)$$

where

$$L(\cdot) = -y_d^{(n)} + \lambda_1 \dot{e} + \lambda_2 \ddot{e} + \cdots + \lambda_{n-1}e^{(n-1)}$$

is a computable function.

Choosing the Lyapunov function candidate as $V_1 = \frac{1}{2}s^2$, the derivative of V_1 along (5.25) yields:

$$\dot{V}_1 = s\dot{s} = s(f + gu + L). \qquad (5.26)$$

For the nonlinear term $s(f + L)$, it follows that $sf \leq |s|a\phi$ and $sL \leq |s||L|$, thus one has:

$$s(f + L) \leq |s|a\phi + |s||L| \leq \underline{g}\theta|s|\Phi,$$

where $\theta = \max\left\{\frac{a}{\underline{g}}, \frac{1}{\underline{g}}\right\}$ represents an unknown virtual parameter, and

$$\Phi(\cdot) = \phi + L^2 + \frac{1}{4} > 0$$

is a computable scalar function.

By using Theorem 2.11, one has:

$$s(f + L) \leq \underline{g}\theta|s|\Phi \leq \underline{g}\theta\varepsilon(t) + \frac{\underline{g}\theta s^2\Phi^2}{\sqrt{s^2\Phi^2 + \varepsilon^2(t)}}, \qquad (5.27)$$

where $\varepsilon(t)$ is some positive function that satisfies $\int_0^t \varepsilon(\tau)d\tau \leq \bar{\varepsilon} < \infty$, with $\bar{\varepsilon}$ being a positive constant.

Substituting (5.27) into (5.26), it is readily derived that:

$$\dot{V}_1 \leq \underline{g}su + \underline{g}\theta\varepsilon(t) + \frac{\underline{g}\theta s^2\Phi^2}{\sqrt{s^2\Phi^2 + \varepsilon^2(t)}}. \qquad (5.28)$$

Now we illustrate the following theorem.

5.3.4 THEOREM AND STABILITY ANALYSIS

Theorem 5.3 *Consider the n-order normal-form nonlinear system as described in (5.18) under Assumptions 5.6-5.9. If the adaptive controller with the adaptive law is designed as:*

$$\begin{cases} u = -cs - \dfrac{\hat{\theta}s\Phi^2}{\sqrt{s^2\Phi^2+\varepsilon^2(t)}}, \\[4mm] \dot{\hat{\theta}} = \dfrac{\gamma s^2\Phi^2}{\sqrt{s^2\Phi^2+\varepsilon^2(t)}}, \quad \hat{\theta}(0) \geq 0, \end{cases} \tag{5.29}$$

where $c > 0$ and $\gamma > 0$ are design parameters, $\hat{\theta}$ is the estimate of the virtual parameter θ, then the proposed control scheme not only guarantees that all signals in the closed-loop system are bounded, but also ensures that the tracking error and its derivatives up to $(n-1)$-th converge to zero asymptotically.

Proof. Choosing the Lyapunov function candidate as $V = V_1 + \frac{g}{2\gamma}\tilde{\theta}^2$, where $\tilde{\theta} = \theta - \hat{\theta}$ denotes the estimation error. Then the derivative of V along (5.28) yields:

$$\dot{V} \leq gsu + g\theta\varepsilon(t) + \frac{g\theta s^2\Phi^2}{\sqrt{s^2\Phi^2+\varepsilon^2(t)}} - \frac{g}{\gamma}\tilde{\theta}\dot{\hat{\theta}}. \tag{5.30}$$

Substituting the control law as shown in (5.29) into gsu, one obtains:

$$gsu = -gcs^2 - \frac{g\hat{\theta}s^2\Phi^2}{\sqrt{s^2\Phi^2+\varepsilon^2(t)}}.$$

Noting that $\hat{\theta}(t) \geq 0$, then it is readily verified that:

$$gsu = -gcs^2 - \frac{g\hat{\theta}s^2\Phi^2}{\sqrt{s^2\Phi^2+\varepsilon^2(t)}} \leq -\underline{g}cs^2 - \frac{g\hat{\theta}s^2\Phi^2}{\sqrt{s^2\Phi^2+\varepsilon^2(t)}},$$

with $0 < \underline{g} < g < \infty$. Therefore, (5.30) can be further expressed as:

$$\dot{V} \leq -\underline{g}cs^2 - \frac{g\hat{\theta}s^2\Phi^2}{\sqrt{s^2\Phi^2+\varepsilon^2(t)}} + \frac{g\theta s^2\Phi^2}{\sqrt{s^2\Phi^2+\varepsilon^2(t)}} - \frac{g}{\gamma}\tilde{\theta}\dot{\hat{\theta}} + g\theta\varepsilon(t)$$

$$= -\underline{g}cs^2 + \frac{g\tilde{\theta}s^2\Phi^2}{\sqrt{s^2\Phi^2+\varepsilon^2(t)}} - \frac{g}{\gamma}\tilde{\theta}\dot{\hat{\theta}} + g\theta\varepsilon(t). \tag{5.31}$$

Substituting the adaptive law as given in (5.29) into (5.31), we obtain:

$$\dot{V} \leq -\underline{g}cs^2 + g\theta\varepsilon(t). \tag{5.32}$$

Integrating (5.32) on $[0,t]$, one has:

$$V(t) + \underline{g}c\int_0^t s^2(\tau)d\tau \leq V(0) + g\theta\bar{\varepsilon}. \tag{5.33}$$

By following the analysis procedure in Section 5.2, it is seen that $\int_0^t s^2(\tau)d\tau \in \mathscr{L}_\infty$, $V \in \mathscr{L}_\infty$, $\hat{\theta} \in \mathscr{L}_\infty$, and $s \in \mathscr{L}_\infty$. Upon using Lemma 5.1, it follows that $e^{(k)}$, $k = 0,1,\cdots,n-1$, is bounded. According to Assumption 5.7, it is shown that $x^{(k)}$ is bounded, which implies that $\phi \in \mathscr{L}_\infty$, $L(\cdot) \in \mathscr{L}_\infty$, $\Phi(\cdot) \in \mathscr{L}_\infty$, $u \in \mathscr{L}_\infty$, $\dot{\hat{\theta}} \in \mathscr{L}_\infty$, and $\dot{s} \in \mathscr{L}_\infty$. Therefore, all signals in the closed-loop system are bounded.

Note that $s \in \mathscr{L}_\infty$, $\dot{s} \in \mathscr{L}_\infty$, and $\int_0^t s^2(\tau)d\tau \in \mathscr{L}_\infty$, by applying Barbalat's Lemma, we have $\lim_{t\to\infty} s(t) = 0$, then it follows from Lemma 5.1 that $\lim_{t\to\infty} e^{(k)} = 0$ for $k = 0,1,\cdots,n-1$, which implies that the tracking error and its derivatives up to $(n-1)$-th converge to zero asymptotically. The proof is completed.

5.3.5 NUMERICAL SIMULATION

To verify the effectiveness of the proposed control, we consider the normal-form nonlinear system (5.18) with $n = 3$, the detailed expressions of the system model are $f(X,p) = x\dot{x} + \exp(-p\dot{x}^2)$, and $g(X) = 1 + 0.1\cos(x\dot{x})$.

In the simulation, the system parameter is given as $p = 2$, the reference trajectory is $y_d = \sin(t)$, the initial conditions of the system states and parameter estimate are taken as $x(0) = 1$, $\dot{x}(0) = 0.5$, $\ddot{x}(0) = 0.2$, $\hat{\theta}(0) = 0$, the design parameters are selected as $c = 1$, $\lambda_1 = 1$, $\lambda_2 = 2.5$, $\gamma = 0.12$, and $\varepsilon(t) = \exp(-0.7t)$. Furthermore, it is seen from the expression of $f(X,p)$ that:

$$|f(X,p)| \le |x||\dot{x}| + 1 \le \frac{1}{2}x^2 + \frac{1}{2}\dot{x}^2 + 1 \le a\phi(X),$$

where $a = 1$ and $\phi(X) = \frac{1}{2}x^2 + \frac{1}{2}\dot{x}^2 + 1$.

Under the proposed control, the simulation results are shown in Fig. 5.7–Fig. 5.10. The evolutions of the system state x and reference signal y_d are plotted in Fig. 5.7. The evolution of the tracking error e is shown in Fig. 5.8, from which it is observed that the tracking error asymptotically converges to zero, which is consistent with the theoretical analysis. Furthermore, the evolutions of the control input u and parameter estimate $\hat{\theta}$ are plotted in Fig. 5.9 and Fig. 5.10, respectively.

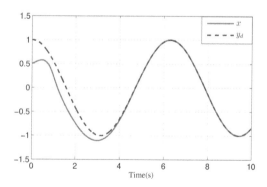

FIGURE 5.7 The evolutions of x and y_d.

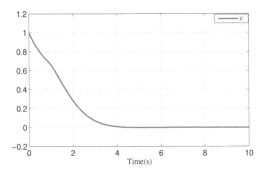

FIGURE 5.8   The evolution of the tracking error e.

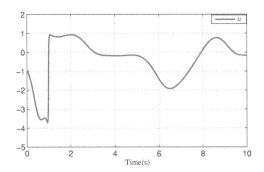

FIGURE 5.9   The evolution of the control input u.

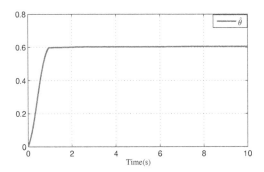

FIGURE 5.10   The evolution of the parameter estimate $\hat{\theta}$.

5.4 NOTES

This chapter presents the control design method for normal-form nonlinear systems with uncertainty. Firstly, by introducing a filtered error, we convert an arbitrary relative degree system into a relative degree one system. Secondly, we develop an important lemma to ensure that the boundedness of the filtered error is able to guarantee the boundedness of the tracking error and its derivatives. Finally, based on the filtered error, we design some adaptive control schemes for normal-form nonlinear systems and the simulation results verify the effectiveness of the proposed approaches.

6 Control of Strict-feedback Systems

The control design of first-order systems and n-order normal-form systems was investigated in Chapter 4 and Chapter 5, respectively. In this chapter, we focus on introducing the control design of strict-feedback nonlinear systems.

6.1 BACKSTEPPING DESIGN WITH UNCERTAINTY

In Section 3.4, we illustrated the main idea and detailed design procedure of the backstepping method for strict-feedback nonlinear systems without uncertainty. The task of control design is much more challenging if the uncertainty is included in the nonlinear systems [9]. In this section, we introduce the adaptive backstepping design in the presence of matched/extended matched/unmatched uncertainty, respectively.

6.1.1 ADAPTIVE BACKSTEPPING WITH MATCHED UNCERTAINTY

Here we briefly introduce the backstepping method with the following second-order strict-feedback nonlinear system in the presence of matched uncertainty:

$$\begin{cases} \dot{x}_1 = x_2 + \varphi_1(x_1), \\ \dot{x}_2 = u + \theta \varphi_2(\bar{x}_2), \\ y = x_1, \end{cases} \tag{6.1}$$

where $x_1 \in \mathbb{R}$ and $x_2 \in \mathbb{R}$ are the system states, $\varphi_1 \in \mathbb{R}$ and $\varphi_2 \in \mathbb{R}$ are the smooth and known functions, $\theta \in \mathbb{R}$ denotes an unknown parameter, $u \in \mathbb{R}$ and $y \in \mathbb{R}$ are the control input and system output, respectively. It should be noted that the "matched uncertainty" means that the uncertainty appears in the same equation as control and uncertainty can be canceled by the designed controller.

Define the coordinate transformation as:

$$\begin{cases} z_1 = x_1, \\ z_2 = x_2 - \alpha_1, \end{cases} \tag{6.2}$$

where α_1 is the virtual controller.

Now we carry out the control design.

Step 1. The derivative of z_1 is:

$$\dot{z}_1 = \dot{x}_1 = x_2 + \varphi_1. \tag{6.3}$$

Noting that $x_2 = z_2 + \alpha_1$, then (6.3) becomes:

$$\dot{z}_1 = z_2 + \alpha_1 + \varphi_1. \tag{6.4}$$

DOI: 10.1201/9781003474364-6

The virtual controller α_1 is designed as:

$$\alpha_1 = -c_1 z_1 - \varphi_1, \tag{6.5}$$

with $c_1 > 0$.

Then (6.4) can be further rewritten as:

$$\dot{z}_1 = z_2 - c_1 z_1, \tag{6.6}$$

where the term z_2 will be handled in Step 2.

Step 2. According to $z_2 = x_2 - \alpha_1$, one has:

$$\dot{z}_2 = \dot{x}_2 - \dot{\alpha}_1 = \theta \varphi_2 + u - \frac{\partial \alpha_1}{\partial x_1}(x_2 + \varphi_1). \tag{6.7}$$

If θ were known for the control design, the actual control u is:

$$u = -c_2 z_2 - \theta \varphi_2 + \frac{\partial \alpha_1}{\partial x_1}(x_2 + \varphi_1) - z_1, \tag{6.8}$$

with $c_2 > 0$ being a design parameter.

Since θ is unknown, we again replace it with its estimate $\hat{\theta}$ in (6.8) to obtain the adaptive control law:

$$u = -c_2 z_2 - \hat{\theta} \varphi_2 + \frac{\partial \alpha_1}{\partial x_1}(x_2 + \varphi_1) - z_1, \tag{6.9}$$

then (6.7) becomes:

$$\dot{z}_2 = \tilde{\theta} \varphi_2 - c_2 z_2 - z_1, \tag{6.10}$$

with $\tilde{\theta} = \theta - \hat{\theta}$ being the parameter estimate error.

Choosing the Lyapunov function candidate as:

$$V = \frac{1}{2} z_1^2 + \frac{1}{2} z_2^2 + \frac{1}{2\gamma} \tilde{\theta}^2,$$

then the derivation of V along (6.6) and (6.10) is:

$$\dot{V} = z_1 \dot{z}_1 + z_2 \dot{z}_2 - \frac{1}{\gamma} \tilde{\theta} \dot{\hat{\theta}} = -c_1 z_1^2 - c_2 z_2^2 + \tilde{\theta}\left(z_2 \varphi_2 - \frac{1}{\gamma}\dot{\hat{\theta}}\right). \tag{6.11}$$

The choice of the update law

$$\dot{\hat{\theta}} = \gamma z_2 \varphi_2$$

eliminates the $\tilde{\theta}$−term in (6.11) and renders the derivative of the Lyapunov function (6.11) non-positive:

$$\dot{V} = -c_1 z_1^2 - c_2 z_2^2 \leq 0.$$

Then using the standard Lyapunov theorem and Barbalat's Lemma, it is ensured that all signals in the closed-loop system are bounded and $\lim_{t \to \infty} z_k(t) = 0$, $k = 1, 2$.

Remark 6.1 *For the following nth-order strict-feedback nonlinear systems with matched uncertainty:*

$$\begin{cases} \dot{x}_k = x_{k+1} + \varphi_k(\bar{x}_k), & k = 1, 2, \cdots, n-1 \\ \dot{x}_n = u + \theta \varphi_n(\bar{x}_n), \\ y = x_1. \end{cases}$$

Let

$$\begin{cases} z_1 = x_1, \\ z_k = x_k - \alpha_{k-1}, & k = 2, 3, \cdots, n, \end{cases} \tag{6.12}$$

be the virtual error, the virtual/actual controllers with the adaptive law can be designed as:

$$\begin{cases} \alpha_1 = -c_1 z_1 - \varphi_1, & c_1 > 0, \\ \alpha_i = -z_{i-1} - c_i z_i + \sum_{k=1}^{i-1} \frac{\partial \alpha_{i-1}}{\partial x_k}(x_{k+1} + \varphi_k), & c_i > 0, \\ u = -z_{n-1} - c_n z_n + \sum_{k=1}^{n-1} \frac{\partial \alpha_{n-1}}{\partial x_k}(x_{k+1} + \varphi_k) - \hat{\theta}\varphi_n, & c_n > 0, \\ \dot{\hat{\theta}} = \gamma z_n \varphi_n, & \gamma > 0, \end{cases}$$

for $i = 2, 3, \cdots, n-1$. The detailed proof is omitted.

6.1.2 ADAPTIVE BACKSTEPPING WITH EXTENDED MATCHED UNCERTAINTY

The adaptive design in Section 6.1.1 is simple because the uncertainty is matched (i.e., the parametric uncertainty is in the spall of control). We now move to the more general case of extended matching (i.e., the parametric uncertainty enters the system one integrator before control)[1]:

$$\begin{cases} \dot{x}_1 = x_2 + \theta \varphi(x_1), \\ \dot{x}_2 = u, \\ y = x_1, \end{cases} \tag{6.13}$$

where $x_1 \in \mathbb{R}$ and $x_2 \in \mathbb{R}$ are the system states, $\varphi \in \mathbb{R}$ is a smooth and known function, $\theta \in \mathbb{R}$ denotes an unknown parameter, $u \in \mathbb{R}$ and $y \in \mathbb{R}$ are the control input and system output, respectively.

If θ were known, upon using the coordinate transformation (6.2) and the Lyapunov function candidate $V = \frac{1}{2}\sum_{k=1}^{2} z_k^2$, the virtual controller and the actual controller are given as:

$$\begin{cases} \alpha_1 = -c_1 z_1 - \theta \varphi, \\ u = -c_2 z_2 - z_1 + \frac{\partial \alpha_1}{\partial x_1}(x_2 + \theta \varphi), \end{cases} \tag{6.14}$$

with $c_k > 0$ $(k = 1, 2)$ being design parameters.

---

[1]Here we take the second-order system as an example.

Since θ is unknown and appears in one equation before the control does, then the virtual control as shown in (6.14) cannot be implementable. Now we can utilize adaptive backstepping to design the controller.

Step 1. The derivative of z_1 is:

$$\dot{z}_1 = \dot{x}_1 = x_2 + \theta\varphi = z_2 + \alpha_1 + \theta\varphi.$$

The virtual controller α_1 is designed as:

$$\alpha_1 = -c_1 z_1 - \vartheta_1 \varphi,$$

where ϑ_1 is the estimate of θ and is updated by:

$$\dot{\vartheta}_1 = \gamma_1 z_1 \varphi, \tag{6.15}$$

where $c_1 > 0$ and $\gamma_1 > 0$ are design parameters.

The $\dot{z}_1$-equation becomes:

$$\dot{z}_1 = -c_1 z_1 + z_2 + (\theta - \vartheta_1)\varphi, \tag{6.16}$$

and the derivative of $V_1 = \frac{1}{2}z_1^2 + \frac{1}{2\gamma_1}(\theta - \vartheta_1)^2$ along (6.16) yields:

$$\dot{V}_1 = z_1 \dot{z}_1 - \frac{1}{\gamma_1}(\theta - \vartheta_1)\dot{\vartheta}_1$$

$$= -c_1 z_1^2 + z_1 z_2 + z_1(\theta - \vartheta_1)\varphi - \frac{1}{\gamma_1}(\theta - \vartheta_1)\dot{\vartheta}_1$$

$$= -c_1 z_1^2 + z_1 z_2, \tag{6.17}$$

where the term $z_1 z_2$ will be coped with in the next step.

Step 2. The derivative of $z_2 = x_2 - \alpha_1$ is:

$$\dot{z}_2 = \dot{x}_2 - \dot{\alpha}_1 = u - \frac{\partial\alpha_1}{\partial x_1}(x_2 + \theta\varphi) - \frac{\partial\alpha_1}{\partial\vartheta_1}\dot{\vartheta}_1$$

$$= u - \frac{\partial\alpha_1}{\partial x_1}(x_2 + \theta\varphi) - \frac{\partial\alpha_1}{\partial\vartheta_1}\gamma_1 z_1 \varphi. \tag{6.18}$$

At this point, we need to select a Lyapunov function and design u to render its derivative non-positive. As we have designed the adaptive law for the parameter estimate in Step 1, then we attempt to select the following Lyapunov function candidate

$$V_2 = V_1 + \frac{1}{2}z_2^2, \tag{6.19}$$

then its derivative along (6.17) and (6.18) is:

$$\dot{V}_2 = \dot{V}_1 + z_2 \dot{z}_2 = -c_1 z_1^2 + z_2\left[u + z_1 - \frac{\partial\alpha_1}{\partial x_1}(x_2 + \theta\varphi) - \frac{\partial\alpha_1}{\partial\vartheta_1}\gamma_1 z_1 \varphi\right].$$

The control u should now be able to cancel the indefinite terms in $\dot{V}_2$. To deal with the terms containing the unknown parameter θ, we will try to employ the existing estimate ϑ_1:

$$u = -z_1 - c_2 z_2 + \frac{\partial \alpha_1}{\partial x_1} x_2 + \frac{\partial \alpha_1}{\partial x_1} \vartheta_1 \varphi + \frac{\partial \alpha_1}{\partial \vartheta_1} \gamma_1 z_1 \varphi,$$

where $c_2 > 0$ is a design parameter.

Therefore, the derivative, $\dot{V}_2$, becomes

$$\dot{V}_2 = -c_1 z_1^2 - c_2 z_2^2 - z_2 \varphi \frac{\partial \alpha_1}{\partial x_1}(\theta - \vartheta_1). \tag{6.20}$$

It is seen from (6.20) that we have no design freedom left to cancel the $(\theta - \vartheta_1)$-term. To overcome this difficulty, we replace the original parameter estimate ϑ_1 in the expression for u with a new estimate ϑ_2, i.e., the new adaptive controller u can be designed as:

$$u = -z_1 - c_2 z_2 + \frac{\partial \alpha_1}{\partial x_1} x_2 + \frac{\partial \alpha_1}{\partial x_1} \vartheta_2 \varphi + \frac{\partial \alpha_1}{\partial \vartheta_1} \gamma_1 z_1 \varphi. \tag{6.21}$$

With the choice (6.21), the z_2-equation becomes:

$$\dot{z}_2 = -c_2 z_2 - z_1 - (\theta - \vartheta_2) \frac{\partial \alpha_1}{\partial x_1} \varphi. \tag{6.22}$$

Due to the presence of the new parameter estimate ϑ_2, the Lyapunov function candidate V_2 as shown in (6.19) must be changed as:

$$V_2 = V_1 + \frac{1}{2} z_2^2 + \frac{1}{2\gamma_2}(\theta - \vartheta_2)^2, \tag{6.23}$$

where $\gamma_2 > 0$ is a design parameter.

The derivative of V_2 defined in (6.23) is:

$$\dot{V}_2 = \dot{V}_1 + z_2 \dot{z}_2 - \frac{1}{\gamma_2}(\theta - \vartheta_2)\dot{\vartheta}_2$$

$$= -c_1 z_1^2 + z_2 \left[u + z_1 - \frac{\partial \alpha_1}{\partial x_1}(x_2 + \theta \varphi) - \frac{\partial \alpha_1}{\partial \vartheta_1} \gamma_1 z_1 \varphi \right] - \frac{1}{\gamma_2}(\theta - \vartheta_2)\dot{\vartheta}_2. \tag{6.24}$$

Substituting the actual control law (6.21) into (6.24), one has:

$$\dot{V}_2 = -\sum_{k=1}^{2} c_k z_k^2 - z_2 \frac{\partial \alpha_1}{\partial x_1}(\theta - \vartheta_2)\varphi - \frac{1}{\gamma_2}(\theta - \vartheta_2)\dot{\vartheta}_2$$

$$= -\sum_{k=1}^{2} c_k z_k^2 - (\theta - \vartheta_2)\left[z_2 \frac{\partial \alpha_1}{\partial x_1}\varphi + \frac{1}{\gamma_2}\dot{\vartheta}_2 \right].$$

Now the $(\theta - \vartheta_2)$-term can be eliminated with the update law

$$\dot{\vartheta}_2 = -\gamma_2 z_2 \frac{\partial \alpha_1}{\partial x_1} \varphi, \tag{6.25}$$

which yields $\dot{V} = -\sum_{k=1}^{2} c_k z_k^2 \leq 0$.

The equations (6.16) and (6.22) along with (6.15) and (6.25) form the error system representation of the resulting closed-loop adaptive system:

$$\begin{cases} \dot{z}_1 = -c_1 z_1 + z_2 + (\theta - \vartheta_1)\varphi, \\ \dot{z}_2 = -c_2 z_2 - z_1 - (\theta - \vartheta_2)\frac{\partial \alpha_1}{\partial x_1}\varphi, \\ \dot{\vartheta}_1 = \gamma_1 z_1 \varphi, \\ \dot{\vartheta}_2 = -\gamma_2 z_2 \frac{\partial \alpha_1}{\partial x_1}\varphi. \end{cases}$$

Using the standard Lyapunov theorem in reference [9] and Barbalat's Lemma, it is ensured that all signals in the closed-loop system are bounded and $\lim_{t \to \infty} z_k(t) = 0$, $k = 1, 2$.

6.1.3   ADAPTIVE BACKSTEPPING WITH UNMATCHED UNCERTAINTY

In this section, we consider the following form of nonlinear systems, which is called "parametric strict-feedback form":

$$\begin{cases} \dot{x}_1 = x_2 + \theta^\top \varphi_1(x_1), \\ \dot{x}_2 = u + \theta^\top \varphi_2(\bar{x}_2), \\ y = x_1, \end{cases} \tag{6.26}$$

where $\varphi_k \in \mathbb{R}^r$, $k = 1, 2$, is a known and smooth nonlinear function, $\theta \in \mathbb{R}^r$ is an unknown constant parameter vector. Such a form of uncertainty is called the unmatched uncertainty.

To carry out the control design, we still employ the coordinate transformation (6.2) by utilizing the backstepping method.

Step 1. The derivative of $z_1 = x_1$ is:

$$\dot{z}_1 = \dot{x}_1 = z_2 + \alpha_1 + \theta^\top \varphi_1. \tag{6.27}$$

The virtual controller α_1 is designed as:

$$\alpha_1 = -c_1 z_1 - \vartheta_1^\top \varphi_1, \tag{6.28}$$

where ϑ_1 is the parameter estimate of θ, and $c_1 > 0$ is a design parameter.

Substituting (6.28) into (6.27) yields:

$$\dot{z}_1 = z_2 - c_1 z_1 + (\theta - \vartheta_1)^\top \omega_1,$$

with $\omega_1 = \varphi_1$.

The derivative of the quadratic function $V_1 = \frac{1}{2}z_1^2 + \frac{1}{2}(\theta - \vartheta_1)^\top \Gamma_1^{-1}(\theta - \vartheta_1)$ is:

$$\dot{V}_1 = z_1 z_2 - c_1 z_1^2 + z_1(\theta - \vartheta_1)^\top \omega_1 - (\theta - \vartheta_1)^\top \Gamma_1^{-1} \dot{\vartheta}_1.$$

By choosing the following adaptive law for $\dot{\vartheta}_1$ as:

$$\dot{\vartheta}_1 = \Gamma_1 z_1 \omega_1,$$

with $\Gamma_1 = \Gamma_1^\top > 0$ being an adaption gain matrix, $\dot{V}_1$ becomes:

$$\dot{V}_1 = z_1 z_2 - c_1 z_1^2.$$

Step 2. As α_1 is the function of variables x_1 and ϑ_1, then the derivative of z_2 is:

$$\dot{z}_2 = \dot{x}_2 - \dot{\alpha}_1 = u + \theta^\top \varphi_2 - \frac{\partial \alpha_1}{\partial x_1}(x_2 + \theta^\top \varphi_1) - \frac{\partial \alpha_1}{\partial \vartheta_1}\dot{\vartheta}_1$$

$$= u + \theta^\top \omega_2 - \frac{\partial \alpha_1}{\partial x_1}x_2 - \frac{\partial \alpha_1}{\partial \vartheta_1}\dot{\vartheta}_1, \qquad (6.29)$$

with $\omega_2 = \varphi_2 - \frac{\partial \alpha_1}{\partial x_1}\varphi_1$.

The actual control law u is:

$$u = -z_1 - c_2 z_2 - \vartheta_2^\top \omega_2 + \frac{\partial \alpha_1}{\partial x_1}x_2 + \frac{\partial \alpha_1}{\partial \vartheta_1}\dot{\vartheta}_1, \qquad (6.30)$$

where c_2 is a positive design parameter, ϑ_2 denotes the parameter estimate of θ in Step 2 and is updated by:

$$\dot{\vartheta}_2 = \Gamma_2 z_2 \omega_2, \qquad (6.31)$$

with $\Gamma_2 = \Gamma_2^\top > 0$ being an adaption gain matrix.

Substituting the actual control law as shown in (6.30) into (6.29), it is checked that:

$$\dot{z}_2 = -z_1 - c_2 z_2 + \omega_2^\top (\theta - \vartheta_2). \qquad (6.32)$$

Choosing the whole Lyapunov function candidate as $V_2 = V_1 + \frac{1}{2}z_2^2 + \frac{1}{2}(\theta - \vartheta_2)^\top \Gamma_2^{-1}(\theta - \vartheta_2)$, then, with the aid of (6.31), the derivative of V_2 along (6.32) yields:

$$\dot{V}_2 = -\sum_{k=1}^{2} c_k z_k^2 + z_2 \omega_2^\top (\theta - \vartheta_2) - (\theta - \vartheta_2)^\top \Gamma_2^{-1}\dot{\vartheta}_2 = -\sum_{k=1}^{2} c_k z_k^2 \leq 0.$$

With the similar stability analysis in the aforementioned sections, it is not difficult to get that: 1) all signals in the closed-loop systems are bounded; and 2) $\lim_{t \to \infty} z_k(t) = 0$, $k = 1, 2$, is guaranteed.

Remark 6.2 *By employing the design procedure in Section 6.1.3, for the n-order parametric strict-feedback nonlinear systems:*

$$\begin{cases} \dot{x}_k = x_{k+1} + \theta^\top \varphi_k(\bar{x}_k), & k = 1, 2, \cdots, n-1, \\ \dot{x}_n = u + \theta^\top \varphi_n(\bar{x}_n), \\ y = x_1, \end{cases} \qquad (6.33)$$

the virtual/actual controllers and the adaptive law are designed as:

$$
\begin{cases}
\alpha_1 = -c_1 z_1 - \vartheta_1^\top \omega_1, \\
\alpha_i = -z_{i-1} - c_i z_i - \vartheta_i^\top \omega_i + \sum_{k=1}^{i-1} \frac{\partial \alpha_{i-1}}{\partial x_k} x_{k+1} + \sum_{k=1}^{i-1} \frac{\partial \alpha_{i-1}}{\partial \vartheta_k} \dot{\vartheta}_k, \\
u = \alpha_n, \\
\dot{\vartheta}_j = \Gamma_j z_j \omega_j, \\
\omega_1 = \varphi_1, \quad \omega_i = \varphi_i - \sum_{k=1}^{i-1} \frac{\partial \alpha_{i-1}}{\partial x_k} \varphi_k,
\end{cases}
$$

for $j = 1, 2, \cdots, n$, $i = 2, 3, \cdots, n$, where $c_j > 0$ denotes a positive design parameter and $\Gamma_j = \Gamma_j^\top > 0$ is an adaption gain matrix, and ϑ_j is the parameter estimate of the unknown constant θ in each step. The detailed proof and stability analysis are omitted. Interested readers may refer to reference [9].

6.2   ADAPTIVE BACKSTEPPING WITH TUNING FUNCTION

For the above backstepping designs of strict-feedback nonlinear systems with extended matched uncertainty or unmatched uncertainty, it is not difficult to see that in each step of backstepping designs, a new parameter estimate must be imposed such that the derivative of the Lyapunov function is semi-negative definite, which may cause the problem of over-parametrization, dramatically increasing the computational burden.

Take the control design of the parametric strict-feedback system in Section 6.1.3 as an example. For systems in the form (6.33) the number of design steps required is equal to the degree n of the system. At each step, an error variable z_i ($i = 1, 2, \cdots, n$), a virtual controller α_i, and a parameter estimate ϑ_i are generated. As a result, if the dimension of the system parameter θ is r, the over-parameterized adaptive controller may employ as many as $n \times r$ parameter estimates. Therefore, to solve the problem of over-parametrization, we introduce an important concept, the tuning function, such that the number of parameter estimates is equal to the number of unknown parameters. This minimum-order design is advantageous not only for implementation but also because it guarantees the strongest achievable stability and convergence properties.

In the tuning function procedure, the parameter update law is designed recursively. At each consecutive step, we design a tuning function as a potential update law. In contrast to the adaptive backstepping design in Section 6.1.3, these intermediate update laws are not implemented. Instead, the controller uses them to compensate for the effects of parameter estimation transients. Only the final tuning function is used as the parameter update law.

6.2.1   CONTROL DESIGN

In this section, we focus on introducing the detailed design procedures of adaptive backstepping design with tuning function. Take the n-order parametric strict-feedback nonlinear systems as an example, we rewrite the expression in what follows:

$$\begin{cases} \dot{x}_k = x_{k+1} + \theta^\top \varphi_k(\bar{x}_k), \quad k = 1, 2, \cdots, n-1, \\ \dot{x}_n = u + \theta^\top \varphi_n(\bar{x}_n), \\ y = x_1, \end{cases} \tag{6.34}$$

where $x_i \in \mathbb{R}$ represents the system state and $\bar{x}_i = [x_1, x_2, \cdots, x_i]^\top$, $i = 1, 2, \cdots, n$, $\varphi_i \in \mathbb{R}^r$ is a smooth and known nonlinear function, $\theta \in \mathbb{R}^r$ denotes the unknown constant parameter, $u \in \mathbb{R}$ and $y \in \mathbb{R}$ are the control input and system output, respectively.

The objective of this section is to design an adaptive stabilization controller such that:

(i)   all signals in the closed-loop system are bounded;
(ii)   the system output converges to zero asymptotically;
(iii) the number of the parameter estimate is equal to the number of the parameter θ.

For n-order parametric strict-feedback nonlinear systems, n steps are required. The first $n-1$ steps are to construct the virtual controllers, and the actual control law is given in step n.

Now we carry out the control design.

Step 1. The derivative of $z_1 = x_1$ w.r.t. time is:

$$\dot{z}_1 = \dot{x}_1 = \theta^\top \varphi_1 + x_2 = \theta^\top \varphi_1 + z_2 + \alpha_1. \tag{6.35}$$

It should be noted that as the system parameter θ is unknown, then to design the virtual/actual controller, an adaptive control scheme can be employed to estimate the unknown parameter θ. The virtual controller α_1 is selected as:

$$\alpha_1 = -c_1 z_1 - \hat{\theta}^\top \varphi_1, \tag{6.36}$$

where $c_1 > 0$ denotes a design parameter, and $\hat{\theta}$ is the estimate of the unknown parameter θ.

Substituting (6.36) into (6.35), one has:

$$\dot{z}_1 = z_2 - c_1 z_1 + \tilde{\theta}^\top \varphi_1, \tag{6.37}$$

where $\tilde{\theta} = \theta - \hat{\theta}$ denotes the parameter estimation error.

Choosing the following quadratic function $V_1 = \frac{1}{2} z_1^2 + \frac{1}{2} \tilde{\theta}^\top \Gamma^{-1} \tilde{\theta}$, where $\Gamma = \Gamma^\top > 0$ is a design matrix, thus, the time derivative of V_1 along (6.37) yields:

$$\dot{V}_1 = z_1 \left(z_2 - c_1 z_1 + \tilde{\theta}^\top \varphi_1 \right) - \tilde{\theta}^\top \Gamma^{-1} \dot{\hat{\theta}}$$

$$= z_1 z_2 - c_1 z_1^2 + \tilde{\theta}^\top \Gamma^{-1} \left(\Gamma \tau_1 - \dot{\hat{\theta}} \right), \tag{6.38}$$

where $\tau_1 = z_1 \varphi_1$, the term $z_1 z_2$ will be handled in Step 2.

Step 2. Since α_1 is the function of x_1 and $\hat{\theta}$, the derivative of the virtual error z_2 as defined in (6.12) w.r.t. time is:

$$\dot{z}_2 = z_3 + \alpha_2 + \theta^\top \varphi_2 - \frac{\partial \alpha_1}{\partial x_1}\left(x_2 + \theta^\top \varphi_1\right) - \frac{\partial \alpha_1}{\partial \hat{\theta}}\dot{\hat{\theta}}$$

$$= z_3 + \alpha_2 + \hat{\theta}^\top \omega_2 + \tilde{\theta}^\top \omega_2 - \frac{\partial \alpha_1}{\partial x_1}x_2 - \frac{\partial \alpha_1}{\partial \hat{\theta}}\dot{\hat{\theta}}, \qquad (6.39)$$

where $\omega_2 = \varphi_2 - \frac{\partial \alpha_1}{\partial x_1}\varphi_1$ and $x_3 = z_3 + \alpha_2$.

Choosing the virtual controller α_2 as:

$$\alpha_2 = -c_2 z_2 - z_1 - \hat{\theta}^\top \omega_2 + \frac{\partial \alpha_1}{\partial x_1}x_2 + \frac{\partial \alpha_1}{\partial \hat{\theta}}\Gamma \tau_2, \qquad (6.40)$$

with

$$\tau_2 = \tau_1 + z_2 \omega_2,$$

where $c_2 > 0$ is a design parameter.

Substituting the virtual controller as shown in (6.40) into (6.39), we have:

$$\dot{z}_2 = z_3 - c_2 z_2 - z_1 + \tilde{\theta}^\top \omega_2 + \frac{\partial \alpha_1}{\partial \hat{\theta}}\left(\Gamma \tau_2 - \dot{\hat{\theta}}\right).$$

Choosing the quadratic function V_2 as $V_2 = V_1 + \frac{1}{2}z_2^2$, the derivative of V_2 is:

$$\dot{V}_2 = -\sum_{k=1}^{2}c_k z_k^2 + z_2 z_3 + z_2 \hat{\theta}^\top \omega_2 + \tilde{\theta}^\top \Gamma^{-1}\left(\Gamma \tau_1 - \dot{\hat{\theta}}\right) + z_2 \frac{\partial \alpha_1}{\partial \hat{\theta}}\left(\Gamma \tau_2 - \dot{\hat{\theta}}\right). \quad (6.41)$$

Note that $\Gamma \tau_1 - \dot{\hat{\theta}}$ can be decomposed into:

$$\Gamma \tau_1 - \dot{\hat{\theta}} = \Gamma \tau_1 - \Gamma \tau_2 + \Gamma \tau_2 - \dot{\hat{\theta}} = -\Gamma z_2 \omega_2 + \Gamma \tau_2 - \dot{\hat{\theta}}.$$

Therefore, (6.41) can be rewritten as:

$$\dot{V}_2 = -\sum_{k=1}^{2}c_k z_k^2 + z_2 z_3 + \left(z_2 \frac{\partial \alpha_1}{\partial \hat{\theta}} + \tilde{\theta}^\top \Gamma^{-1}\right)\left(\Gamma \tau_2 - \dot{\hat{\theta}}\right).$$

Step 3. To clarify the idea of the tuning-function-based backstepping design more clearly, it is necessary to introduce the design skill of the virtual controller in this step. Since $x_4 = z_4 + \alpha_3$, then $\dot{x}_3 = x_4 + \varphi_3^\top \theta$ can be rewritten as:

$$\dot{z}_3 = x_4 + \theta^\top \varphi_3 - \dot{\alpha}_2 = z_4 + \alpha_3 + \hat{\theta}^\top \omega_3 + \tilde{\theta}^\top \omega_3 - \sum_{k=1}^{2}\frac{\partial \alpha_2}{\partial x_k}x_{k+1} - \frac{\partial \alpha_2}{\partial \hat{\theta}}\dot{\hat{\theta}}, \quad (6.42)$$

where $\omega_3 = \varphi_3 - \sum_{k=1}^{2}\frac{\partial \alpha_2}{\partial x_k}\varphi_k$.

Designing the virtual controller α_3 as:

$$
\begin{cases}
\alpha_3 = -c_3 z_3 - z_2 - \hat{\theta}^\top \omega_3 + \sum_{k=1}^{2} \frac{\partial \alpha_2}{\partial x_k} x_{k+1} + \frac{\partial \alpha_2}{\partial \hat{\theta}} \Gamma \tau_3 + z_2 \frac{\partial \alpha_1}{\partial \hat{\theta}} \Gamma \omega_3, \\
\tau_3 = \tau_2 + \omega_3 z_3,
\end{cases}
\tag{6.43}
$$

where $c_3 > 0$.

Substituting the expression of the virtual controller α_3 into (6.42), one has:

$$
\dot{z}_3 = z_4 - c_3 z_3 - z_2 + \tilde{\theta}^\top \omega_3 + \frac{\partial \alpha_2}{\partial \hat{\theta}} \left(\Gamma \tau_3 - \dot{\hat{\theta}} \right) + z_2 \frac{\partial \alpha_1}{\partial \hat{\theta}} \Gamma \omega_3.
\tag{6.44}
$$

Choosing the quadratic function as $V_3 = V_2 + \frac{1}{2} z_3^2$, the derivative of V_3 can be expressed as:

$$
\dot{V}_3 = -\sum_{k=1}^{3} c_k z_k^2 + z_3 z_4 + z_2 z_3 \frac{\partial \alpha_1}{\partial \hat{\theta}} \Gamma \omega_3 + z_3 \tilde{\theta}^\top \omega_3
$$
$$
+ \left(z_2 \frac{\partial \alpha_1}{\partial \hat{\theta}} + \tilde{\theta}^\top \Gamma^{-1} \right) \left(\Gamma \tau_2 - \dot{\hat{\theta}} \right) + z_3 \frac{\partial \alpha_2}{\partial \hat{\theta}} \left(\Gamma \tau_3 - \dot{\hat{\theta}} \right).
\tag{6.45}
$$

Noting that $\tau_3 = \tau_2 + \omega_3 z_3$, then $\Gamma \tau_2 - \dot{\hat{\theta}} = \Gamma \tau_2 - \Gamma \tau_3 + \Gamma \tau_3 - \dot{\hat{\theta}} = -\Gamma \omega_3 z_3 + \Gamma \tau_3 - \dot{\hat{\theta}}$.
Therefore,

$$
\left(z_2 \frac{\partial \alpha_1}{\partial \hat{\theta}} + \tilde{\theta}^\top \Gamma^{-1} \right) \left(\Gamma \tau_2 - \dot{\hat{\theta}} \right) = \left(z_2 \frac{\partial \alpha_1}{\partial \hat{\theta}} + \tilde{\theta}^\top \Gamma^{-1} \right) \left(-\Gamma \omega_3 z_3 + \Gamma \tau_3 - \dot{\hat{\theta}} \right)
$$
$$
= -z_2 \frac{\partial \alpha_1}{\partial \hat{\theta}} \Gamma \omega_3 z_3 - \tilde{\theta}^\top \omega_3 z_3 + \left(z_2 \frac{\partial \alpha_1}{\partial \hat{\theta}} + \tilde{\theta}^\top \Gamma^{-1} \right) \left(\Gamma \tau_3 - \dot{\hat{\theta}} \right).
\tag{6.46}
$$

The first and second terms in the right-hand side of (6.46) are able to cancel the third and fourth terms in the right-hand side of (6.45), thus (6.45) can be simplified to:

$$
\dot{V}_3 = -\sum_{k=1}^{3} c_k z_k^2 + z_3 z_4 + \left(z_2 \frac{\partial \alpha_1}{\partial \hat{\theta}} + z_3 \frac{\partial \alpha_2}{\partial \hat{\theta}} + \tilde{\theta}^\top \Gamma^{-1} \right) \left(\Gamma \tau_3 - \dot{\hat{\theta}} \right).
\tag{6.47}
$$

Remark 6.3 *The major difference in the virtual controller design between Step 3 and the first two steps is the term $z_2 \frac{\partial \alpha_1}{\partial \hat{\theta}} \Gamma \omega_3$, leading to the term $z_2 z_3 \frac{\partial \alpha_1}{\partial \hat{\theta}} \Gamma \omega_3$ appears in the derivative of V_3 [as shown in (6.45)], the purpose of employing such a term is used to cancel the first term $-z_2 \frac{\partial \alpha_1}{\partial \hat{\theta}} \Gamma \omega_3 z_3$ in the right-hand side of (6.46).*

Step $j \, (j = 4, 5, \cdots, n)$. According to (6.12), we have:

$$
\dot{z}_j = \dot{x}_j - \dot{\alpha}_{j-1} = z_{j+1} + \alpha_j + \theta^\top \varphi_j - \sum_{k=1}^{j-1} \frac{\partial \alpha_{j-1}}{\partial x_k} \left(x_{k+1} + \theta^\top \varphi_k \right) - \frac{\partial \alpha_{j-1}}{\partial \hat{\theta}} \dot{\hat{\theta}}.
$$

Therefore, the virtual controller, the actual controller, and the adaptive law are designed as:

$$\alpha_j = -c_j z_j - z_{j-1} - \hat{\theta}^\top \omega_j + \sum_{k=1}^{j-1} \frac{\partial \alpha_{j-1}}{\partial x_k} x_{k+1}$$

$$+ \frac{\partial \alpha_{j-1}}{\partial \hat{\theta}} \Gamma \tau_j + \sum_{k=2}^{j-1} \frac{\partial \alpha_{k-1}}{\partial \hat{\theta}} \Gamma \omega_j z_k, \tag{6.48}$$

$$\tau_j = \tau_{j-1} + \omega_j z_j, \quad \omega_j = \varphi_j - \sum_{k=1}^{j-1} \frac{\partial \alpha_{j-1}}{\partial x_k} \varphi_k, \tag{6.49}$$

$$u = \alpha_n, \tag{6.50}$$

$$\dot{\hat{\theta}} = \Gamma \tau_n, \tag{6.51}$$

with $c_j > 0$.

The stability analysis of the closed-loop system under the controller (6.50) is given in the following theorem.

6.2.2 THEOREM AND STABILITY ANALYSIS

Theorem 6.1 *For the strict-feedback nonlinear system (6.34), if the adaptive controller (6.50) with the adaptive law (6.51) is applied, the proposed control scheme ensures that the objectives (i)-(iii) are achieved.*

Proof. According to the virtual controller (6.48) and actual controller (6.50), the following closed-loop system is obtained:

$$\dot{z}_1 = z_2 - c_1 z_1 + \tilde{\theta}^\top \varphi_1, \tag{6.52}$$

$$\dot{z}_2 = z_3 - c_2 z_2 - z_1 + \tilde{\theta}^\top \omega_2 + \frac{\partial \alpha_1}{\partial \hat{\theta}} \left(\Gamma \tau_2 - \dot{\hat{\theta}} \right), \tag{6.53}$$

$$\dot{z}_3 = z_4 - c_3 z_3 - z_2 + \tilde{\theta}^\top \omega_3 + \frac{\partial \alpha_2}{\partial \hat{\theta}} \left(\Gamma \tau_3 - \dot{\hat{\theta}} \right) + z_2 \frac{\partial \alpha_1}{\partial \hat{\theta}} \Gamma \omega_3, \tag{6.54}$$

$$\dot{z}_j = z_{j+1} - c_j z_j - z_{j-1} + \tilde{\theta}^\top \omega_j + \frac{\partial \alpha_{j-1}}{\partial \hat{\theta}} \left(\Gamma \tau_j - \dot{\hat{\theta}} \right) + \sum_{k=1}^{j-2} z_{k+1} \frac{\partial \alpha_k}{\partial \hat{\theta}} \Gamma \omega_j, \tag{6.55}$$

$$\dot{z}_n = -c_n z_n - z_{n-1} + \tilde{\theta}^\top \omega_n + \frac{\partial \alpha_{n-1}}{\partial \hat{\theta}} \left(\Gamma \tau_n - \dot{\hat{\theta}} \right) + \sum_{k=1}^{n-2} z_{k+1} \frac{\partial \alpha_k}{\partial \hat{\theta}} \Gamma \omega_n, \tag{6.56}$$

for $j = 4, 5, \cdots, n-1$.

Defining the Lyapunov function candidate as $V = \frac{1}{2} \sum_{k=1}^n z_k^2 + \frac{1}{2} \tilde{\theta}^\top \Gamma^{-1} \tilde{\theta}$. With the aid of (6.52)-(6.56), the derivative of the Lyapunov function candidate is:

$$\dot{V} = -\sum_{k=1}^n c_k z_k^2 \leq 0. \tag{6.57}$$

Integrating (6.57) on $[0,t]$, one has:

$$V(t) + c_1 \int_0^t z_1^2(\tau) d\tau + \cdots + c_n \int_0^t z_n^2(\tau) d\tau = V(0). \qquad (6.58)$$

Now, we first prove that all signals in the closed-loop system are bounded. According to (6.58), it is seen that $V(t)$ is bounded, then the error z_k $(k = 1, 2, \cdots, n)$ and the parameter estimation error $\tilde{\theta}$ are bounded. Since $\tilde{\theta} = \theta - \hat{\theta}$ and θ is an unknown yet bounded constant, then $\hat{\theta} \in \mathscr{L}_\infty$. Note that $z_1 = x_1$, it shows that $x_1 \in \mathscr{L}_\infty$ and $\varphi_1 \in \mathscr{L}_\infty$, then it is not difficult to get the boundedness of the virtual controller α_1 and $\dot{z}_1$. As z_2 and α_1 are bounded, then it follows that the system state x_2 is bounded, thus $\varphi_2(\bar{x}_2) \in \mathscr{L}_\infty$. In addition, due to the smoothness and boundedness of the virtual controller α_1, it is shown that $\frac{\partial \alpha_1}{\partial x_1}$ and $\frac{\partial \alpha_1}{\partial \hat{\theta}}$ are bounded. Thus, $\omega_2 \in \mathscr{L}_\infty$, $\tau_2 \in \mathscr{L}_\infty$, and $\alpha_2 \in \mathscr{L}_\infty$. Utilizing the similar analysis in Step 2, it is guaranteed that the system states x_i $(i = 3, 4, \cdots, n)$, the smooth functions φ_i, the partial derivatives of the virtual controllers, the virtual controllers, the actual controller u, the adaptive law $\dot{\hat{\theta}}$, and the derivative of the virtual error $\dot{z}_i$ are bounded. Therefore, all signals of the closed-loop systems are bounded.

Secondly, we prove that the virtual error converges to zero asymptotically. According to (6.58), it is seen that $z_k \in \mathscr{L}_2$ $(k = 1, 2, \cdots, n)$. Note that $z_k \in \mathscr{L}_\infty$ and $\dot{z}_k \in \mathscr{L}_\infty$, thus, by using Barbalat's Lemma, one has $\lim\limits_{t \to \infty} z_k(t) = 0$, $k = 1, 2, \cdots, n$. The proof is completed.

Remark 6.4 *The backstepping design with the tuning function for tracking is only a minor modification of the stabilization design procedure, to achieve the goal of asymptotic tracking, we define the following coordinate transformation:*

$$\begin{cases} z_1 = x_1 - y_d(t), \\ z_k = x_k - \alpha_{k-1}, \quad k = 2, 3, \cdots, n, \end{cases}$$

with $y_d(t)$ being the reference signal and $y_d^{(k)}(t)$, $k = 0, 1, \cdots, n$, being assumed to be known, bounded, and piecewise continuous. The virtual and actual controllers with the adaptive law should be modified as:

$$\alpha_1 = -c_1 z_1 - \hat{\theta}^\top \varphi_1 + \dot{y}_d,$$

$$\alpha_2 = -c_2 z_2 - z_1 - \hat{\theta}^\top \omega_2 + \frac{\partial \alpha_1}{\partial x_1} x_2 + \sum_{k=0}^{1} \frac{\partial \alpha_1}{\partial y_d^{(k)}} y_d^{(k+1)} + \frac{\partial \alpha_1}{\partial \hat{\theta}} \Gamma \tau_2,$$

$$\alpha_j = -c_j z_j - z_{j-1} - \omega_j^\top \hat{\theta} + \sum_{k=1}^{j-1} \frac{\partial \alpha_{j-1}}{\partial x_k} x_{k+1} + \sum_{k=0}^{j-1} \frac{\partial \alpha_{j-1}}{\partial y_d^{(k)}} y_d^{(k+1)}$$

$$+ \frac{\partial \alpha_{j-1}}{\partial \hat{\theta}} \Gamma \tau_j + \sum_{k=2}^{j-1} \frac{\partial \alpha_{k-1}}{\partial \hat{\theta}} \Gamma \omega_j z_k,$$

$$\tau_j = \tau_{j-1} + \omega_j z_j, \quad \omega_j = \varphi_j - \sum_{k=1}^{j-1}$$

$$\frac{\partial \alpha_{j-1}}{\partial x_k} \varphi_k,$$

$$u = \alpha_n,$$

$$\dot{\hat{\theta}} = \Gamma \tau_n,$$

where $\Gamma = \Gamma^\top > 0$ is an adaptation gain matrix. The detailed proof and stability analysis are similar to those in Section 6.1.3, thus they are omitted.

6.2.3 NUMERICAL SIMULATION

To demonstrate the effectiveness of the adaptive control scheme proposed in Section 6.2.1, the following second-order nonlinear system is employed:

$$\begin{cases} \dot{x}_1 = x_2 + \varphi_1(x_1)\theta, \\ \dot{x}_2 = u, \end{cases} \tag{6.59}$$

where $\varphi_1(x_1) = x_1^2$, $\theta = 1$.

In the simulation, the reference signal is given as $y_d = 0.2\sin(t)$. The design parameters are chosen as: $c_1 = 1$, $c_2 = 5$, and $\Gamma = 1$. The initial values of the system states and parameter estimate are selected as: $x_1(0) = 1$, $x_2(0) = -1$, and $\hat{\theta}(0) = 0$. Under the control scheme in Remark 6.4, the simulation results are shown in Fig. 6.1-Fig. 6.4. The evolutions of the system output x_1 and desired signal $y_d(t)$ are shown in Fig. 6.1. The evolution of the tracking error z_1 is plotted in Fig. 6.2, from which it is seen that the tracking error converges to zero asymptotically, which verifies the effectiveness of the adaptive algorithm and confirms the theoretical analysis. In addition, the evolutions of the adaptive control law and parameter estimate are shown in Fig. 6.3 and Fig. 6.4, respectively, which are bounded for all time.

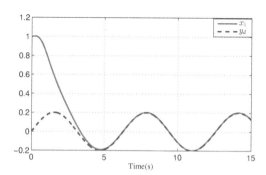

FIGURE 6.1 The evolutions of x_1 and y_d.

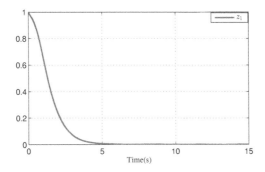

FIGURE 6.2   The evolution of the tracking error e.

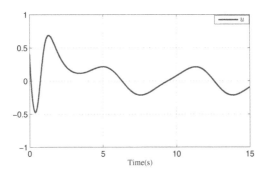

FIGURE 6.3   The evolution of the control law u.

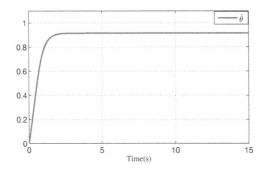

FIGURE 6.4   The evolution of the parameter estimate $\hat{\theta}$.

6.3 ROBUST ADAPTIVE CONTROL OF NON-PARAMETRIC STRICT-FEEDBACK SYSTEMS

6.3.1 PROBLEM FORMULATION

In Section 6.2, the precondition of the corresponding control algorithm is that the nonlinear function f_k ($k = 1, \cdots, n$) in (6.34) must satisfy the parameterized decomposition condition, i.e., $f_k(\bar{x}_k, \theta) = \theta^\top \varphi_k(\bar{x}_k)$. If the nonlinear term does not satisfy such a condition, the corresponding control scheme will fail. Therefore, how to deal with the case of non-parametric uncertainty in strict-feedback nonlinear systems is a difficult yet challenging work.

Moreover, the system model considered in Section 6.2 is simple, which is mainly reflected in:

(i) the parameters in each subsystem are all equal to θ;
(ii) the control gain is 1, which greatly reduces the difficulty of controller design.

However, in practical engineering systems, the parameter in each subsystem is normally in different values and the control gain is unknown or even time-varying. To reflect the above two points, the following strict-feedback nonlinear system is considered:

$$\begin{cases} \dot{x}_k = f_k(\bar{x}_k, \theta_k) + g_k(\bar{x}_k)x_{k+1}, \ k = 1, 2, \cdots, n-1, \\ \dot{x}_n = f_n(\bar{x}_n, \theta_n) + g_n(\bar{x}_n)u, \\ y = x_1, \end{cases} \qquad (6.60)$$

where $x_1 \in \mathbb{R}, \cdots, x_n \in \mathbb{R}$ denote the system states and $\bar{x}_i = [x_1, \cdots, x_i]^\top$, $i = 1, 2, \cdots, n$, $f_k \in \mathbb{R}$ denotes a smooth nonlinear function, $\theta_k \in \mathbb{R}^{r_k}$ is an unknown parameter vector, $g_k(\bar{x}_k) \in \mathbb{R}$ is the control coefficient; $u \in \mathbb{R}$ and $y \in \mathbb{R}$ are the control input and system output, respectively.

If the nonlinear function f_k does not satisfy the parameterized decomposition condition, the form shown in (6.60) is called the "non-parametric strict-feedback system". Obviously, the system (6.34) that satisfies the parameterized decomposition condition is a special case of (6.60).

In this section, the control objective for the non-parametric strict-feedback nonlinear system (6.60) is to design a robust adaptive controller such that:

(i) all signals in the closed-loop systems are bounded;
(ii) the system output $y = x_1$ tracks the reference signal closely and the tracking error converges to a small compact set around zero.

To this end, the following assumptions are introduced.

Assumption 6.1 *The system states* $x_1, x_2, \cdots, x_n$ *are measurable.*

Assumption 6.2 *The control gain* $g_k(\bar{x}_k), k = 1, 2, \cdots, n$ *is unknown and time-varying. In addition, there exist unknown constants* $\underline{g}_k$ *and* $\bar{g}_k$ *such that* $0 < \underline{g}_k \le g_k(\bar{x}_k) \le \bar{g}_k < \infty$.

Assumption 6.3 *The reference signal $y_d(t)$ and its derivatives $y_d^{(k)}(t)$, $k = 1, 2, \cdots, n$, are known, bounded, and piecewise continuous.*

Assumption 6.4 *For the unknown nonlinear function $f_k(\bar{x}_k, \theta_k)$, there exist an unknown constant $a_k > 0$ and a known smooth function $\phi_k(\bar{x}_k) \geq 0$ such that:*

$$|f_k(\bar{x}_k, \theta_k)| \leq a_k \phi_k(\bar{x}_k), \quad k = 1, 2, \cdots, n. \tag{6.61}$$

If $\bar{x}_k$ is bounded, so do f_k and ϕ_k.

6.3.2 CONTROL DESIGN

To develop a robust adaptive control scheme for the non-parametric strict-feedback nonlinear system (6.60), we define the following coordinate transformation:

$$\begin{cases} z_1 = x_1 - y_d, \\ z_k = x_k - \alpha_{k-1}, \quad k = 2, 3, \cdots, n, \end{cases} \tag{6.62}$$

where z_1 is the actual tracking error, z_k $(k = 2, 3, \cdots, n)$ is the virtual error, α_{k-1} is the virtual controller that will be designed later.

Now we carry out the robust adaptive control design.

Step 1. According to the coordinate transformation (6.62), the derivative of the tracking error z_1 w.r.t. time is:

$$\dot{z}_1 = f_1 + g_1 z_2 + g_1 \alpha_1 - \dot{y}_d. \tag{6.63}$$

Noting that the nonlinear function f_1 does not satisfy the parameterized decomposition condition and the control gain g_1 is unavailable for control design, thus it is impossible to design the virtual controller directly. To describe the design idea of the robust adaptive method clearly, the derivative of $V_{11} = \frac{1}{2} z_1^2$ along (6.63) is:

$$\dot{V}_{11} = z_1 \dot{z}_1 = z_1 (f_1 + g_1 z_2 + g_1 \alpha_1 - \dot{y}_d) = g_1 z_1 \alpha_1 + g_1 z_1 z_2 + \Xi_1, \tag{6.64}$$

where $\Xi_1 = z_1 (f_1 - \dot{y}_d)$ is the lumped uncertainty.

In this section, as the control gain g_k is time-varying and unknown, how to deal with the problem of unmatched non-parametric uncertainty f_1 is a challenging topic. As such, Assumption 6.4 in terms of core function is employed. It is seen from (6.61) that this expression indeed is an equality condition, which implies that, compared with the tuning function-based adaptive control in Section 6.2, some new technologies must be imposed to handle the non-parametric uncertainty f_1.

Now we utilize Young's inequality to cope with the non-parametric uncertainty, it follows that:

$$z_1 f_1 \leq |z_1| a_1 \phi_1 \leq \underline{g}_1 a_1^2 z_1^2 \phi_1^2 + \frac{1}{4\underline{g}_1}, \quad -z_1 \dot{y}_d \leq \underline{g}_1 z_1^2 \dot{y}_d^2 + \frac{1}{4\underline{g}_1}.$$

Thus, the lumped uncertainty term Ξ_1 can be upper bounded by:

$$\Xi_1 \leq \underline{g}_1 b_1 z_1^2 \Phi_1 + \frac{1}{2\underline{g}_1},$$

with $b_1 = \max\{1, a_1^2\}$ denoting an unknown constant virtual parameter and

$$\Phi_1 = \phi_1^2 + \dot{y}_d^2 \geq 0$$

being an available function.

Therefore, (6.64) can be rewritten as:

$$\dot{V}_{11} \leq g_1 z_1 \alpha_1 + g_1 z_1 z_2 + \underline{g}_1 b_1 z_1^2 \Phi_1 + \frac{1}{2\underline{g}_1}. \tag{6.65}$$

Designing the virtual controller α_1 as:

$$\alpha_1 = -c_1 z_1 - \hat{b}_1 z_1 \Phi_1, \tag{6.66}$$

where $c_1 > 0$ is the design parameter, $\hat{b}_1$ is the estimate of the virtual parameter b_1, which satisfies the following condition:

$$\hat{b}_1(t) \geq 0, \quad \text{for } t \geq 0. \tag{6.67}$$

Substituting the virtual controller as given in (6.66) into the term $g_1 z_1 \alpha_1$, one has:

$$g_1 z_1 \alpha_1 = -g_1 c_1 z_1^2 - g_1 \hat{b}_1 z_1^2 \Phi_1. \tag{6.68}$$

According to Assumption 6.2, one has $-g_1 \leq -\underline{g}_1$. Thus, (6.68) becomes:

$$g_1 z_1 \alpha_1 \leq -\underline{g}_1 c_1 z_1^2 - \underline{g}_1 \hat{b}_1 z_1^2 \Phi_1, \tag{6.69}$$

then (6.65) can be further expressed as:

$$\dot{V}_{11} \leq g_1 z_1 z_2 - \underline{g}_1 c_1 z_1^2 + \underline{g}_1 \tilde{b}_1 z_1^2 \Phi_1 + \frac{1}{2\underline{g}_1}, \tag{6.70}$$

where $\tilde{b}_1 = b_1 - \hat{b}_1$ is the parameter estimation error. The term $g_1 z_1 z_2$ will be handled in Step 2.

To deal with the term $\underline{g}_1 \tilde{b}_1 z_1^2 \Phi_1$, a quadratic function is introduced as $V_{12} = \frac{\underline{g}_1}{2\gamma_1} \tilde{b}_1^2$, where $\gamma_1 > 0$ denotes the design parameter. According to (6.70), the derivative of $V_1 = V_{11} + V_{12}$ along (6.70) is:

$$\dot{V}_1 \leq g_1 z_1 z_2 - \underline{g}_1 c_1 z_1^2 + \underline{g}_1 \tilde{b}_1 z_1^2 \Phi_1 + \frac{1}{2\underline{g}_1} + \frac{\underline{g}_1}{\gamma_1} \tilde{b}_1 \dot{\tilde{b}}_1. \tag{6.71}$$

Since b_1 is an unknown constant and $\tilde{b}_1 = b_1 - \hat{b}_1$, one has $\dot{\tilde{b}}_1 = -\dot{\hat{b}}_1$, then (6.71) can be further rewritten as:

$$\dot{V}_1 \leq g_1 z_1 z_2 - \underline{g}_1 c_1 z_1^2 + \underline{g}_1 \tilde{b}_1 z_1^2 \Phi_1 + \frac{1}{2\underline{g}_1} - \frac{\underline{g}_1}{\gamma_1} \tilde{b}_1 \dot{\hat{b}}_1. \tag{6.72}$$

To ensure that the condition on parameter estimate as given in (6.67) holds for all time, the adaptive law is designed as:

$$\dot{\hat{b}}_1 = \gamma_1 z_1^2 \Phi_1 - \sigma_1 \hat{b}_1, \quad \hat{b}_1(0) \geq 0, \tag{6.73}$$

where γ_1 and σ_1 are positive design parameters, $\hat{b}_1(0)$ is the initial condition of the parameter estimate $\hat{b}_1(t)$. According to Theorem 2.12, it is not difficult to get that $\hat{b}_1(t) \geq 0$ holds for $t \geq 0$.

Substituting the expression of the adaptive law as defined in (6.73) into $-\frac{g_1}{\gamma_1}\tilde{b}_1\dot{\hat{b}}_1$, it follows that:

$$-\frac{g_1}{\gamma_1}\tilde{b}_1\dot{\hat{b}}_1 = -\frac{g_1}{\gamma_1}\tilde{b}_1\left(\gamma_1 z_1^2 \Phi_1 - \sigma_1 \hat{b}_1\right) = -\underline{g}_1\tilde{b}_1 z_1^2 \Phi_1 + \frac{g_1\sigma_1}{\gamma_1}\tilde{b}_1\hat{b}_1.$$

Therefore, (6.72) becomes:

$$\dot{V}_1 \leq g_1 z_1 z_2 - \underline{g}_1 c_1 z_1^2 + \frac{1}{2g_1} + \frac{g_1\sigma_1}{\gamma_1}\tilde{b}_1\hat{b}_1. \tag{6.74}$$

Since

$$\frac{g_1\sigma_1}{\gamma_1}\tilde{b}_1\hat{b}_1 = \frac{g_1\sigma_1}{\gamma_1}\left(b_1\tilde{b}_1 - \tilde{b}_1^2\right) \leq \frac{g_1\sigma_1}{\gamma_1}\left(\frac{1}{2}b_1^2 + \frac{1}{2}\tilde{b}_1^2 - \tilde{b}_1^2\right),$$

then (6.74) can be simplified as:

$$\dot{V}_1 \leq g_1 z_1 z_2 - \underline{g}_1 c_1 z_1^2 - \frac{g_1\sigma_1}{2\gamma_1}\tilde{b}_1^2 + \Delta_1, \tag{6.75}$$

where $\Delta_1 = \frac{1}{2g_1} + \frac{g_1\sigma_1}{2\gamma_1}b_1^2 > 0$ is an unknown constant, and $g_1 z_1 z_2$ will be handled in Step 2.

Step 2. According to the coordinate transformation (6.62), it follows that $x_3 = z_3 + \alpha_2$, then the derivative of the virtual error z_2 w.r.t. time is:

$$\dot{z}_2 = \dot{x}_2 - \dot{\alpha}_1 = f_2 + g_2(z_3 + \alpha_2) - \dot{\alpha}_1. \tag{6.76}$$

As the virtual controller α_1 is the function of $x_1, y_d, \dot{y}_d$, and the parameter estimate $\hat{b}_1$, then one has:

$$\dot{\alpha}_1 = \frac{\partial \alpha_1}{\partial x_1}\dot{x}_1 + \sum_{k=0}^{1}\frac{\partial \alpha_1}{\partial y_d^{(k)}}y_d^{(k+1)} + \frac{\partial \alpha_1}{\partial \hat{b}_1}\dot{\hat{b}}_1.$$

Substituting the first subsystem of the controlled system into (6.76), we have:

$$\dot{z}_2 = f_2 + g_2(z_3 + \alpha_2) - \frac{\partial \alpha_1}{\partial x_1}(g_1 x_2 + f_1) + \ell_2,$$

where $\ell_2 = -\sum_{k=0}^{1}\frac{\partial \alpha_1}{\partial y_d^{(k)}}y_d^{(k+1)} - \frac{\partial \alpha_1}{\partial \hat{b}_1}\dot{\hat{b}}_1$ is computable, which leads to

$$z_2\dot{z}_2 = g_2 z_2(z_3 + \alpha_2) + \Xi_2, \tag{6.77}$$

with

$$\Xi_2 = z_2 \left(f_2 - \frac{\partial \alpha_1}{\partial x_1} (g_1 x_2 + f_1) + \ell_2 \right)$$

being an uncertain function.

Define the quadratic function as $V_{21} = V_1 + \frac{1}{2} z_2^2$, with the aid of (6.75) and (6.77), one has:

$$\dot{V}_{21} \le -\underline{g}_1 c_1 z_1^2 - \frac{g_1 \sigma_1}{2\gamma_1} \tilde{b}_1^2 + \Delta_1 + g_2 z_2 (z_3 + \alpha_2) + \Xi_2', \tag{6.78}$$

where $\Xi_2' = g_1 z_1 z_2 + \Xi_2$ is the lumped uncertainty.

Upon using Young's inequality, one has:

$$z_2 f_2 \le |z_2| a_2 \phi_2 \le \underline{g}_2 a_2^2 z_2^2 \phi_2^2 + \frac{1}{4\underline{g}_2}, \tag{6.79}$$

$$-z_2 \frac{\partial \alpha_1}{\partial x_1} f_1 \le |z_2| \left| \frac{\partial \alpha_1}{\partial x_1} \right| a_1 \phi_1 \le \underline{g}_2 z_2^2 \left(\frac{\partial \alpha_1}{\partial x_1} \right)^2 a_1^2 \phi_1^2 + \frac{1}{4\underline{g}_2}, \tag{6.80}$$

$$-z_2 \frac{\partial \alpha_1}{\partial x_1} g_1 x_2 \le |z_2| \left| \frac{\partial \alpha_1}{\partial x_1} x_2 \right| \bar{g}_1 \le \underline{g}_2 z_2^2 \left(\frac{\partial \alpha_1}{\partial x_1} x_2 \right)^2 \bar{g}_1^2 + \frac{1}{4\underline{g}_2}, \tag{6.81}$$

$$z_2 \ell_2 \le \underline{g}_2 z_2^2 \ell_2^2 + \frac{1}{4\underline{g}_2}, \tag{6.82}$$

$$g_1 z_1 z_2 \le \underline{g}_2 z_1^2 z_2^2 + \frac{\bar{g}_1^2}{4\underline{g}_2}. \tag{6.83}$$

Then Ξ_2' can be upper bounded by:

$$\Xi_2' \le \underline{g}_2 b_2 z_2^2 \Phi_2 + \frac{4 + \bar{g}_1^2}{4\underline{g}_2},$$

where

$$b_2 = \max \left\{ 1, a_1^2, a_2^2, \bar{g}_1^2 \right\} > 0$$

is the virtual unknown parameter, and

$$\Phi_2 = \phi_2^2 + \left(\frac{\partial \alpha_1}{\partial x_1} \right)^2 \phi_1^2 + \left(\frac{\partial \alpha_1}{\partial x_1} x_2 \right)^2 + \ell_2^2 + z_1^2 \ge 0$$

is the computable function.

Thus, (6.78) can be further rewritten as:

$$\dot{V}_{21} \le -\underline{g}_1 c_1 z_1^2 - \frac{g_1 \sigma_1}{2\gamma_1} \tilde{b}_1^2 + \Delta_1 + g_2 z_2 (z_3 + \alpha_2) + \underline{g}_2 b_2 z_2^2 \Phi_2 + \frac{4 + \bar{g}_1^2}{4\underline{g}_2}. \tag{6.84}$$

The virtual controller α_2 is designed as:

$$\alpha_2 = -c_2 z_2 - \hat{b}_2 z_2 \Phi_2, \tag{6.85}$$

where $c_2 > 0$ is a design parameter, $\hat{b}_2$ is the estimate of the virtual parameter b_2, which satisfies the following adaptive law:

$$\dot{\hat{b}}_2 = \gamma_2 z_2^2 \Phi_2 - \sigma_2 \hat{b}_2, \quad \hat{b}_2(0) \geq 0, \tag{6.86}$$

where γ_2 and σ_2 are positive design parameters, $\hat{b}_2(0)$ is the initial value of the parameter estimate $\hat{b}_2(t)$. It is worth noting that, according to (6.86) and the initial value of the parameter estimate, it is not difficult to get that $\hat{b}_2(t) \geq 0$ holds for $t \geq 0$.

Invoking the virtual controller as shown in (6.85) into the term $g_2 z_2 \alpha_2$, one obtains:

$$g_2 z_2 \alpha_2 \leq -\underline{g}_2 c_2 z_2^2 - \underline{g}_2 \hat{b}_2 z_2^2 \Phi_2,$$

where the fact that $-g_2 \leq -\underline{g}_2$ is used. Then (6.84) can be expressed as:

$$\dot{V}_{21} \leq g_2 z_2 z_3 - \sum_{k=1}^{2} \underline{g}_k c_k z_k^2 - \frac{\underline{g}_1 \sigma_1}{2\gamma_1} \tilde{b}_1^2 + \Delta_1 + \underline{g}_2 \tilde{b}_2 z_2^2 \Phi_2 + \frac{4 + \bar{g}_1^2}{4\underline{g}_2}, \tag{6.87}$$

where $\tilde{b}_2 = b_2 - \hat{b}_2$ is the parameter estimation error.

To deal with the term $\underline{g}_2 \tilde{b}_2 z_2^2 \Phi_2$, we introduce a quadratic function $V_2 = V_{21} + \frac{\underline{g}_2}{2\gamma_2} \tilde{b}_2^2$. Together with (6.87), the derivative of quadratic function V_2 is:

$$\dot{V}_2 \leq g_2 z_2 z_3 - \sum_{k=1}^{2} \underline{g}_k c_k z_k^2 - \frac{\underline{g}_1 \sigma_1}{2\gamma_1} \tilde{b}_1^2 + \Delta_1 + \underline{g}_2 \tilde{b}_2 z_2^2 \Phi_2 + \frac{4 + \bar{g}_1^2}{4\underline{g}_2} - \frac{\underline{g}_2}{\gamma_2} \tilde{b}_2 \dot{\hat{b}}_2. \tag{6.88}$$

Substituting the adaptive law as given in (6.86) into $-\frac{\underline{g}_2}{\gamma_2} \tilde{b}_2 \dot{\hat{b}}_2$, one has:

$$-\frac{\underline{g}_2}{\gamma_2} \tilde{b}_2 \dot{\hat{b}}_2 = -\frac{\underline{g}_2}{\gamma_2} \tilde{b}_2 \left(\gamma_2 z_2^2 \Phi_2 - \sigma_2 \hat{b}_2 \right) = -\underline{g}_2 \tilde{b}_2 z_2^2 \Phi_2 + \frac{\underline{g}_2 \sigma_2}{\gamma_2} \tilde{b}_2 \hat{b}_2.$$

Since $\frac{\underline{g}_2 \sigma_2}{\gamma_2} \tilde{b}_2 \hat{b}_2 \leq \frac{\underline{g}_2 \sigma_2}{\gamma_2} \left(\frac{1}{2} b_2^2 - \frac{1}{2} \tilde{b}_2^2 \right)$, then (6.88) becomes:

$$\dot{V}_2 \leq g_2 z_2 z_3 - \sum_{k=1}^{2} \underline{g}_k c_k z_k^2 - \sum_{k=1}^{2} \frac{\underline{g}_2 \sigma_2}{2\gamma_2} \tilde{b}_2^2 + \Delta_2, \tag{6.89}$$

where $\Delta_2 = \Delta_1 + \frac{4 + \bar{g}_1^2}{4\underline{g}_2} + \frac{\underline{g}_2 \sigma_2}{2\gamma_2} b_2^2 > 0$ is an unknown constant, and the term $g_2 z_2 z_3$ will be handled in Step 3.

Step i $(i = 3, 4, \cdots, n)$. To describe the subsequent controller design procedures, the following virtual parameters are defined:

$$b_i = \max\left\{ 1, a_1^2, \cdots, a_{i-1}^2, a_i^2, \bar{g}_1^2, \cdots, \bar{g}_{i-1}^2 \right\}. \tag{6.90}$$

According to the coordinate transformation (6.62), we have $x_{i+1} = z_{i+1} + \alpha_i$. It should be noted that if $i = n$, $x_{n+1} = \alpha_n = u$ and $z_{n+1} = 0$, then the derivative of the virtual error z_i is:

$$\dot{z}_i = \dot{x}_i - \dot{\alpha}_{i-1} = f_i + g_i(z_{i+1} + \alpha_i) - \dot{\alpha}_{i-1}.$$

As the virtual controller α_{i-1} is the function of states $x_1, \cdots, x_{i-1}$, reference signals $y_d, \dot{y}_d, \cdots, y_d^{(i-1)}$, and the parameter estimation $\hat{b}_1, \cdots, \hat{b}_{i-1}$, then one has:

$$\dot{\alpha}_{i-1} = \sum_{k=1}^{i-1} \frac{\partial \alpha_{i-1}}{\partial x_k}(f_k + g_k x_{k+1}) + \sum_{k=0}^{i-1} \frac{\partial \alpha_1}{\partial y_d^{(k)}} y_d^{(k+1)} + \sum_{k=1}^{i-1} \frac{\partial \alpha_{i-1}}{\partial \hat{b}_k}\dot{\hat{b}}_k,$$

therefore,

$$\dot{z}_i = f_i + g_i(z_{i+1} + \alpha_i) - \sum_{k=1}^{i-1} \frac{\partial \alpha_{i-1}}{\partial x_k}(f_k + g_k x_{k+1}) + \ell_i,$$

where $\ell_i = -\sum_{k=0}^{i-1} \frac{\partial \alpha_1}{\partial y_d^{(k)}} y_d^{(k+1)} - \sum_{k=1}^{i-1} \frac{\partial \alpha_{i-1}}{\partial \hat{b}_k}\dot{\hat{b}}_k$ is available for control design, then $\frac{d}{dt}\left(\frac{1}{2}z_i^2\right)$ can be written as:

$$z_i\dot{z}_i = g_i z_i(z_{i+1} + \alpha_i) + \Xi_i,$$

where $\Xi_i = z_i\left(f_i - \sum_{k=1}^{i-1} \frac{\partial \alpha_{i-1}}{\partial x_k}(f_k + g_k x_{k+1}) + \ell_i\right)$ is an uncertain nonlinear function.

Choosing the quadratic function as $V_i = V_{i-1} + \frac{1}{2}z_i^2 + \frac{g_i}{2\gamma_i}\tilde{b}_i^2$, where $\tilde{b}_i = b_i - \hat{b}_i$ is the parameter estimate error with $\hat{b}_i$ being the estimate of the unknown virtual parameter b_i, γ_i is the positive design parameter, then the derivative of the quadratic function V_i is:

$$\dot{V}_i \leq -\sum_{k=1}^{i-1} g_k c_k z_k^2 - \sum_{k=1}^{i-1} \frac{g_k \sigma_k}{2\gamma_k}\tilde{b}_k^2 + \Delta_{i-1} + g_i z_i(z_{i+1} + \alpha_i) + \Xi_i' - \frac{g_i}{\gamma_i}\tilde{b}_i\dot{\hat{b}}_i, \quad (6.91)$$

where $\Xi_i' = \Xi_i + g_{i-1}z_{i-1}z_i$ is the lumped uncertainties.

Similar to (6.79)-(6.83), it follows that:

$$z_i f_i \leq \underline{g}_i a_i^2 z_i^2 \phi_i^2 + \frac{1}{4\underline{g}_i},$$

$$-z_i \sum_{k=1}^{i-1} \frac{\partial \alpha_{i-1}}{\partial x_k}f_k \leq \underline{g}_i z_i^2 \sum_{k=1}^{i-1}\left(\frac{\partial \alpha_{i-1}}{\partial x_k}\right)^2 a_k^2 \phi_k^2 + \frac{i-1}{4\underline{g}_i},$$

$$-z_i \sum_{k=1}^{i-1} \frac{\partial \alpha_{i-1}}{\partial x_k}g_k x_{k+1} \leq \underline{g}_i z_i^2 \sum_{k=1}^{i-1}\left(\frac{\partial \alpha_{i-1}}{\partial x_k}x_{k+1}\right)^2 \bar{g}_k^2 + \frac{i-1}{4\underline{g}_i},$$

$$z_i \ell_i \leq \underline{g}_i z_i^2 \ell_i^2 + \frac{1}{4\underline{g}_i},$$

$$g_{i-1}z_{i-1}z_i \leq \underline{g}_i z_{i-1}^2 z_i^2 + \frac{\bar{g}_{i-1}^2}{4\underline{g}_i}.$$

Then Ξ_i' can be upper bounded by:

$$\Xi_i' \leq \underline{g}_i b_i z_i^2 \Phi_i + \frac{2i + \bar{g}_{i-1}^2}{4\underline{g}_i},$$

where $b_i > 0$ is given in (6.90), and

$$\Phi_i = \phi_i^2 + \sum_{k=1}^{i-1}\left(\frac{\partial \alpha_{i-1}}{\partial x_k}\right)^2 \phi_k^2 + \sum_{k=1}^{i-1}\left(\frac{\partial \alpha_{i-1}}{\partial x_k}x_{k+1}\right)^2 + \ell_i^2 + z_{i-1}^2 \geq 0.$$

Therefore, (6.91) can be further rewritten as:

$$\dot{V}_i \leq -\sum_{k=1}^{i-1}\underline{g}_k c_k z_k^2 - \sum_{k=1}^{i-1}\frac{\underline{g}_k \sigma_k}{2\gamma_k}\tilde{b}_k^2 + \Delta_{i-1} + g_i z_i(z_{i+1} + \alpha_i)$$
$$+ \underline{g}_i b_i z_i^2 \Phi_i + \frac{2i + \bar{g}_{i-1}^2}{4\underline{g}_i} - \frac{g_i}{\gamma_i}\tilde{b}_i\dot{\hat{b}}_i. \tag{6.92}$$

The virtual/actual controller α_i with the adaptive law is designed as:

$$\begin{cases} \alpha_i = -c_i z_i - \hat{b}_i z_i \Phi_i, \\ u = \alpha_n, \\ \dot{\hat{b}}_i = \gamma_i z_i^2 \Phi_i - \sigma_i \hat{b}_i, \quad \hat{b}_i(0) \geq 0, \end{cases} \tag{6.93}$$

where c_i, γ_i, and σ_i are positive design parameters, $\hat{b}_i(0)$ are the initial values of the parameter estimates $\hat{b}_i(t)$.

Substituting the virtual/actual controller as given in (6.93) into $g_i z_i \alpha_i$, it follows that:

$$g_i z_i \alpha_i \leq -\underline{g}_i c_i z_i^2 - \underline{g}_i \hat{b}_i z_i^2 \Phi_i. \tag{6.94}$$

Using (6.94) and the adaptive law as shown in (6.93), (6.92) becomes:

$$\dot{V}_i \leq -\sum_{k=1}^{i}\underline{g}_k c_k z_k^2 - \sum_{k=1}^{i}\frac{\underline{g}_k \sigma_k}{2\gamma_k}\tilde{b}_k^2 + \Delta_i + g_i z_i z_{i+1}, \tag{6.95}$$

where $\Delta_i = \Delta_{i-1} + \frac{2i + \bar{g}_{i-1}^2}{4\underline{g}_i} + \frac{\underline{g}_i \sigma_i}{2\gamma_i}b_i^2 > 0$ is an unknown constant.

Note that $g_i z_i z_{i+1} = 0$ as $i = n$, then (6.95) can be simplified as:

$$\dot{V}_n \leq -\sum_{k=1}^{n}\underline{g}_k c_k z_k^2 - \sum_{k=1}^{n}\frac{\underline{g}_k \sigma_k}{2\gamma_k}\tilde{b}_k^2 + \Delta_n. \tag{6.96}$$

6.3.3 THEOREM AND STABILITY ANALYSIS

Theorem 6.2 *For the non-parametric strict-feedback nonlinear system (6.60), under Assumptions 6.1-6.4, if the robust adaptive controller (6.93) with the adaptive law is applied, then the objectives (i)-(ii) can be ensured.*

Proof. Choosing the Lyapunov function candidate as $V = V_n$. According to (6.96), one has:

$$\dot{V} \leq -l_1 V + l_2, \tag{6.97}$$

where $l_1 = \min\left\{2\underline{g}_k c_k, \sigma_k\right\} > 0$, $l_2 = \Delta_n$, $k = 1, 2, \cdots, n$.

Firstly, we prove that all signals in the closed-loop system are bounded. According to (6.97), it is seen that $V(t)$ is bounded, thus $z_k \in \mathscr{L}_\infty$, $\tilde{b}_k \in \mathscr{L}_\infty$, $k = 1, 2, \cdots, n$. As $\tilde{b}_k = b_k - \hat{b}_k$ and b_k is a constant, then it follows that $\hat{b}_k \in \mathscr{L}_\infty$. Noting that $z_1 = x_1 - y_d$ and $y_d \in \mathscr{L}_\infty$, then $x_1 \in \mathscr{L}_\infty$, it further follows from Assumption 5.8 that f_1 and ϕ_1 are bounded, which implies that $\Phi_1 \in \mathscr{L}_\infty$. Therefore, the virtual controller α_1 and the adaptive law $\dot{\hat{b}}_1$ are bounded. Using the similar analysis in step 1, we get that the system states x_i $(i = 2, \cdots, n)$, virtual controllers α_j $(j = 2, \cdots, n-1)$, actual controller u, and adaptive laws $\dot{\hat{b}}_i$ are all bounded. Therefore, all signals in the closed-loop systems are bounded.

Secondly, we analyze the tracking performance. (6.97) can be written as:

$$\dot{V} \le -\underline{g}_1 c_1 z_1^2 + l_2.$$

If $|z_1| > \sqrt{\dfrac{l_2 + v}{\underline{g}_1 c_1}}$ with $v > 0$ being a small constant, then $\dot{V}$ becomes negative and $|z_1|$ enters and remains in the compact set

$$\Omega_{z1} = \left\{ z_1 \in \mathbb{R} \,\middle|\, |z_1| \le \sqrt{\dfrac{l_2 + v}{\underline{g}_1 c_1}} \right\},$$

which implies that the tracking performance can be improved by increasing the design parameter c_1. The proof is completed.

6.3.4   NUMERICAL SIMULATION

To verify the effectiveness of the robust adaptive algorithm in Section 6.3.2, the following non-parametric second-order nonlinear system is considered:

$$\begin{cases} \dot{x}_1 = g_1(x_1)x_2 + f_1(x_1, \theta_1), \\ \dot{x}_2 = g_2(\bar{x}_2)u, \end{cases}$$

where $f_1(x_1, \theta_1) = \theta_{11}x_1 + \exp\left(-\theta_{12}x_1^2\right)$ with $\theta = [\theta_{11}, \theta_{12}]^\top = [2, 1]^\top$, $g_1(x_1) = 3 + 0.2\cos(x_1)$, and $g_2(\bar{x}_2) = 2 + 0.1\sin(x_1 x_2)$.

According to the expression of f_1, it is shown that f_1 does not satisfy the parameter decomposition condition, however, it is not difficult to verify that f_1 satisfies Assumption 6.4, therefore the core function can be chosen as $\phi_1 = x_1^2 + 1$. Furthermore, as $f_2 = 0$, then $\phi_2 = 0$. In the simulation, the reference signal is given as $y_d = 0.5\sin(t)$. The design parameters are selected as: $c_1 = 6$, $c_2 = 6$, $\gamma_1 = 0.5$, $\gamma_2 = 0.2$, $\sigma_1 = 0.5$, and $\sigma_2 = 0.5$. The initial values of the system states and parameter estimates are: $x_1(0) = 0.1$, $x_2(0) = -0.5$, and $\hat{b}_1(0) = \hat{b}_2(0) = 0$. Under the developed robust adaptive control scheme in Section 6.3.2, the simulation results are shown in Fig. 6.5-Fig. 6.7. The evolutions of the system output x_1 and reference signal $y_d(t)$ are plotted in Fig. 6.5, from which it is seen that the system output tracks the desired signal well. Fig. 6.6 shows the evolution of the actual controller u. The evolutions of the parameter estimates $\hat{b}_1$ and $\hat{b}_2$ are plotted in Fig. 6.7. It is seen that all signals in the closed-loop systems are bounded, verifying the effectiveness of the proposed control algorithm in Section 6.3.2.

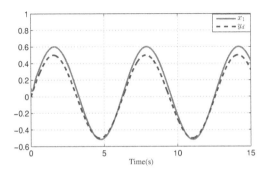

FIGURE 6.5   The evolutions of x_1 and y_d.

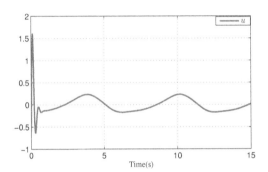

FIGURE 6.6   The evolution of the control input u.

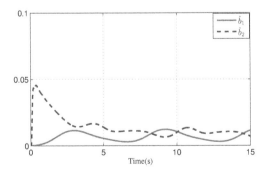

FIGURE 6.7   The evolutions of the parameter estimates $\hat{b}_1$ and $\hat{b}_2$.

6.4 NOTES

In this chapter, we illustrate the basic procedures of backstepping design. The descriptions and the writing of adaptive backstepping design are influenced by the clear presentation in references [9, 38, 86]. Readers may also consult references [39, 41, 87, 88, 89], which contain many meaningful results in backstepping design.

7 Control of MIMO Systems

In the previous chapters, we introduced the control design of SISO nonlinear systems. However, many practical engineering systems are not in the SISO form, but in the MIMO form with complicated structures, high nonlinearities, and strong dynamic couplings [90, 91, 92, 93, 94, 95, 96, 97], which can be expressed by the following nonlinear equation:

$$\begin{cases} \dot{x} = F(x,p,u), \\ y = h(x,u), \end{cases}$$

where $x = [x_1, x_2, \cdots, x_m]^\top \in \mathbb{R}^m$ is the system state, $u = [u_1, u_2, \cdots, u_n]^\top \in \mathbb{R}^n$ denotes the control input, $y = [y_1, y_2, \cdots, y_l]^\top \in \mathbb{R}^l$ is the system output, $F \in \mathbb{R}^m$ and $h \in \mathbb{R}^l$ are nonlinear functions, as shown in Fig. 7.1.

For MIMO systems, the control design is rather difficult compared with the control of SISO systems. The main challenges come from the following aspects.

(1) For SISO systems, only the scalar computation is involved in the control design and implementation. However, a large number of matrix operations are unavoidable for MIMO systems, such as matrix inversion operations, matrix decomposition operations, etc., making the computational burden dramatically increased and the corresponding controls of SISO systems not applicable any more.

(2) In most practical engineering systems, strong coupling is embedded within the MIMO systems so that the control gain of MIMO systems is in the matrix form, which extremely increases the difficulty of control design.

Therefore, in this chapter, we mainly focus on illustrating the control design methods for MIMO nonlinear systems.

7.1 CONTROL OF MIMO NORMAL-FORM SYSTEMS

7.1.1 PROBLEM FORMULATION

In this section, we consider the following MIMO normal-form nonlinear systems:

$$\begin{cases} \dot{x}_k = x_{k+1}, \ k = 1, 2, \cdots, n-1, \\ \dot{x}_n = G(\bar{x}_n)u + F(\bar{x}_n, \theta), \\ y = x_1, \end{cases} \tag{7.1}$$

where $x_i = [x_{i1}, x_{i2}, \cdots, x_{im}]^\top \in \mathbb{R}^m$, $i = 1, 2, \cdots, n$, is the system state and $\bar{x}_n = [x_1^\top, \cdots, x_n^\top]^\top$, $u \in \mathbb{R}^m$ and $y \in \mathbb{R}^m$ are the control input and system output, respectively, $G(\bar{x}_n) \in \mathbb{R}^{m \times m}$ is a smooth control gain matrix, $F(\bar{x}_n, \theta) \in \mathbb{R}^m$ denotes a smooth function vector, and $\theta \in \mathbb{R}^r$ denotes a parameter vector.

To facilitate the control design, we define the tracking error as $E = x_1 - y_d = [e_{11}, \cdots, e_{1m}]^\top \in \mathbb{R}^m$ with $y_d \in \mathbb{R}^m$ being the reference signal.

DOI: 10.1201/9781003474364-7

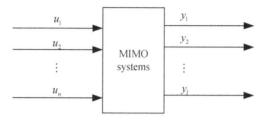

FIGURE 7.1 The schematic diagram of MIMO systems.

7.1.2 MODEL-DEPENDENT CONTROL DESIGN

The control objective of this section is to design a model-dependent control scheme for the MIMO normal-form nonlinear system (7.1) such that:

(*i*)   all signals in the closed-loop system are bounded;
(*ii*)   the tracking error $E = x_1 - y_d$ converges to zero asymptotically.

To this end, the following assumptions are imposed.

Assumption 7.1 *The system states $x_1, x_2, \cdots, x_n$ are measurable.*

Assumption 7.2 *The reference signal $y_d = [y_{d1}, \cdots, y_{dm}]^\top$ and its derivatives $y_d^{(k)}$, $k = 1, 2, \cdots, n$, are known, bounded, and piecewise continuous.*

Assumption 7.3 *The control gain matrix $G(\cdot)$ is symmetric and positive definite, and there exist some constants $\underline{g}$ and $\bar{g}$ satisfying $0 < \underline{g} < \bar{g} < \infty$ such that for any $X \in \mathbb{R}^m$,*

$$\underline{g}\|X\|^2 \le X^\top GX \le \bar{g}\|X\|^2$$

Furthermore, $F(\cdot)$ is bounded if $\cdot$ is bounded.

Assumption 7.4 *The control gain matrix $G(\cdot)$ and the nonlinear function vector $F(\cdot)$ are available for control design.*

Remark 7.1 *According to Assumption 7.3, it is seen that the inverse of the control gain matrix, $G^{-1} = M$, is also symmetric and positive definite.*

Upon using the tracking error, the filtered error is defined as:

$$s = \lambda_1 E + \lambda_2 \dot{E} + \cdots + \lambda_{n-1} E^{(n-2)} + E^{(n-1)}, \tag{7.2}$$

where $\lambda_1, \cdots, \lambda_{n-1}$ are design parameters so that the polynomial $l^{n-1} + \lambda_{n-1} l^{n-2} + \cdots + \lambda_1$ is Hurwitz. Then the derivative of s is:

$$\dot{s} = \lambda_1 \dot{E} + \lambda_2 E^{(2)} + \cdots + \lambda_{n-1} E^{(n-1)} + E^{(n)}. \tag{7.3}$$

Since $E = x_1 - y_d$, together with the system model (7.1), (7.3) becomes:

$$\dot{s} = \lambda_1 \dot{E} + \lambda_2 E^{(2)} + \cdots + \lambda_{n-1} E^{(n-1)} + F + Gu - y_d^{(n)} = F + Gu + L, \qquad (7.4)$$

where $L = \lambda_1 \dot{E} + \lambda_2 E^{(2)} + \cdots + \lambda_{n-1} E^{(n-1)} - y_d^{(n)}$ is a computable function.
  Now we state the following theorem.

Theorem 7.1 *Consider the MIMO normal-formal nonlinear system (7.1). Under Assumptions 7.1-7.4, if the model-dependent controller u is designed as:*

$$u = G^{-1}(-cs - F - L), \qquad (7.5)$$

where $c > 0$ is a design parameter, then the objectives (i)-(ii) can be ensured.

  Proof. Substituting the model-dependent controller (7.5) into (7.4), one has $\dot{s} = -cs$. As the stability analysis is similar to that in Section 5.1, then the detailed steps are omitted. The proof is completed.

7.1.3 PARAMETER-DECOMPOSITION-BASED CONTROL DESIGN

In Section 7.1.2, we introduced the model-dependent control scheme for MIMO normal-form nonlinear systems. However, for most practical applications, the system model is unknown and the system parameters are difficult (or even impossible) to be obtained. Therefore, for MIMO nonlinear systems with unknown parameters, how to design an advanced controller and analyze the stability of the closed-loop system is a difficult yet challenging issue in the control community. In this section, we focus on developing a model-independent control scheme for parametric MIMO normal-form nonlinear systems.
  Noting that (7.4) can be rewritten as:

$$M\dot{s} = MF + u + ML. \qquad (7.6)$$

Since there exist some uncertainties in the system model, it indicates that the matrix M and the function vector F are unavailable for control design. To address this issue, except for Assumptions 7.1–7.3, we also impose the following assumption.

Assumption 7.5 *The uncertain functions $MF + ML$ and $\dot{M}s$ satisfy the following condition:*

$$MF + ML = \varphi_1(\bar{x}_n) p, \quad \dot{M}s = \varphi_2(\bar{x}_n) p,$$

where $\varphi_1 \in \mathbb{R}^{m \times p}$ and $\varphi_2 \in \mathbb{R}^{m \times p}$ are smooth available functions, and $p \in \mathbb{R}^p$ is an unknown parameter vector.

  Constructing the quadratic function as:

$$V_1 = \frac{1}{2} s^\top M s,$$

then its derivative along (7.6) yields:

$$\dot{V}_1 = s^\top M \dot{s} + \frac{1}{2} s^\top \dot{M} s = s^\top (MF + ML + u) + \frac{1}{2} s^\top \dot{M} s. \tag{7.7}$$

According to Assumption 7.5, (7.7) becomes:

$$\dot{V}_1 = s^\top (\varphi_1 p + u) + \frac{1}{2} s^\top \varphi_2 p = s^\top \Phi p + s^\top u, \tag{7.8}$$

where $\Phi = \varphi_1 + \frac{1}{2} \varphi_2 \in \mathbb{R}^{m \times p}$ is a known smooth function.

Noting that adaptive technique is a powerful tool for uncertain parametric systems, then with the aid of the standard adaptive control method [9], the adaptive controller is designed as:

$$u = -cs - \Phi \hat{p}, \tag{7.9}$$

where $c > 0$ is a design parameter, and $\hat{p}$ is the estimate of the unknown parameter vector p, which is updated by:

$$\dot{\hat{p}} = \Gamma \Phi^\top s, \tag{7.10}$$

with $\Gamma = \Gamma^\top > 0$ being a constant matrix.

Now we state the following theorem.

Theorem 7.2 *Consider the MIMO normal-form nonlinear system (7.1) in the presence of parametric uncertainty. Under Assumptions 7.1–7.3, and 7.5, if the adaptive controller (7.9) with the adaptive law (7.10) is applied, then it is guaranteed that all signals in the closed-loop systems are bounded, and the tracking error converges to zero asymptotically.*

Proof. Substituting the control law as shown in (7.9) into (7.8), one obtains:

$$\dot{V}_1 = s^\top \Phi \tilde{p} - cs^\top s.$$

The Lyapunov function candidate is selected as $V = V_1 + \frac{1}{2} \tilde{p}^\top \Gamma^{-1} \tilde{p}$, then its derivative w.r.t. time is:

$$\dot{V} = s^\top \Phi \tilde{p} - cs^\top s - \tilde{p}^\top \Gamma^{-1} \dot{\hat{p}}. \tag{7.11}$$

Involving the adaptive law (7.10) into (7.11),

$$\dot{V} = -cs^\top s \le 0. \tag{7.12}$$

It follows from (7.12) that $V(t) \in \mathcal{L}_\infty$, $s \in \mathcal{L}_\infty$, $\tilde{p} \in \mathcal{L}_\infty$. According to Lemma 5.1, it is seen that $E, \dot{E}, \cdots, E^{(n-1)}$ are bounded. Since the reference signal and its derivatives up to n-th are bounded, then the system states $x_1, \cdots, x_n$ are bounded, it further follows that φ_1 and φ_2 are bounded. Since $\tilde{p} = p - \hat{p}$ and p is a constant, then $\hat{p}$ is bounded. According to the expressions of the control law and adaptive law, we have $u \in \mathcal{L}_\infty$, $\dot{\hat{p}} \in \mathcal{L}_\infty$ and $\dot{s} \in \mathcal{L}_\infty$. In addition, upon using (7.12), one has $s \in \mathcal{L}_2$, therefore, by utilizing Barbalat's Lemma, it is not difficult to get that $\lim_{t \to \infty} s(t) = 0$, which further implies from Lemma 5.1 that $\lim_{t \to \infty} e(t) = 0$.

7.1.4   CORE FUNCTION-BASED CONTROL DESIGN

If the nonlinear terms $MF + ML$ and $\dot{M}s$ don't satisfy the parameter decomposition condition, the corresponding adaptive control scheme in Section 7.1.3 is no longer applicable. Therefore, in this section, we utilize the core function-based control algorithm in Section 4.3.1 to handle this uncertainty.

To facilitate the control design, the following assumption is introduced.

Assumption 7.6 *For the unknown nonlinear function $F(\cdot)$, there exist an unknown constant $a \geq 0$ and an available function $\phi(\bar{x}_n) \geq 0$ such that $\|F(\cdot)\| \leq a\phi(\bar{x}_n)$. In addition, F and ϕ are bounded if $\bar{x}_n$ is bounded.*

With the aid of (7.4), the derivative of the quadratic function $\frac{1}{2}s^\top s$ w.r.t. time is:

$$s^\top \dot{s} = s^\top Gu + \Xi, \tag{7.13}$$

where $\Xi = s^\top (F + L)$ denotes the lumped uncertainty.

According to Young's inequality, it is obtained that:

$$s^\top F \leq \|s\| a\phi \leq \underline{\lambda}\|s\|^2 a^2\phi^2 + \frac{1}{4\lambda}, \tag{7.14}$$

$$s^\top L \leq \|s\| \|L\| \leq \underline{\lambda}\|s\|^2\|L\|^2 + \frac{1}{4\lambda}. \tag{7.15}$$

Then Ξ can be upper bounded by:

$$\Xi \leq \underline{\lambda}\|s\|^2 b\Phi + \frac{1}{2\lambda}, \tag{7.16}$$

where $b = \max\{a^2, 1\}$ is an unknown constant, and $\Phi = \phi^2 + \|L\|^2$ denotes a computable function.

Combining (7.14)-(7.16), (7.13) becomes:

$$s^\top \dot{s} \leq s^\top Gu + \underline{\lambda}\|s\|^2 b\Phi + \frac{1}{2\lambda}. \tag{7.17}$$

Designing the robust adaptive controller as:

$$u = -cs - \hat{b}s\Phi, \tag{7.18}$$

where $c > 0$ is a design parameter, $\hat{b}$ denotes the estimate of the unknown parameter b, which is updated by:

$$\dot{\hat{b}} = \gamma\|s\|^2\Phi - \sigma\hat{b}, \;\; \hat{b}(0) \geq 0, \tag{7.19}$$

where $\sigma > 0$ and $\gamma > 0$ are design parameters, and $\hat{b}(0)$ is the initial value of the parameter estimate $\hat{b}(t)$. According to (7.19), it is worth noting that $\hat{b}(t) \geq 0$ holds for $\hat{b}(0) \geq 0$.

Now we give the following theorem.

Theorem 7.3 *For the non-parametric MIMO normal-form nonlinear systems (7.1) under Assumptions 7.1-7.3 and 7.6, the proposed control algorithms (7.18)-(7.19) ensure that all signals in the closed-loop systems are bounded.*

Proof. Substituting the control law u as defined in (7.18) into the term $s^\top Gu$, we have:

$$s^\top Gu = -cs^\top Gs - \hat{b}\Phi s^\top Gs.$$

Upon utilizing Assumption 7.3, the term $s^\top Gu$ can be rewritten as:

$$s^\top Gu \leq -c\underline{\lambda}s^\top s - \underline{\lambda}\hat{b}\Phi s^\top s,$$

where the fact that $\hat{b}(t) \geq 0$ is used.

Therefore, (7.17) becomes:

$$s^\top \dot{s} \leq -c\underline{\lambda}s^\top s + \underline{\lambda}\tilde{b}\Phi s^\top s + \frac{1}{2\underline{\lambda}}. \tag{7.20}$$

Constructing the Lyapunov function candidate as $V = \frac{1}{2}s^\top s + \frac{\lambda}{2\gamma}\tilde{b}^2$ with $\gamma > 0$ being a design parameter, then the derivative of V along (7.20) yields:

$$\dot{V} \leq -c\underline{\lambda}\|s\|^2 + \underline{\lambda}\|s\|^2\tilde{b}\Phi - \frac{\underline{\lambda}}{\gamma}\tilde{b}\dot{\hat{b}} + \frac{1}{2\underline{\lambda}}. \tag{7.21}$$

Substituting the update law $\dot{\hat{b}}$ as given in (7.19) into (7.21), one has:

$$\dot{V} \leq -c\underline{\lambda}\|s\|^2 + \frac{\sigma\underline{\lambda}}{\gamma}\tilde{b}\hat{b} + \frac{1}{2\underline{\lambda}}.$$

Noting that $\frac{\sigma\underline{\lambda}}{\gamma}\tilde{b}\hat{b} \leq \frac{\sigma\underline{\lambda}}{\gamma}\tilde{b}(b - \tilde{b}) \leq -\frac{\sigma\underline{\lambda}}{2\gamma}\tilde{b}^2 + \frac{\sigma\underline{\lambda}}{2\gamma}b^2$, then it further follows that:

$$\dot{V} \leq -c\underline{\lambda}\|s\|^2 - \frac{\sigma\underline{\lambda}}{2\gamma}\tilde{b}^2 + \frac{\sigma\underline{\lambda}}{2\gamma}b^2 + \frac{1}{2\underline{\lambda}} \leq -l_1 V + l_2, \tag{7.22}$$

where $l_1 = \min\{2c\underline{\lambda}, \sigma\}$, $l_2 = \frac{\sigma\underline{\lambda}}{2\gamma}b^2 + \frac{1}{2\underline{\lambda}}$.

Now we prove the boundedness of all signals in the closed-loop system. According to (7.22), we have $V \in \mathcal{L}_\infty$, which implies that $s \in \mathcal{L}_\infty$ and $\hat{b} \in \mathcal{L}_\infty$. It is seen from Lemma 5.1 that $E \in \mathcal{L}_\infty$, $\dot{E} \in \mathcal{L}_\infty$, $\cdots$, $E^{(n-1)} \in \mathcal{L}_\infty$. Noting that s is bounded, then it further follows that L is bounded. Since $E = x_1 - y_d$, $\dot{E} = x_2 - \dot{y}_d$, $\cdots$, $E^{(n-1)} = x_n - y_d^{(n-1)}$, and $y_d, \cdots, y_d^{(n-1)}$ are bounded, then $x_1, x_2, \cdots, x_n$ are all bounded, which implies that $F \in \mathcal{L}_\infty$, $\phi \in \mathcal{L}_\infty$, $u \in \mathcal{L}_\infty$, and $\dot{\hat{b}} \in \mathcal{L}_\infty$. Therefore, all signals in the closed-loop system are bounded. The proof is completed.

7.2 CONTROL OF ROBOTIC SYSTEMS

Since robotic manipulators are able to carry out sophisticated jobs efficiently in various kinds of industrial systems, then, in the past decades, the control design of robotic systems has attracted more and more attention from scientists and engineers [16, 17, 32, 98, 99, 100, 101, 102, 103]. As the robotic system belongs to a typical class of MIMO normal-form nonlinear systems, in this section we focus on introducing the control design for robotic systems.

7.2.1 PROBLEM FORMULATION

We first recall the system model of the rigid robotic systems that was introduced in Chapter 1, which can be described by the following dynamics:

$$H(q,p)\ddot{q}+N_g(q,\dot{q},p)\dot{q}+G_g(q,p)+\tau_d(\dot{q},p)=u, \qquad (7.23)$$

where $q=[q_1,\cdots,q_m]^\top \in \mathbb{R}^m$ denotes the link position, $H(q,p) \in \mathbb{R}^{m\times m}$ denotes the inertia matrix, $p \in \mathbb{R}^l$ denotes the unknown parameter vector, $N_g(q,\dot{q},p) \in \mathbb{R}^{m\times m}$ denotes the centripetal-Coriolis matrix, $G_g(q,p) \in \mathbb{R}^m$ represents the gravitation vector, $\tau_d(\dot{q},p) \in \mathbb{R}^m$ denotes the frictional and disturbing force, and $u \in \mathbb{R}^m$ is the control input.

Let $q=x_1$ and $\dot{q}=x_2$, the robotic system (7.23) can be converted into the following from:

$$\begin{cases} \dot{x}_1 = x_2, \\ \dot{x}_2 = F(x,p)+G(x,p)u+D(x,p), \end{cases} \qquad (7.24)$$

where $F(x,p)=H^{-1}(-N_g x_2 - G_g)$, $D(x,p)=-H^{-1}\tau_d$, $G=H^{-1}$.

Let $E=x_1-y_d=[e_{11},\cdots,e_{1m}]^\top$ be the tracking error, where $y_d=[y_{d1},\cdots,y_{dm}]^\top$ is the reference signal. The control objective of this section is to design a robust adaptive control method for the robotic system (7.23) such that all signals in the closed-loop systems are bounded.

To this end, the following assumptions are imposed.

Assumption 7.7 *The system states x_1 and x_2 are measurable.*

Assumption 7.8 *The reference signal y_d and its derivatives $y_d^{(k)}$, $k=1,2$, are known, bounded, and piecewise continuous.*

Assumption 7.9 *For the frictional and disturbing force τ_d, there exists an unknown constant γ_τ such that $\|\tau_d\| \leq \gamma_\tau(1+\|x_2\|)$.*

7.2.2 CONTROL DESIGN

We define the tracking error-based filtered error as:

$$s = \lambda_1 E + \dot{E}, \qquad (7.25)$$

where $\lambda_1 > 0$ denotes a design parameter.

Taking the derivative of filtered error (7.25) w.r.t. time yields:

$$\dot{s} = \dot{x}_2 - \ddot{y}_d + \lambda_1 \dot{E}. \tag{7.26}$$

Substituting the second equation of (7.24) into (7.26), it follows that:

$$\dot{s} = Gu + F + D - \ddot{y}_d + \lambda_1 \dot{E}, \tag{7.27}$$

then the derivative of the quadratic function $\frac{1}{2}s^\top s$ along (7.27) yields:

$$s^\top \dot{s} = s^\top Gu + s^\top \left(F + D - \ddot{y}_d + \lambda_1 \dot{E} \right). \tag{7.28}$$

According to the properties of robotic systems in Chapter 1, together with Assumption 7.9, one has:

$$\|F\| \leq \|G\| \left(\|N_g\| \|x_2\| + \|G_g\| \right) \leq \overline{\lambda} \left(\gamma_N \|x_2\|^2 + \gamma_G \right)$$
$$\leq \max \left\{ \overline{\lambda} \gamma_N, \overline{\lambda} \gamma_G \right\} \left(\|x_2\|^2 + 1 \right),$$
$$\|D\| \leq \gamma_\tau \|G\| \left(1 + \|x_2\| \right) \leq \gamma_\tau \overline{\lambda} \left(1 + \|x_2\| \right).$$

By employing Young's inequality, we have:

$$s^\top F \leq \underline{\lambda} \max \left\{ \overline{\lambda} \gamma_N, \overline{\lambda} \gamma_G \right\}^2 \|s\|^2 \left(\|x_2\|^2 + 1 \right)^2 + \frac{1}{4\underline{\lambda}},$$
$$s^\top D \leq \underline{\lambda} \|s\|^2 \gamma_\tau^2 \overline{\lambda}^2 \left(1 + \|x_2\|^2 \right)^2 + \frac{1}{4\underline{\lambda}},$$
$$s^\top \left(\lambda_1 \dot{E} - \ddot{y}_d \right) \leq \underline{\lambda} \|s\|^2 \left(\lambda_1 \dot{E} - \ddot{y}_d \right)^2 + \frac{1}{4\underline{\lambda}}.$$

Therefore, we further have:

$$s^\top \left(F + D - \ddot{y}_d + \lambda_1 \dot{E} \right) \leq \underline{\lambda} a \|s\|^2 \Phi + \frac{3}{4\underline{\lambda}},$$

where $a = \max \left\{ \max \left\{ \overline{\lambda} \gamma_N, \overline{\lambda} \gamma_G \right\}^2, \gamma_\tau^2 \overline{\lambda}^2 \right\}$ is an unknown parameter, and $\Phi = 2 \left(\|x_2\|^2 + 1 \right)^2 + \left(\lambda_1 \dot{E} - \ddot{y}_d \right)^2$ is an available function for control design.

The robust adaptive controller is designed as:

$$u = -(c + \hat{a}\Phi) s, \tag{7.29}$$

where $c > 0$ is a design parameter, $\hat{a}$ is the estimate of the unknown parameter a, which is updated by:

$$\dot{\hat{a}} = \gamma \|s\|^2 \Phi - \sigma \hat{a}, \quad \hat{a}(0) \geq 0, \tag{7.30}$$

where $\sigma > 0$ and $\gamma > 0$ are design parameters.

For the nonlinear robotic system (7.23), we give the following theorem.

Theorem 7.4 *For the nonlinear robotic system (7.23) under Assumptions 7.7-7.9, the proposed robust adaptive control schemes (7.29) and (7.30) guarantee that all signals in the closed-loop system are bounded.*

Proof. The detailed proof is similar to that in Section 7.1.4, so it is omitted.

7.2.3 NUMERICAL SIMULATION

To validate the effectiveness of the proposed method for robotic systems in Section 7.2.2, we consider the following 2-link robotic system and the detailed information of $H(\cdot)$, $N_g(\cdot)$ and $G_g(\cdot)$ are expressed as follows:

$$H(q,p) = \begin{bmatrix} p_1 + p_2 + 2p_3\cos(q_2) & p_2 + p_3\cos(q_2) \\ p_2 + p_3\cos(q_2) & p_2 \end{bmatrix},$$

$$N_g(q,\dot{q},p) = \begin{bmatrix} -p_3\dot{q}_2\sin(q_2) & -p_3(\dot{q}_1 + \dot{q}_2)\sin(q_2) \\ p_3\dot{q}_1\sin(q_2) & 0 \end{bmatrix},$$

$$G_g(q,p) = \begin{bmatrix} p_4 g\cos(q_1) + p_5 g\cos(q_1 + q_2) \\ p_5 g\cos(q_1 + q_2) \end{bmatrix},$$

$$\tau_d = p_6(\tanh(p_7\dot{q}) - \tanh(p_8\dot{q})) + p_9\tanh(p_{10}\dot{q}) + p_{11}\dot{q},$$

where the system parameters are given as: $p_1 = 2.9$, $p_2 = 0.76$, $p_3 = 0.87$, $p_4 = 3.04$, $p_5 = 0.87$, $p_6 = 0.5$, $p_7 = 0.6$, $p_8 = 0.8$, $p_9 = 0.4$, $p_{10} = 0.1$, and $p_{11} = 0.1$.
   In order to facilitate the description, let $q = x_1$ and $\dot{q} = x_2$. In the simulation, the initial conditions of the system are selected as: $x_1(0) = [1.5, 1]^T$, $x_2(0) = [-1, -1]^T$, and $\hat{b}_2(0) = 0$, the reference signal is $y_d = [\sin(t), \sin(t)]^T$, the design parameters are given as: $c = 800$, $\lambda_1 = 2$, $\gamma = 0.1$, and $\sigma = 0.8$. Under the proposed control scheme, the simulation results are shown in Fig. 7.2–Fig. 7.6. The evolutions of the tracking errors, e_{11} and e_{12}, are shown in Fig. 7.2 and Fig. 7.3, respectively. The evolutions of the control inputs and parameter estimate are plotted in Fig. 7.4–Fig. 7.6, which indicate that all signals in the closed-loop system are bounded.

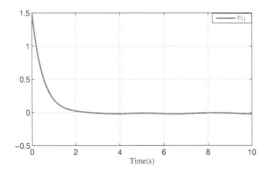

FIGURE 7.2 The evolution of the tracking error e_{11}.

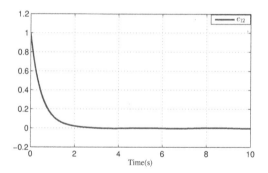

FIGURE 7.3 The evolution of the tracking error e_{12}.

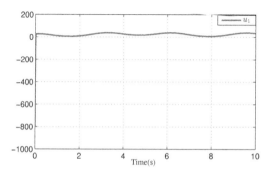

FIGURE 7.4 The evolution of the control input u_1.

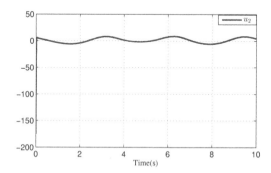

FIGURE 7.5 The evolution of the control input u_2.

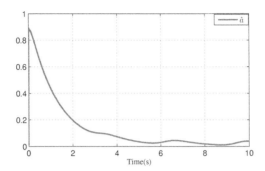

FIGURE 7.6 The evolution of the parameter estimate $\hat{a}$.

7.3 MODEL-DEPENDENT CONTROL OF MIMO STRICT-FEEDBACK SYSTEMS

In Sections 7.1 and 7.2, only the special case with MIMO normal-form nonlinear systems is considered, we will study the control design of more general MIMO strict-feedback nonlinear systems in this section.

7.3.1 PROBLEM FORMULATION

We consider the following three-order MIMO strict-feedback nonlinear systems:

$$\begin{cases} \dot{x}_1 = F_1(x_1) + G_1(x_1)x_2, \\ \dot{x}_2 = F_2(\bar{x}_2) + G_2(\bar{x}_2)x_3, \\ \dot{x}_3 = F_3(\bar{x}_3) + G_3(\bar{x}_3)u, \\ y = x_1, \end{cases} \tag{7.31}$$

where $x_i = [x_{i1}, x_{i2}, \cdots, x_{im}]^\top \in \mathbb{R}^m$, $i = 1, 2, 3$, is the system state and $\bar{x}_i = [x_1^\top, x_2^\top, x_3^\top]^\top$, $u \in \mathbb{R}^m$ represents the control input, $y = x_1 = [x_{11}, x_{12}, \cdots, x_{1m}]^\top \in \mathbb{R}^m$ is the system output, $G_i(\cdot) \in \mathbb{R}^{m \times m}$ represents the control gain matrix, and $F_i(\cdot) \in \mathbb{R}^m$ denotes a smooth nonlinear function.

The control objective of this section is to design a model-dependent control so that:

(*i*)   all signals in the closed-loop system are bounded;
(*ii*) the tracking error converges to zero asymptotically.

To this end, the following assumptions are introduced.

Assumption 7.10 *The system states x_1, x_2, and x_3 are measurable.*

Assumption 7.11 *The reference signal $y_d = [y_{d1}, \cdots, y_{dm}]^\top$ and its derivatives $y_d^{(k)}$, $k = 1, 2, 3$, are known, and bounded, and piecewise continuous.*

Assumption 7.12 *The system model is known, i.e., G_k and F_k, $k = 1, 2, 3$, are available for control design.*

Assumption 7.13 *The control gain matrix G_k is symmetric and positive definite, so does $M_k = G_k^{-1}$. Furthermore, there exist some positive constants $\underline{g}_k$ and $\overline{g}_k$ such that $\underline{g}_k \leq \|M_k\| \leq \overline{g}_k$.*

7.3.2 CONTROL DESIGN

In this section, we employ the backstepping technique to design the model-dependent controller. Define the tracking/virtual error as:

$$\begin{cases} z_1 = x_1 - y_d, \\ z_k = x_k - \alpha_{k-1}, \quad k = 2, 3, \end{cases} \tag{7.32}$$

where $z_1 = [z_{11}, \cdots, z_{1m}]^\top \in \mathbb{R}^m$ is the tracking error, $z_k = [z_{k1}, \cdots, z_{km}]^\top \in \mathbb{R}^m$ represents the virtual error, and α_{i-1} represents the virtual control law.

Step 1. Taking the derivative of z_1 w.r.t. time yields:

$$\dot{z}_1 = \dot{x}_1 - \dot{y}_d = F_1 + G_1 x_2 - \dot{y}_d. \tag{7.33}$$

According to (7.32), one has $x_2 = z_2 + \alpha_1$, then (7.33) can be rewritten as:

$$\dot{z}_1 = F_1 + G_1 z_2 + G_1 \alpha_1 - \dot{y}_d. \tag{7.34}$$

Choosing the quadratic function as $V_1 = \frac{1}{2} z_1^\top z_1$, then the derivative of V_1 along (7.34) yields:

$$\dot{V}_1 = z_1^\top \dot{z}_1 = z_1^\top [F_1 + G_1 (z_2 + \alpha_1) - \dot{y}_d]. \tag{7.35}$$

It is seen from Assumption 7.12 that the nonlinear function vector F_1 and control gain matrix G_1 are available for control design, thus, the virtual controller α_1 is designed as:

$$\alpha_1 = G_1^{-1}(-c_1 z_1 - F_1 + \dot{y}_d), \tag{7.36}$$

with $c_1 > 0$.

Substituting (7.36) into (7.35), we have:

$$\dot{V}_1 = z_1^\top(-c_1 z_1 + G_1 z_2) = -c_1 \|z_1\|^2 + z_1^\top G_1 z_2, \tag{7.37}$$

where the term $z_1^\top G_1 z_2$ will be handled in Step 2.

Step 2. The derivative of z_2 is:

$$\dot{z}_2 = \dot{x}_2 - \dot{\alpha}_1 = F_2 + G_2 x_3 - \dot{\alpha}_1. \tag{7.38}$$

Noting that

$$\dot{\alpha}_1 = \frac{\partial \alpha_1}{\partial x_1} \dot{x}_1 + \sum_{k=0}^{1} \frac{\partial \alpha_1}{\partial y_d^{(k)}} y_d^{(k+1)} = \frac{\partial \alpha_1}{\partial x_1}(G_1 x_2 + F_1) + \sum_{k=0}^{1} \frac{\partial \alpha_1}{\partial y_d^{(k)}} y_d^{(k+1)}.$$

As $x_3 = z_3 + \alpha_2$, (7.38) can be rewritten as:

$$\dot{z}_2 = F_2 + G_2 z_3 + G_2 \alpha_2 - \frac{\partial \alpha_1}{\partial x_1}(G_1 x_2 + F_1) - \sum_{k=0}^{1} \frac{\partial \alpha_1}{\partial y_d^{(k)}} y_d^{(k+1)}. \tag{7.39}$$

Constructing the quadratic function as $V_2 = \frac{1}{2}z_2^\top z_2 + V_1$, then the derivative of V_2 along (7.37) and (7.39) yields:

$$\dot{V}_2 = z_2^\top \left[F_2 + G_2 z_3 + G_2 \alpha_2 - \frac{\partial \alpha_1}{\partial x_1}(G_1 x_2 + F_1) - \sum_{k=0}^{1} \frac{\partial \alpha_1}{\partial y_d^{(k)}} y_d^{(k+1)} \right]$$
$$- c_1 \|z_1\|^2 + G_1 z_1^\top z_2. \tag{7.40}$$

As the nonlinear function vectors F_1, F_2, control gain matrices G_1, G_2, and y_d, $\dot{y}_d$ are known, then the virtual controller α_2 is designed as:

$$\alpha_2 = G_2^{-1}\left(-c_2 z_2 - F_2 + \frac{\partial \alpha_1}{\partial x_1}(G_1 x_2 + F_1) + \sum_{k=0}^{1} \frac{\partial \alpha_1}{\partial y_d^{(k)}} y_d^{(k+1)} - G_1 z_1 \right), \tag{7.41}$$

where $c_2 > 0$ is a design parameter.

Substituting the virtual controller α_2 as defined in (7.41) into (7.40), it is not difficult to get:

$$\dot{V}_2 = -c_1 \|z_1\|^2 - c_2 \|z_2\|^2 + z_1^\top G_1 z_2 + z_2^\top G_2 z_3 - z_2^\top G_1 z_1. \tag{7.42}$$

Noting that $\left(z_1^\top G_1 z_2\right)^\top = z_2^\top \left(z_1^\top G_1\right)^\top = z_2^\top G_1^\top z_1 = z_1^\top G_1 z_2$, then (7.42) becomes:

$$\dot{V}_2 = -c_1 \|z_1\|^2 - c_2 \|z_2\|^2 + z_2^\top G_2 z_3, \tag{7.43}$$

where $z_2^\top G_2 z_3$ will be handled in Step 3.

Step 3. The derivative of the virtual error z_3 is:

$$\dot{z}_3 = F_3 + G_3 u - \sum_{k=1}^{2} \frac{\partial \alpha_2}{\partial x_k}(G_k x_{k+1} + F_k) - \sum_{k=0}^{2} \frac{\partial \alpha_2}{\partial y_d^{(k)}} y_d^{(k+1)}. \tag{7.44}$$

Choosing the Lyapunov function candidate as $V_3 = \frac{1}{2}z_3^\top z_3 + V_2$, its derivative is:

$$\dot{V}_3 = z_3^\top \left[F_3 + G_3 u - \sum_{k=1}^{2} \frac{\partial \alpha_2}{\partial x_k}(G_k x_{k+1} + F_k) - \sum_{k=0}^{2} \frac{\partial \alpha_2}{\partial y_d^{(k)}} y_d^{(k+1)} \right]$$
$$- c_1 \|z_1\|^2 - c_2 \|z_2\|^2 + z_2^\top G_2 z_3. \tag{7.45}$$

The actual controller u is designed as:

$$u = G_3^{-1}\left(-c_3 z_3 - F_3 + \sum_{k=1}^{2} \frac{\partial \alpha_2}{\partial x_k}(G_k x_{k+1} + F_k) + \sum_{k=0}^{2} \frac{\partial \alpha_2}{\partial y_d^{(k)}} y_d^{(k+1)} - G_2 z_2 \right), \tag{7.46}$$

where $c_3 > 0$ is a design parameter.

For the MIMO strict-feedback nonlinear systems described in (7.31), we give the following theorem.

7.3.3   THEOREM AND STABILITY ANALYSIS

Theorem 7.5 *Considering the MIMO strict-feedback nonlinear system (7.31), the virtual controllers α_1, α_2, and actual controller u. Under Assumptions 7.10-7.13, the proposed control scheme ensures that objectives (i) and (ii) are achieved.*

Proof. Substituting (7.46) into (7.45), we have:

$$\dot{V}_3 = -c_1 z_1^\top z_1 - c_2 z_2^\top z_2 - c_3 z_3^\top z_3 \leq 0. \tag{7.47}$$

Integrating (7.47) on $[0,t]$, one has:

$$V_3(t) + \sum_{k=1}^{3} \int_0^t c_k z_k^\top(\tau) z_k(\tau) d\tau = V_3(0). \tag{7.48}$$

Firstly, we prove that all signals in the closed-loop system are bounded. According to (7.48), it is shown that $V_3(t) \in \mathscr{L}_\infty$ and $z_k \in \mathscr{L}_2$. By using the definition of the Lyapunov function, one has $z_k \in \mathscr{L}_\infty$, $k = 1, 2, 3$. Due to $z_1 = x_1 - y_d$ and y_d are bounded, then it follows that $x_1 \in \mathscr{L}_\infty$. Since F_1 is smooth, then F_1 is also bounded. Upon using the definition of the virtual controller α_1, we have $\alpha_1 \in \mathscr{L}_\infty$, $\frac{\partial \alpha_1}{\partial x_1} \in \mathscr{L}_\infty$, $\frac{\partial \alpha_1}{\partial y_d} \in \mathscr{L}_\infty$, $\frac{\partial \alpha_1}{\partial \dot{y}_d} \in \mathscr{L}_\infty$. As $z_2 = x_2 - \alpha_1$, then the boundedness of x_2 is ensured. Owing to the smoothness of the nonlinear function F_2, one has $F_2 \in \mathscr{L}_\infty$. With the similar analysis processes in the above steps, the virtual control law α_2, system state x_3, nonlinear function F_3, actual controller u, and $\dot{z}_k$ are bounded for all time.

Secondly, we prove that the tracking error converges to zero asymptotically, i.e., $\lim_{t \to \infty} z_1(t) = 0$. Since $z_1 \in \mathscr{L}_\infty \cap \mathscr{L}_2$, $\dot{z}_1 \in \mathscr{L}_\infty$, then it is seen from Barbalat's Lemma that $\lim_{t \to \infty} z_1(t) = 0$. The proof is completed.

7.4   MODEL-INDEPENDENT CONTROL OF MIMO STRICT-FEEDBACK SYSTEMS

In Section 7.3, we considered a model-dependent control algorithm for third-order MIMO strict-feedback nonlinear systems, whereas in practical applications, the system model and system parameters are difficult (or even impossible) to obtain. Therefore, in this section, we introduce a model-independent control algorithm for a class of non-parametric MIMO strict-feedback nonlinear systems.

7.4.1   PROBLEM FORMULATION

The following MIMO strict-feedback nonlinear systems are considered:

$$\begin{cases} \dot{x}_k = F_k(\bar{x}_k, \theta_k) + G_k(\bar{x}_k) x_{k+1}, & k = 1, 2, \cdots, n-1, \\ \dot{x}_n = F_n(\bar{x}_n, \theta_n) + G_n(\bar{x}_n) u, \\ y = x_1, \end{cases} \tag{7.49}$$

where $x_i = [x_{i1}, \cdots, x_{i1}]^\top \in \mathbb{R}^m$, $i = 1, 2, \cdots, n$, denotes the system state and $\bar{x}_i = [x_1^\top, \cdots, x_i^\top]^\top \in \mathbb{R}^{mi}$, $G_i(\cdot) \in \mathbb{R}^{m \times m}$ is a smooth but unknown control gain matrix, $F_i(\cdot) \in \mathbb{R}^m$ is a smooth but uncertain nonlinear function, $\theta_i \in R^{r_i}$ denotes an unknown parameter vector, $u \in \mathbb{R}^m$ is the control input, and $y = x_1 = [x_{11}, \cdots, x_{1m}] \in \mathbb{R}^m$ denotes the system output.

The tracking error is defined as $E = x_1 - y_d = [e_{11}, \cdots, e_{1m}]^\top$, where $y_d = [y_{d1}, \cdots, y_{dm}]^\top \in \mathbb{R}^m$ is the reference signal. The control objective of this section is to design a robust adaptive control scheme for MIMO strict-feedback nonlinear systems (7.49) such that:

(*i*)   all signals in the closed-loop system are bounded;
(*ii*) the tracking error converges to a small compact set around zero.

To this end, the following assumptions are imposed.

Assumption 7.14 *The system states $x_1, \cdots, x_n$ are measurable.*

Assumption 7.15 *The reference trajectory y_d and its derivatives $y_d^{(k)}$, $k = 1, 2, \cdots, n$, are known, bounded, and piecewise continuous.*

Assumption 7.16 *The unknown control gain matrix $G_i(\cdot) \in \mathbb{R}^{m \times m}$ is symmetric and positive definite, and there exist some unknown constants $\underline{g}_i$ and $\bar{g}_i$ such that $0 < \underline{g}_i < \min \{eig(G_i)\}$ and $\|G_i\| \le \bar{g}_i$.*

Assumption 7.17 *For the unknown nonlinear function vector $F_i(\cdot)$, there exist an unknown constant $a_i \ge 0$ and a smooth function $\phi_i(\bar{x}_i) \ge 0$ such that $\|F_i(\cdot)\| \le a_i \phi_i(\bar{x}_i)$ for $\forall t \ge 0$. In addition, F_i and ϕ_i are bounded if $\bar{x}_i$ is bounded.*

7.4.2 CONTROL DESIGN

Define the coordinate transformation as:

$$\begin{cases} z_1 = x_1 - y_d, \\ z_k = x_k - \alpha_{k-1}, \quad k = 2, 3, \cdots, n, \end{cases} \tag{7.50}$$

where $z_1 = [z_{11}, z_{12}, \cdots, z_{1m}]^\top \in \mathbb{R}^m$ is the tracking error, $z_k = [z_{k1}, z_{k2}, \cdots, z_{km}]^\top \in \mathbb{R}^m$ is the virtual error, and α_{k-1} is the virtual control law which will be given later.

Now we carry out the control design.

Step 1. According to (7.49) and (7.50), the derivative of z_1 w.r.t. time is $\dot{z}_1 = F_1 + G_1 x_2 - \dot{y}_d$. Since $x_2 = z_2 + \alpha_1$, it follows that:

$$z_1^\top \dot{z}_1 = z_1^\top G_1 \alpha_1 + \Xi_1 + z_1^\top G_1 z_2, \tag{7.51}$$

where $\Xi_1 = z_1^\top (F_1 - \dot{y}_d)$ is the lumped uncertainty.

According to Assumption 7.17, with the aid of Young's inequality, one has:

$$z_1^\top F_1 \leq \|z_1\| a_1 \phi_1 \leq \underline{g}_1 a_1^2 \|z_1\|^2 \phi_1^2 + \frac{1}{4\underline{g}_1}, \qquad (7.52)$$

$$-z_1^\top \dot{y}_d \leq \|z_1\| \|\dot{y}_d\| \leq \underline{g}_1 \|z_1\|^2 \|\dot{y}_d\|^2 + \frac{1}{4\underline{g}_1}. \qquad (7.53)$$

Therefore, Ξ_1 can be upper bounded by:

$$\Xi_1 \leq \underline{g}_1 \|z_1\|^2 b_1 \Phi_1 + \frac{1}{2\underline{g}_1},$$

with $b_1 = \max\{1, a_1^2\}$ being an unknown virtual constant and $\Phi_1 = \phi_1^2 + \|\dot{y}_d\|^2$ being available for the control design.

According to (7.52) and (7.53), (7.51) becomes:

$$z_1^\top \dot{z}_1 \leq z_1^\top G_1 \alpha_1 + \underline{g}_1 b_1 \|z_1\|^2 \Phi_1 + \frac{1}{2\underline{g}_1} + z_1^\top G_1 z_2.$$

Choosing the Lyapunov function candidate as:

$$V_1 = \frac{1}{2} z_1^\top z_1 + \frac{\underline{g}_1}{2\gamma_1} \tilde{b}_1^2,$$

where $\gamma_1 > 0$ is a design parameter, $\tilde{b}_1 = b_1 - \hat{b}_1$ is the estimate error with $\hat{b}_1$ being the estimate of the unknown parameter b_1. Then the derivative of V_1 w.r.t. time yields:

$$\dot{V}_1 = z_1^\top \dot{z}_1 - \frac{\underline{g}_1}{\gamma_1} \tilde{b}_1 \dot{\hat{b}}_1 \leq z_1^\top G_1 \alpha_1 + \underline{g}_1 b_1 \|z_1\|^2 \Phi_1 + \frac{1}{2\underline{g}_1} + z_1^\top G_1 z_2 - \frac{\underline{g}_1}{\gamma_1} \tilde{b}_1 \dot{\hat{b}}_1. \quad (7.54)$$

Designing the virtual controller α_1 and adaptive law $\hat{b}_1$ as:

$$\begin{cases} \alpha_1 = -(c_1 + \hat{b}_1 \Phi_1) z_1, \\ \dot{\hat{b}}_1 = \gamma_1 \|z_1\|^2 \Phi_1 - \sigma_1 \hat{b}_1, \quad \hat{b}_1(0) \geq 0, \end{cases} \qquad (7.55)$$

where $c_1 > 0$, $\gamma_1 > 0$, and $\sigma_1 > 0$ denote the design parameters, $\hat{b}_1(0) \geq 0$ is an arbitrary initial value. It is worth noting that for any initial condition $\hat{b}_1(0) \geq 0$, $\hat{b}_1(t) \geq 0$ always holds for $\forall t \geq 0$ as $\gamma_1 \|z_1\|^2 \Phi_1 \geq 0$.

Substituting the virtual controller α_1 as defined in (7.55) into $z_1^\top G_1 \alpha_1$, we have:

$$z_1^\top G_1 \alpha_1 = -(c_1 + \hat{b}_1 \Phi_1) z_1^\top G_1 z_1.$$

With the aid of Assumption 7.16, it follows that $-z_1^\top G_1 z_1 \leq -\underline{g}_1 \|z_1\|^2$, which leads to:

$$z_1^\top G_1 \alpha_1 \leq -\underline{g}_1 (c_1 + \hat{b}_1 \Phi_1) \|z_1\|^2. \qquad (7.56)$$

Substituting (7.56) into (7.54), one has:

$$\dot{V}_1 \leq -\underline{g}_1 c_1 \|z_1\|^2 + \underline{g}_1 \|z_1\|^2 \tilde{b}_1 \Phi_1 + \frac{1}{2\underline{g}_1} + z_1^\top G_1 z_2 - \frac{\underline{g}_1}{\gamma_1} \tilde{b}_1 \dot{\hat{b}}_1. \qquad (7.57)$$

Invoking the adaptive law $\dot{\hat{b}}_1$ as given in (7.55) into (7.57), one obtains:

$$\dot{V}_1 \leq -\underline{g}_1 c_1 \|z_1\|^2 + \underline{g}_1 \|z_1\|^2 \tilde{b}_1 \Phi_1 + z_1^\top G_1 z_2 + \frac{1}{2\underline{g}_1} + \frac{\sigma_1 g_1}{\gamma_1} \tilde{b}_1 \hat{b}_1 - \underline{g}_1 \|z_1\|^2 \tilde{b}_1 \Phi_1$$

$$\leq -\underline{g}_1 c_1 \|z_1\|^2 + z_1^\top G_1 z_2 + \frac{1}{2\underline{g}_1} + \frac{\sigma_1 g_1}{\gamma_1} \tilde{b}_1 \hat{b}_1. \tag{7.58}$$

As $\frac{\sigma_1 g_1}{\gamma_1} \tilde{b}_1 \hat{b}_1 \leq -\frac{\sigma_1 g_1}{2\gamma_1} \tilde{b}_1^2 + \frac{\sigma_1 g_1}{2\gamma_1} b_1^2$, (7.58) becomes:

$$\dot{V}_1 \leq -\underline{g}_1 c_1 \|z_1\|^2 - \frac{\sigma_1 g_1}{2\gamma_1} \tilde{b}_1^2 + z_1^\top G_1 z_2 + \Delta_1, \tag{7.59}$$

where $\Delta_1 = \frac{1}{2\underline{g}_1} + \frac{\sigma_1 g_1}{2\gamma_1} b_1^2$ denotes a positive constant, the term $z_1^\top G_1 z_2$ will be handled in Step 2.

Step 2. According to (7.49) and (7.50), the derivative of z_2 w.r.t. time is:

$$\dot{z}_2 = \dot{x}_2 - \dot{\alpha}_1 = F_2 + G_2 (z_3 + \alpha_2) - \frac{\partial \alpha_1}{\partial x_1} (F_1 + G_1 x_2) + \ell_1,$$

where $\ell_1 = -\sum_{k=0}^{1} \frac{\partial \alpha_1}{\partial y_d^{(k)}} y_d^{(k+1)} - \frac{\partial \alpha_1}{\partial \hat{b}_1} \dot{\hat{b}}_1$, then the derivative of $z_2^\top z_2$ is:

$$z_2^\top \dot{z}_2 = z_2^\top G_2 z_3 + z_2^\top G_2 \alpha_2 + z_2^\top \left[F_2 - \frac{\partial \alpha_1}{\partial x_1} (F_1 + G_1 x_2) + \ell_1 \right].$$

With (7.59), the derivative of $V_{21} = V_1 + \frac{1}{2} z_2^\top z_2$ is:

$$\dot{V}_{21} \leq -\underline{g}_1 c_1 \|z_1\|^2 - \frac{\sigma_1 g_1}{2\gamma_1} \tilde{b}_1^2 + z_2^\top G_2 \alpha_2 + \Delta_1 + z_2^\top G_2 z_3 + \Xi_2, \tag{7.60}$$

where $\Xi_2 = z_1^\top G_1 z_2 + z_2^\top \left[F_2 - \frac{\partial \alpha_1}{\partial x_1} (F_1 + G_1 x_2) + \ell_1 \right]$ is the lumped nonlinear term.
By utilizing Young's inequality, one has:

$$z_2^\top F_2 \leq \underline{g}_2 \|z_2\|^2 a_2^2 \phi_2^2 + \frac{1}{4\underline{g}_2},$$

$$z_1^\top G_1 z_2 \leq \underline{g}_2 \|z_1\|^2 \|z_2\|^2 + \frac{\bar{g}_1^2}{4\underline{g}_2},$$

$$-z_2^\top \frac{\partial \alpha_1}{\partial x_1} G_1 x_2 \leq \underline{g}_2 \left\| \frac{\partial \alpha_1}{\partial x_1} \right\|^2 \|x_2\|^2 \|z_2\|^2 + \frac{\bar{g}_1^2}{4\underline{g}_2},$$

$$-z_2^\top \frac{\partial \alpha_1}{\partial x_1} F_1 \leq \underline{g}_2 \|z_2\|^2 \left\| \frac{\partial \alpha_1}{\partial x_1} \right\|^2 a_1^2 \phi_1^2 + \frac{1}{4\underline{g}_2},$$

$$z_2^\top \ell_1 \leq \underline{g}_2 \|z_2\|^2 \|\ell_1\|^2 + \frac{1}{4\underline{g}_2}.$$

Therefore, Ξ_2 can be upper bounded by:

$$\Xi_2 \leq \underline{g}_2 b_2 \|z_2\|^2 \Phi_2 + \frac{3}{4\underline{g}_2} + \frac{\bar{g}_1^2}{2\underline{g}_2},$$

where $b_2 = \max\{1, a_1^2, a_2^2\}$ is an unknown virtual parameter, and

$$\Phi_2 = \|z_1\|^2 + \phi_2^2 + \left\|\frac{\partial\alpha_1}{\partial x_1}\right\|^2 \phi_1^2 + \left\|\frac{\partial\alpha_1}{\partial x_1}\right\|^2 \|x_2\|^2 + \|\ell_1\|^2$$

is a computable function, then (7.60) becomes:

$$\dot{V}_{21} \leq -\underline{g}_1 c_1 \|z_1\|^2 - \frac{\sigma_1 \underline{g}_1}{2\gamma_1}\tilde{b}_1^2 + z_2^\top G_2\alpha_2 + \Delta_1 + z_2^\top G_2 z_3 + \underline{g}_2 b_2 \|z_2\|^2 \Phi_2$$

$$+ \frac{3}{4\underline{g}_2} + \frac{\bar{g}_1^2}{2\underline{g}_2}. \tag{7.61}$$

Choosing the Lyapunov function candidate as $V_2 = V_{21} + \frac{g_2}{2\gamma_2}\tilde{b}_2^2$, where $\gamma_2 > 0$ is a design parameter, $\tilde{b}_2 = b_2 - \hat{b}_2$ is an estimate error with $\hat{b}_2$ being the estimate of the unknown parameter b_2, then the derivative of V_2 along (7.61) is:

$$\dot{V}_2 \leq -\underline{g}_1 c_1 \|z_1\|^2 - \frac{\sigma_1 \underline{g}_1}{2\gamma_1}\tilde{b}_1^2 + z_2^\top G_2\alpha_2 + z_2^\top G_2 z_3$$

$$+ \underline{g}_2 b_2 \|z_2\|^2 \Phi_2 + \frac{3}{4\underline{g}_2} + \frac{\bar{g}_1^2}{2\underline{g}_2} - \frac{g_2}{\gamma_2}\tilde{b}_2\dot{\hat{b}}_2 + \Delta_1. \tag{7.62}$$

Designing the virtual controller α_2 and adaptive law $\dot{\hat{b}}_2$ as:

$$\begin{cases} \alpha_2 = -(c_2 + \hat{b}_2\Phi_2)z_2, \\ \dot{\hat{b}}_2 = \gamma_2\|z_2\|^2\Phi_2 - \sigma_2\hat{b}_2, \quad \hat{b}_2(0) \geq 0, \end{cases} \tag{7.63}$$

where $c_2 > 0$, $\sigma_2 > 0$, and $\gamma_2 > 0$ are design parameters, and $\hat{b}_2(0) \geq 0$ is any arbitrary initial value.

Substituting the virtual controller α_2 into $z_2^\top G_2\alpha_2$, one has

$$z_2^\top G_2\alpha_2 \leq -\underline{g}_2(c_2 + \hat{b}_2\Phi_2)\|z_2\|^2.$$

Therefore, (7.62) can be rewritten as:

$$\dot{V}_2 \leq -\sum_{k=1}^{2} \underline{g}_k c_k \|z_k\|^2 - \frac{\sigma_1 \underline{g}_1}{2\gamma_1}\tilde{b}_1^2 + \Delta_1 + z_2^\top G_2 z_3$$

$$+ \underline{g}_2\tilde{b}_2\|z_2\|^2\Phi_2 + \frac{3}{4\underline{g}_2} + \frac{\bar{g}_1^2}{2\underline{g}_2} - \frac{g_2}{\gamma_2}\tilde{b}_2\dot{\hat{b}}_2. \tag{7.64}$$

Substituting the adaptive law $\dot{\hat{b}}_2$ as defined in (7.63) into (7.64), it is checked that:

$$\dot{V}_2 \leq -\sum_{k=1}^{2} \underline{g}_k c_k \|z_k\|^2 - \sum_{k=1}^{2} \frac{\sigma_k \underline{g}_k}{2\gamma_k} \tilde{b}_k^2 + \Delta_2 + z_2^\top G_2 z_3, \qquad (7.65)$$

where $\Delta_2 = \Delta_1 + \frac{3}{4g_2} + \frac{\bar{g}_1^2}{2g_2}$ denotes a positive constant, the term $z_2^\top G_2 z_3$ will be handled in the following step.

Step $i\,(i = 3, 4, \cdots, n)$. Before designing the virtual controller and actual controller, the following parameters are given as follows:

$$b_i = \max\left\{a_1^2, \cdots, a_i^2, 1\right\}, \quad \tilde{b}_i = b_i - \hat{b}_i, \qquad (7.66)$$

where $\hat{b}_i$ is the estimate of the parameter b_i and $\tilde{b}_i$ is the estimation error.

Upon using the coordinate transformation (7.50), the derivative of z_i w.r.t. time is:

$$\dot{z}_i = \dot{x}_i - \dot{\alpha}_{i-1} = F_i + G_i z_{i+1} + G_i \alpha_i - \dot{\alpha}_{i-1}.$$

Note that, as $i = n$, one has $\alpha_n = u$ and $z_{n+1} = 0$. In addition, as α_{i-1} is the function of variables x_k, $\hat{b}_k$ $(k = 1, 2, \cdots, i-1)$, and $y_d^{(j)}$ $(j = 0, 1, \cdots, i-1)$, then

$$\dot{\alpha}_{i-1} = \sum_{k=1}^{i-1} \frac{\partial \alpha_{i-1}}{\partial x_k} (F_k + G_k x_{k+1}) + \sum_{k=0}^{i-1} \frac{\partial \alpha_{i-1}}{\partial y_d^{(k)}} y_d^{(k+1)} + \sum_{k=1}^{i-1} \frac{\partial \alpha_{i-1}}{\partial \hat{b}_k} \dot{\hat{b}}_k$$

$$= \sum_{k=1}^{i-1} \frac{\partial \alpha_{i-1}}{\partial x_k} (F_k + G_k x_{k+1}) - \ell_{i-1},$$

where $\ell_{i-1} = -\sum_{k=0}^{i-1} \frac{\partial \alpha_{i-1}}{\partial y_d^{(k)}} y_d^{(k+1)} - \sum_{k=1}^{i-1} \frac{\partial \alpha_{i-1}}{\partial \hat{b}_k} \dot{\hat{b}}_k$ denotes the computable function, which leads to:

$$z_i^\top \dot{z}_i = z_i^\top F_i + z_i^\top G_i z_{i+1} + z_i^\top G_i \alpha_i - z_i^\top \sum_{k=1}^{i-1} \frac{\partial \alpha_{i-1}}{\partial x_k} (F_k + G_k x_{k+1}) + z_i^\top \ell_{i-1}. \qquad (7.67)$$

Constructing the Lyapunov function candidate as:

$$V_i = V_{i-1} + \frac{1}{2} z_i^\top z_i + \frac{g_i}{2\gamma_i} \tilde{b}_i^2,$$

where $\gamma_i > 0$ is a design parameter. Combining (7.65) with (7.67), the derivative of V_i is:

$$\dot{V}_i \leq -\sum_{k=1}^{i-1} \underline{g}_k c_k \|z_k\|^2 - \sum_{k=1}^{i-1} \frac{\sigma_k \underline{g}_k}{2\gamma_k} \tilde{b}_k^2 + \Delta_{i-1} + z_i^\top G_i z_{i+1} + z_i^\top G_i \alpha_i + \Xi_i - \frac{g_i}{\gamma_i} \tilde{b}_i \dot{\hat{b}}_i, \qquad (7.68)$$

where

$$\Xi_i = z_{i-1}^\top G_{i-1} z_i + z_i^\top F_i - z_i^\top \sum_{k=1}^{i-1} \frac{\partial \alpha_{i-1}}{\partial x_k} (F_k + G_k x_{k+1}) + z_i^\top \ell_{i-1}$$

is the lumped nonlinear term.

Upon using Young's inequality for Ξ_i, we have:

$$z_i^\top F_i \leq \underline{g}_i \|z_i\|^2 a_i^2 \phi_i^2 + \frac{1}{4\underline{g}_i},$$

$$z_{i-1}^\top G_{i-1} z_i \leq \underline{g}_i \|z_{i-1}\|^2 \|z_i\|^2 + \frac{\bar{g}_{i-1}^2}{4\underline{g}_i},$$

$$-z_i^\top \sum_{k=1}^{i-1} \frac{\partial \alpha_{i-1}}{\partial x_k} F_k \leq \underline{g}_i \|z_i\|^2 \sum_{k=1}^{i-1} \left\|\frac{\partial \alpha_{i-1}}{\partial x_k}\right\|^2 a_k^2 \phi_k^2 + \frac{i-1}{4\underline{g}_i},$$

$$-z_i \sum_{k=1}^{i-1} \frac{\partial \alpha_{i-1}}{\partial x_k} G_k x_{k+1} \leq \underline{g}_i \|z_i\|^2 \sum_{k=1}^{i-1} \left\|\frac{\partial \alpha_{i-1}}{\partial x_k}\right\|^2 \|x_{k+1}\|^2 + \sum_{k=1}^{i-1} \frac{\bar{g}_k^2}{4\underline{g}_i},$$

$$z_i^\top \ell_{i-1} \leq \underline{g}_i \|z_i\|^2 \|\ell_{i-1}\|^2 + \frac{1}{4\underline{g}_i}.$$

Thus Ξ_i can be upper bounded by:

$$\Xi_i \leq \underline{g}_i b_i \|z_i\|^2 \Phi_i + \frac{i+1}{4\underline{g}_i} + \frac{\bar{g}_{i-1}^2}{4\underline{g}_i} + \sum_{k=1}^{i-1} \frac{\bar{g}_k^2}{4\underline{g}_i},$$

where b_i is defined in (7.66), and

$$\Phi_i = \|z_{i-1}\|^2 + \phi_i^2 + \sum_{k=1}^{i-1} \left\|\frac{\partial \alpha_{i-1}}{\partial x_k}\right\|^2 \phi_k^2 + \sum_{k=1}^{i-1} \left\|\frac{\partial \alpha_{i-1}}{\partial x_k}\right\|^2 \|x_{k+1}\|^2 + \|\ell_{i-1}\|^2$$

is available for control design.

Then (7.68) can be rewritten as:

$$V_i \leq -\sum_{k=1}^{i-1} \underline{g}_k c_k \|z_k\|^2 - \sum_{k=1}^{i-1} \frac{\sigma_k \underline{g}_k}{2\gamma_k} \tilde{b}_k^2 + \Delta_{i-1} + z_i^\top G_i z_{i+1} + z_i^\top G_i \alpha_i$$

$$+ \underline{g}_i b_i \|z_i\|^2 \Phi_i + \frac{i+1}{4\underline{g}_i} + \frac{\bar{g}_{i-1}^2}{4\underline{g}_i} + \sum_{k=1}^{i-1} \frac{\bar{g}_k^2}{4\underline{g}_i} - \frac{\underline{g}_i}{\gamma_i} \tilde{b}_i \dot{\hat{b}}_i. \tag{7.69}$$

The virtual controller α_i, adaptive law $\hat{b}_i$, and actual controller u are designed as:

$$\begin{cases} \alpha_i = -(c_i + \hat{b}_i \Phi_i) z_i, \\ \dot{\hat{b}}_i = \gamma_i \|z_i\|^2 \Phi_i - \sigma_i \hat{b}_i, \quad \hat{b}_i(0) \geq 0, \\ u = \alpha_n, \end{cases} \tag{7.70}$$

where $c_i > 0$, $\sigma_i > 0$, and $\gamma_i > 0$ are design parameters, $\hat{b}_i(0) \geq 0$ is any arbitrary initial value.

Substituting the virtual controller α_i as defined in (7.70) into (7.69), one has:

$$\dot{V}_i \leq -\sum_{k=1}^{i} \underline{g}_k c_k \|z_k\|^2 - \sum_{k=1}^{i-1} \frac{\sigma_k \underline{g}_k}{2\gamma_k} \tilde{b}_k^2 + \Delta_{i-1} + z_i^\top G_i z_{i+1}$$

$$+ \underline{g}_i \tilde{b}_i \|z_i\|^2 \Phi_i + \frac{i+1}{4\underline{g}_i} + \frac{\overline{g}_{i-1}^2}{4\underline{g}_i} + \sum_{k=1}^{i-1} \frac{\overline{g}_k^2}{4\underline{g}_i} - \frac{\underline{g}_i}{\gamma_i} \tilde{b}_i \dot{\hat{b}}_i. \qquad (7.71)$$

Inserting the adaptive law $\dot{\hat{b}}_i$ into (7.71), we have:

$$\dot{V}_i \leq -\sum_{k=1}^{i} \underline{g}_k c_k \|z_k\|^2 - \sum_{k=1}^{i} \frac{\sigma_k \underline{g}_k}{2\gamma_k} \tilde{b}_k^2 + \Delta_i + z_i^\top G_i z_{i+1}, \qquad (7.72)$$

where $\Delta_i = \Delta_{i-1} + \frac{i+1}{4\underline{g}_i} + \frac{\overline{g}_{i-1}^2}{4\underline{g}_i} + \sum_{k=1}^{i-1} \frac{\overline{g}_k^2}{4\underline{g}_i} + \frac{\underline{g}_i \sigma_i}{2\gamma_i} b_i^2$ is a positive constant.

Now, we give the following theorem for robust adaptive control of MIMO strict-feedback nonlinear system (7.49).

7.4.3 THEOREM AND STABILITY ANALYSIS

Theorem 7.6 *Consider the MIMO strict-feedback nonlinear system (7.49) under Assumptions 7.14-7.17, if the proposed adaptive control (7.70) is applied, the objectives (i) and (ii) are achieved.*

Proof. For $i = n$, $z_{n+1} = 0$, then (7.72) becomes:

$$\dot{V}_n \leq -\sum_{k=1}^{n} \underline{g}_k c_k \|z_k\|^2 - \sum_{k=1}^{n} \frac{\sigma_k \underline{g}_k}{2\gamma_k} \tilde{b}_k^2 + \Delta_n \leq -l_1 V_n + l_2, \qquad (7.73)$$

where $l_1 = \min\left\{2\underline{g}_k c_k, 2\sigma_k\right\}$, $k = 1, 2, \cdots, n$, $l_2 = \Delta_n$. Firstly, we prove that all signals in the closed-loop system are bounded. According to (7.73), one has $z_i \in \mathcal{L}_\infty$ and $\hat{b}_i \in \mathcal{L}_\infty$, $i = 1, 2, \cdots, n$. Since $z_1 = x_1 - y_d$ and $x_1 \in \mathcal{L}_\infty$, then $\phi_1 \in \mathcal{L}_\infty$, it implies that $F_1(x_1) \in \mathcal{L}_\infty$, and $\Phi_1 \in \mathcal{L}_\infty$, then it follows from (7.55) that the virtual controller α_1 and adaptive law $\hat{b}_1$ are bounded. Furthermore, it is not difficult to get that $\partial \alpha_1 / \partial x_1 \in \mathcal{L}_\infty$, $\partial \alpha_1 / \partial y_d \in \mathcal{L}_\infty$, and $\partial \alpha_1 / \partial \dot{y}_d \in \mathcal{L}_\infty$. Note that $z_2 = x_2 - \alpha_1$, one has $x_2 \in \mathcal{L}_\infty$, then it further follows that $\phi_2 \in \mathcal{L}_\infty$ and $F_2 \in \mathcal{L}_\infty$, which implies that α_2 is bounded. Using the similar analysis, it is shown that the system state x_i, nonlinear function F_i, virtual controller α_{i-1}, actual controller u, and adaptive update law $\hat{b}_i$ are all bounded.

Secondly, we prove that the tracking error converges to a small compact set. Noting that (7.73) can be rewritten as $\dot{V}_n \leq -\underline{g}_1 c_1 \|z_1\|^2 + l_2$, then for $\|z_1\| > \sqrt{\frac{l_2 + v}{\underline{g}_1 c_1}}$,

$\dot{V}_n < 0$, where ν is a small constant, then the tracking error $\|z_1\|$ will enter and re-
main in the compact set $\Omega_{z_1} = \left\{ z_1 \in \mathbb{R}^m \,\middle|\, \|z_1\| \leq \sqrt{\frac{l_2 + \nu}{g_1 c_1}} \right\}$. The proof is completed.

Remark 7.2 *The size of Ω_{z_1} is smaller if c_1 is larger and η_2 is smaller. To make η_2
small, we need small σ_i and large γ_i, $i = 1, 2, \cdots, n$, which, however, might lead to
larger control effort or less effectiveness in preventing parameter estimation drift-
ing. Therefore, certain compromises between control performance and control effort
should be made in practice. It should be noted that as the tracking result is derived
from the Lyapunov stability theorem (resulting in sufficient conditions for tracking
stability), the actual tracking error bound could be smaller than what is theoretically
estimated. How to further reduce the tracking error bound represents an interesting
topic for future research.*

7.4.4 NUMERICAL SIMULATION

To demonstrate the effectiveness of the proposed control scheme in Section 7.4.2, a
numerical example is carried out on a second-order MIMO strict-feedback nonlinear
system described by (7.49) and the detailed expressions for F_i and G_i, $i = 1, 2$, are
given as:

$$F_1(x_1, p_1) = \begin{bmatrix} \theta_{11} x_{11}^2 \\ x_{11} x_{12} \exp(-\theta_{12} t) \end{bmatrix}, F_2(\bar{x}_2) = \begin{bmatrix} x_{11} x_{21} \cos(\theta_{21} t) \\ \theta_{22} x_{11} x_{22}^3 \end{bmatrix},$$

$$G_1(x_1) = \begin{bmatrix} 3 + \sin(x_{11}) & \tanh(x_{12}) \\ \tanh(x_{12}) & 2 + \cos(x_{12}) \end{bmatrix},$$

$$G_2(\bar{x}_2) = \begin{bmatrix} 2 + x_{21}^2 & \arctan(x_{11} x_{21}) \\ \arctan(x_{11} x_{21}) & 3 + x_{22}^2 \end{bmatrix},$$

where $\theta_1 = [\theta_{11}, \theta_{12}]^\top = [1, 0.5]^\top$ and $\theta_2 = [\theta_{21}, \theta_{22}]^\top = [0.1, 0.5]^\top$.

The objective is to ensure that all closed-loop signals are bounded and that the
tracking error converges to a compact set around zero. To implement the control
algorithm, the initial states are chosen as: $x_1(0) = [x_{11}(0), x_{12}(0)]^\top = [0.5, -0.5]^\top$
and $x_2(0) = [x_{21}(0), x_{22}(0)]^\top = [-0.2, 0.2]^\top$. In addition, the initial values of $\hat{\theta}_k$
are given as: $\hat{\theta}_1(0) = \hat{\theta}_2(0) = 0$. The reference signal is $y_d(t) = [y_{d1}(t), y_{d2}(t)]^\top = [0.3 \sin(t), 0.3 \sin(t)]^\top$, which satisfies Assumption 7.15. Furthermore, the simulation model is in strict-feedback form and satisfies Assumptions 7.16 and 7.17, which
indicate that the proposed control can be applied to such a MIMO nonlinear system.

In the simulation, the design parameters are chosen as: $k_1 = 2$, $k_2 = 6$, $\gamma_1 = 0.1$,
$\gamma_2 = 0.2$, $\sigma_1 = 0.1$, and $\sigma_2 = 0.1$. The simulation results are depicted in Fig. 7.7-Fig.
7.10, where Fig. 7.7 and Fig. 7.8 show the evolutions of x_{1k} and x_{2k}, $k = 1, 2$, as
well as the reference signals, which confirm that under the proposed control scheme
(7.70), the system output can track the reference signals well. Moreover, the evolutions of the tracking errors, e_{11} and e_{12}, and control input $u = [u_1, u_2]^\top$ are plotted in
Fig. 7.9 and Fig. 7.10, respectively.

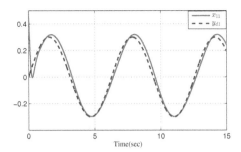

FIGURE 7.7 The evolutions of x_{11} and y_{d1}.

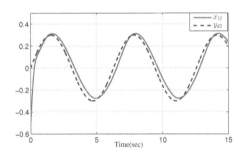

FIGURE 7.8 The evolutions of x_{12} and y_{d2}.

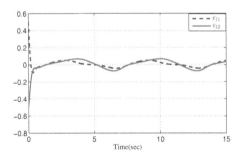

FIGURE 7.9 The evolutions of the tracking errors e_{11} and e_{12}.

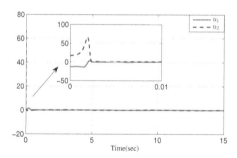

FIGURE 7.10 The evolutions of the control inputs u_1 and u_2.

7.5 NOTES

This chapter illustrates the control design methods for MIMO strict-feedback non-linear systems in the presence of parametric/non-parametric uncertainty. To show the design procedures of MIMO systems, we first present the model-dependent control scheme in Section 7.1.2 and Section 7.3, respectively, for MIMO normal-form non-linear systems and strict-feedback nonlinear systems. Subsequently, some adaptive (or robust adaptive) methods are proposed in Sections 7.1.3, and 7.1.4, and Section 7.4 for handling the uncertainties. The effectiveness of the developed controls is verified via simulations.

Note, however, that in this chapter the control gain matrix $G(\cdot)$ must be symmetric and positive definite. In practical applications, this condition may not be satisfied. For example, if actuator/sensor faults [104, 105] or input nonlinearities [92, 96, 106] occur in the system under consideration, this may introduce an additional gain matrix in the original nonlinear dynamics, which means that the positive definite property of the gain matrix is no longer satisfied. The main difficulties and challenges stem from the following aspects.

(1)   In contrast to nonlinear systems with the SISO form, the control gain in MIMO systems is in the matrix form, so the commutative law does not apply to general matrix multiplication.
(2)   Due to the introduction of additional matrices caused by actuator faults or input nonlinearities, it is difficult to ensure that the new control gain matrix satisfies the property of symmetry and positive definiteness.

Up to now, there are some techniques to address the above issue, which can be divided into the following directions.

(1)   The first direction is to utilize the NN or the fuzzy logic systems to estimate the uncertain matrix [92, 107], whereas extra efforts must be made to avoid the singularity problem when calculating the inverse of the estimated matrix. Moreover, the approximation ability of NN depends on the number of NN

nodes and the network structure. Selecting too many learning neurons bears a heavy computational burden.

(2) The second direction is to impose an assumption on the control gain matrix. For example, it is required that $G_1 = \frac{G\rho + (G\rho)^\top}{2}$ is positive/ negative definite [94, 97, 108], where ρ is the introduced new matrix caused by the actuator faults or input saturation. However, this condition is demanding and difficult to satisfy in practice.

(3) The third direction is to use the matrix factorization technique [109, 110], based on which it may remove the restrictive requirement on the uncertain control gain matrix. Noting that the precondition of utilizing such a method is that the derivative of the gain matrix only relies on the system state and is independent of the system input u, then before using such a method, the designer must check whether such a condition technology is satisfied.

Furthermore, the problem of control directions is also a challenging yet an interesting topic for MIMO nonlinear systems [91, 93]. However, most of the existing Nussbaum gain-based methods [83, 111] are only suitable for SISO nonlinear systems. The main reason is that the normally employed Nussbaum results don't satisfy the superposition principle. One way is to utilize the method in reference [85], by imposing an important lemma, we convert the problem of a gain matrix control direction into the problem of a scalar control direction; the other way is to construct some new Nussbaum functions [112] so that they can handle the problem of the gain matrix control direction directly.

Last but not the least is that the considered nonlinear system is in a square form, how to handle the more general nonsquare MIMO systems [42, 113] represents an interesting topic in future work.

Section III

**Performance/Prescribed-time
Control**

8 Accelerated Adaptive Control

We studied the control methods for nonlinear systems with steady-state performance in the previous chapters. However, with the rapid development of advanced science and technology as well as the increasing demand for high-quality products, the transient performance of a closed-loop system is more and more important for improving the system transient performance, that is, the controlled plant must be responded quickly with high accuracy and with small overshoot, which poses new challenges to the control design of nonlinear systems. Therefore, from Chapters 8 to 13 we will focus on introducing the prescribed performance, exponential convergence, and finite-/prescribed-time controls for improving the transient performance.

In this chapter, we seek for a tracking control scenario, referred to as accelerated tracking control, which not only steers the tracking error to zero (or to a small region of the equilibrium point), but also allows the tracking rate of convergence to be explicitly and arbitrarily specifiable so that the tracking error, before reaching the residual region, can be made decay as fast as exponential or even faster. Clearly, such a feature is of great practical importance in a number of applications. For instance, in spacecraft rendezvous and docking, vehicle self-parking, unmanned ground or aerial vehicle formation, missile interception, and machine part self-assembling, etc., it is often required or desired that the response speed during the startup phase is made faster while the approaching speed to the target is reduced for safe and reliable mission accomplishment. From the point of view of control, this literally involves synthesizing a control scheme capable of "speeding" up the tracking process and "slowing" it down to the normal tracking when entering the steady-state region.

To achieve satisfactory tracking performance for nonlinear systems, we introduce the concepts of "rate function" and "accelerated dynamics" in Section 8.1, which is the key point for achieving accelerated tracking. Different from the traditional adaptive tracking control, the developed accelerated adaptive control in Section 8.2 encompasses a key component that explicitly links the feedback control gain and adaptive law with a rate function through multiplication. As such, the resultant control not only exhibits the mechanism to smoothly adjust its control gain during the tracking process, but also has a faster rate for parameter updating, thus giving rise to improved tracking control performance. Note that, on one hand, the enhanced performance obtained by the proposed method agrees with common sense that speeding up the parameter updating rate and/or cranking up the feedback control gain would logically lead to better control performance, and on the other hand, prior to this study, it is unclear on by how much and in what way to increase the control gain and parameter updating the rate in order to achieve improved tracking performance. As a matter of fact, simply choosing a high constant control gain or simply increasing the

DOI: 10.1201/9781003474364-8

parameter updating rate by a large constant value in the adaptive law does not necessarily produce better performance. As shown in this work, those gains have to be alerted at different paces and with different modes as time goes by, and clear guidelines for adjusting such gains for improving tracking performance are analytically established.

Subsequently, we extend the scaling function transformation-based technique to SISO nonlinear systems and MIMO nonlinear systems with uncertainties in Section 8.3 and 8.4, respectively. The developed accelerated adaptive control (AAC) not only can achieve satisfactory steady-state tracking performance, but also can adjust the transient tracking performance. To demonstrate the feasibility of the proposed methods in practical engineering, we develop an AAC scheme for robotic systems in Section 8.5, and the simulation results in Section 8.6 verify the correction and effectiveness of the theoretical analysis.

8.1 RATE FUNCTION AND ACCELERATED DYNAMICS

In this section, we first introduce a pool of rate function, and then establish the accelerated dynamics.

8.1.1 A POOL OF RATE FUNCTION

By "rate function", it refers to a real function $\kappa(t)$, satisfying the following conditions:

(i) $\kappa(t)$ with $\kappa(0) = 1$ is positive and monotonically decreasing w.r.t. time for $t \geq 0$;
(ii) $\kappa(t)$ and its derivatives $\kappa(t)^{(i)}$, $i = 1, 2, \cdots, \infty$, are known, bounded, and piecewise continuous;
(iii) $\lim\limits_{t \to \infty} (\kappa(t))^{(i)} = 0$ for $i = 0, 1, \cdots, \infty$.

Obviously, immediate examples of such $\kappa(t)$ include (but are not limited to) the following functions:

$$\kappa(t) = \frac{1}{1+t^2},$$
$$\kappa(t) = \exp(-t),$$
$$\kappa(t) = a^{-t},$$
$$\kappa(t) = \frac{1}{(1+t)a^t},$$
$$\vdots$$

where, in all cases, $t \in [0, \infty)$, $a > 1$, $\exp(\cdot)$ denotes the exponential function of argument $\cdot$.

8.1.2   ACCELERATED DYNAMICS

Consider a system with error dynamics described by:

$$\dot{e} = f(e,x), \quad e(0) = e_0, \tag{8.1}$$

where $e = x - y_d \in \mathbb{R}^m$ is the tracking error with $x \in \mathbb{R}^m$ being the plant state vector and $y_d \in \mathbb{R}^m$ being the reference trajectory.

We define a "transformed error" $z \in \mathbb{R}^m$ and introduce the following rate transformation:

$$z = \beta(t)e, \tag{8.2}$$

such that (8.1) is converted into the following new nonlinear system:

$$\dot{z} = \beta^{-1}\dot{\beta}z + \beta f(\beta^{-1}z,x) = h(\beta,\beta^{-1}\dot{\beta},z,x), \tag{8.3}$$

with $z(0) = e(0)$ being the initial condition and $h(\cdot)$ being a new nonlinear function, where $\beta(t) \in \mathbb{R}$ is a positive and bounded time-varying scaling function to be determined later.

Note that the transformed system (8.3) differs from the original one as it carries the transformed error dynamics along with it. It is interesting to note that if the system (8.3) is made asymptotically stable, i.e., $z \to 0$ as $t \to \infty$, according to (8.2), we have $e = \beta^{-1}z$, then by properly designing β, in the original system the tracking error e can be made to not only converge to zero as $t \to \infty$, but also exhibit additional desirable tracking performance, such as prescribed rate of convergence and shaped steady-state response. The key to achieving this is to design the scaling function $\beta(t)$.

8.1.3   VARIABLE DECOMPOSITION AND SCALING FUNCTION

Our motivation for designing β stems from the following interesting observations:

(1) Mathematically, any bounded variable (vector or scalar) z, i.e., $\|z\| \leq z_{\max} < \infty$, can be decomposed into two components, i.e.,

$$z = z_m + z_r,$$

with $z_m = (1-b_f)z$, $z_r = b_fz$, where b_f satisfying $0 < b_f \ll 1$ is a small positive constant.

(2) As confirmed and illustrated in Fig. 8.1, if b_f is sufficiently small, $z_m = (1-b_f)z$ is then sufficiently close to z [i.e., $(1-b_f)z$ is almost identical to z], for this reason, $z_m = (1-b_f)z$ is referred to as the major (primary) component of z; whereas $z_r = b_fz$ is the residual component of z which evolves with time and is strictly bounded by $b_fz_{\max}$ (adjustable via b_f).

The above observations suggest that the dynamic behavior of z_m in terms of transient and steady-state response primarily reflects that of z, hence if z_m is controlled and well-shaped, so is z. This motivates us to choose β^{-1} as:

$$\beta^{-1}(t) = (1-b_f)\kappa(t) + b_f, \tag{8.4}$$

where $0 < b_f \ll 1$ is a free design parameter and $\kappa(t)$ is any function from the *rate function pool* in Section 8.1.1. Consequently,

$$e = \beta^{-1} z = \beta^{-1} z_m + \beta^{-1} z_r = z_m \kappa + z_r,$$

indicating that the tracking error e consists of two parts:

$$z_m \kappa = (1 - b_f) z \kappa, \quad \text{and} \quad z_r = b_f z,$$

both are related to z, then it is further shown that:

(i) since $0 < b_f \ll 1$ and z is bounded by z_{max}, $b_f z$ can be made as small as desired. Furthermore, $b_f z$ is strictly confined within $b_f z_{max}$ (adjustable through b_f) during the entire control process and further shrinks to zero if $z(t)$ converges to zero asymptotically, therefore $b_f z$ represents the "residual component" of the tracking error;

(ii) whereas the term $(1 - b_f) z \kappa$ manifests itself as the "primary (main) component" of the tracking error;

(iii) it is interesting to see that, with such β, the two components of the tracking error are either already in and remain within the residual region or are forced to converge toward the region at the rate governed not only by z, but more importantly, by κ as well, and κ is independent of the initial condition and can be explicitly and arbitrarily pre-specified, rendering the tracking performance well shaped by choosing κ and b_f properly, as shown in Fig. 8.1.

For later control development, it is worth noting that (8.4) can be expressed as:

$$\beta(t) = \frac{1}{(1 - b_f)\kappa(t) + b_f}, \tag{8.5}$$

which has the following properties for any κ as defined in Section 8.1.1:

(i) β monotonically increases from 1 to $1/b_f$ and $\beta \in [1, 1/b_f)$ for all $t \geq 0$;

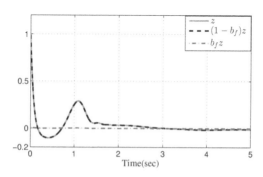

FIGURE 8.1   The schematic illustration of z, $(1 - b_f) z$, and $b_f z$ with $z(0) = 1$, $b_f = 0.01$.

(*ii*) $\beta(t)$ and its derivatives $\beta(t)^{(k)}$, $k = 0, 1, \cdots, \infty$, are known, bounded, and piecewise continuous;

(*iii*) noting that

$$\beta^2 = \frac{1}{\left((1 - b_f)\kappa(t) + b_f\right)^2},$$

$$\beta^{-1}\dot{\beta} = \frac{-(1 - b_f)\dot{\kappa}(t)}{(1 - b_f)\kappa(t) + b_f},$$

$$\beta\dot{\beta} = \frac{-(1 - b_f)\dot{\kappa}(t)}{\left((1 - b_f)\kappa(t) + b_f\right)^3}.$$

Then according to the property about $\kappa(t)$, it is readily computed that β^2, $\beta^{-1}\dot{\beta}$ and $\beta\dot{\beta}$ are bounded, $\lim\limits_{t \to \infty} \left(\beta^{-1}\dot{\beta}\right) = 0$, and $\lim\limits_{t \to \infty} \left(\beta\dot{\beta}\right) = 0$. In addition, $\frac{d}{dt}\left(\beta^{-1}\dot{\beta}\right)$ is also bounded for all $t \geq 0$.

Our focus in this chapter is on the development of the AAC fully equipped with the aforementioned important and unique features.

To help with the understanding of the fundamental idea and technical development of the proposed methods in this chapter, we start with controller design for the first-order systems with parametric uncertainties, followed by the extension to high-order cases with parametric/non-parametric uncertainties.

8.2   ACCELERATED ADAPTIVE CONTROL OF FIRST-ORDER SYSTEMS

We consider the following simple first-order scalar system:

$$\dot{x} = ax^2 + u, \tag{8.6}$$

where $x \in \mathbb{R}$ is the system state assumed to be available for measurement, $u \in \mathbb{R}$ is the control input, and a denotes an unknown constant.

The objective is to design an AAC strategy so that:

(*i*)   not only the tracking error converges to zero at $t \to \infty$, but also the tracking process, before reaching the residual region, is explicitly controllable in that the tracking error converges to the region at an accelerated decay rate determined by κ;

(*ii*) the accelerated decay rate of the tracking error is independent of system initial conditions and can be explicitly and arbitrarily;

(*iii*) all signals in the closed-loop systems are bounded and the control effort is smooth without involving excessively large initial control impulse.

By defining the tracking error as $e = x - y_d$ with y_d being the reference signal, one has:

$$\dot{e} = ax^2 + u - \dot{y}_d = u + \theta^{\top}\phi, \tag{8.7}$$

where $\theta^\top \phi = ax^2 - \dot{y}_d$ with $\theta = [a, 1]^\top$ being an unknown constant vector and $\phi = [x^2, -\dot{y}_d]^\top$ being a computable function vector.

Before designing the AAC, we first give the traditional adaptive control (TAC) method to show the advantages of the AAC method.

8.2.1   TRADITIONAL ADAPTIVE CONTROL

Define $\hat{\theta}$ as the estimate of θ and let $\tilde{\theta} = \theta - \hat{\theta}$ be the parameter estimate error. By utilizing the standard adaptive control [9], the TAC algorithm

$$\begin{cases} u = -c_T e - \hat{\theta}^\top \phi, \\ \dot{\hat{\theta}} = \Gamma \phi e, \end{cases} \tag{8.8}$$

ensures that e tends to zero asymptotically, where $c_T > 0$ is the control gain and $\Gamma = \Gamma^\top > 0$ is an adaptation rate matrix, both of which can be chosen by the designer. The global tracking stability can be readily confirmed by using

$$V_T = \frac{1}{2}e^2 + \frac{1}{2}\tilde{\theta}^\top \Gamma^{-1}\tilde{\theta},$$

leading to $\dot{V}_T = -c_T e^2 \le 0$, and the result follows by using Barbalat's Lemma.

Now we start with the AAC for the first-order scalar system (8.6).

8.2.2   ACCELERATED ADAPTIVE CONTROL

To get more than just asymptotically stable tracking results, by making use of the scaling function β as defined in (8.5), we construct the following AAC:

$$\begin{cases} u = -\left(c_A + \beta^{-1}\dot{\beta}\right)e - \hat{\theta}^\top \phi, \\ \dot{\hat{\theta}} = \Gamma \phi \beta^2 e, \end{cases} \tag{8.9}$$

where $c_A > 0$.

Note that although the structure of (8.9) looks similar to that of the TAC method (8.8) when viewing $c_A + \beta^{-1}\dot{\beta}$ as c_T, the proposed AAC method (8.9) differs from the TAC method (8.8) in several aspects:

(i)   the feedback control gain c_T in (8.8) remains unchanged during the entire control process while the control gain in (8.9) contains a component $\beta^{-1}\dot{\beta}$ that is time-varying during the system operation, which not only avoids the excessively large initial control effort as the TAC with a high constant gain (thus preventing severe actuator wearing), but also reduces energy consumption because the gain here automatically decreases as the tracking error enters the steady-state region, and eventually vanishes, rather than remains at a large value all the time;

(ii)   the adaptive law for the parameter updating, $\hat{\theta}$, in (8.9) involves a scaling factor β^2, the role of which is to increase the updating/learning rate during the critical transient period, thus enhancing the transient performance.

It is those unique properties that lead to the following results.

Theorem 8.1 *Consider the first-order nonlinear system (8.6) with an unknown constant parameter a. Under the assumption that y_d, $\dot{y}_d$, and $\ddot{y}_d$ are known, bounded, and piecewise continuous* [1], *the proposed AAC (8.9) achieves the objectives $(i) - (iii)$.*

Proof. By performing the rate transformation (8.2), we get the following accelerated error dynamics from (8.7),

$$\dot{z} = \dot{\beta}e + \beta(ax^2 + u - \dot{y}_d) = \dot{\beta}e + \beta(u + \theta^\top \phi).$$

With the proposed AAC (8.9), the closed-loop error dynamics becomes:

$$\dot{z} = -c_A z + \beta \tilde{\theta}^\top \phi, \tag{8.10}$$

Consider the Lyapunov function candidate as:

$$V = \frac{1}{2}z^2 + \frac{1}{2}\tilde{\theta}^\top \Gamma^{-1} \tilde{\theta},$$

then the time derivative of V along (8.10) is:

$$\dot{V} = -c_A z^2 + z\beta \tilde{\theta}^\top \phi - \tilde{\theta}^\top \Gamma^{-1} \dot{\hat{\theta}}. \tag{8.11}$$

Upon employing the adaptive law for $\hat{\theta}$ as defined in (8.9), (8.11) becomes:

$$\dot{V} = -c_A z^2 \leq 0, \tag{8.12}$$

from which the following properties can be established.

(1) Boundedness of internal signals. We first show that all signals in the closed-loop systems including z, $\dot{z}$, e, $\dot{e}$, x, $\hat{\theta}$, $\dot{\hat{\theta}}$, u, and $\dot{u}$ are bounded. In fact, from (8.12), it is seen that $z \in \mathscr{L}_\infty \cap \mathscr{L}_2$ and $\hat{\theta} \in \mathscr{L}_\infty$. Note that $e = \beta^{-1}z$, then e is bounded as β^{-1} is bounded, thus x is bounded due to the boundedness of the reference signal y_d, then ϕ is bounded. As β and $\beta^{-1}\dot{\beta}$ are bounded for all $t \geq 0$, it is seen that the control input u and the adaptive law for $\hat{\theta}$ in (8.9) are bounded and implementable. Besides, according to (8.7) and (8.10), $\dot{e} \in \mathscr{L}_\infty$ and $\dot{z} \in \mathscr{L}_\infty$. Moreover, as $\frac{d}{dt}\left(\beta^{-1}\dot{\beta}\right)$ and $\dot{x}$ are bounded, $\frac{\partial \phi}{\partial x}\dot{x}$ and $\frac{\partial \phi}{\partial \dot{y}_d}\ddot{y}_d$ are then bounded, it is readily shown that $\dot{u} \in \mathscr{L}_\infty$ (thus the control action is bounded and continuously differentiable). Therefore, all signals in the closed-loop systems are bounded.

(2) Zero tracking error and accelerated tracking process. As $\dot{z} \in \mathscr{L}_\infty$ and $z \in \mathscr{L}_2 \cap \mathscr{L}_\infty$, one can get the conclusion that $\lim_{t \to \infty} z(t) = 0$ by employing Barbalat's Lemma. Note that $e = \beta^{-1}z$, then it is established that:

$$e(t) = b_f z(t) + (1 - b_f)z(t)\kappa(t), \tag{8.13}$$

---

[1] The boundedness of $\ddot{y}_d$ is imposed to establish the boundedness of the control rate $\dot{u}$.

with $z(t) = 0$ as $t \to \infty$. It is seen that the tracking error e contains two terms:

(i) the first term $b_f z$ is strictly confined within the adjustable residual region $b_f z_{max}$[2] during the entire control process;

(ii) the second term $(1 - b_f)z\kappa$ is explicitly controlled and forced to converge to zero at the rate no less than κ.

Therefore, all the components of the tracking error are either strictly confined within the residual region, or driven to the region and further converge to zero at the rate no less than κ, which can be pre-specified explicitly and arbitrarily. Furthermore, the accelerated rate κ chosen from the pool of rate function is independent of the initial condition and any other parameters, thus can be pre-specified explicitly and arbitrarily. Such features, unique and highly desirable in practice, are not observed in any other existing methods.

(3) Asymptotic constancy of estimate parameters. It is shown that $e \to 0$, $\beta^2 \to 1/b_f^2$, and $\beta^{-1}\dot{\beta} \to 0$ as $t \to \infty$, then from the adaptive law for $\hat{\theta}$, we get that $\dot{\hat{\theta}} \to 0$ as $t \to \infty$, which implies that $\|\hat{\theta}\|$ converges to a constant c_θ. In addition, $c_A + \beta^{-1}\dot{\beta}$ tends c_A as $t \to \infty$. In all, both the transient response and steady-state error are well-shaped and can be adjusted by choosing b_0, b_f, and κ properly. The proof is completed.

8.2.3 NUMERICAL SIMULATION

Compared with the TAC method in Section 8.2.1, to quantitatively verify the superior performance of the proposed AAC in Section 8.2.2, we now present the simulation on the scalar model (8.6). Clearly, the proposed AAC method (8.9) covers the TAC method (8.8) as a special case (i.e., $\kappa = 1$).

In the simulation, the reference signal is $y_d = \sin(t)$ and the unknown parameter as well as initial estimate are set as $a = 2$, $x_0 = 1$, and $\hat{\theta}(0) = [0,0]^\top$.

Four different rate functions for κ are considered, i.e., $\kappa = 1$, $\frac{1}{1+t^2}$, $\exp(-t)$, and $\frac{1}{4^t(1+t^2)}$. Note that, for any control scheme, different control efforts (related to its control gain) could lead to different control precision. Hence, to get a fair comparison between the TAC and the AAC, we first test what control magnitude (effort) is required from each control scheme in order to get the same or similar tracking precision by increasing the control gain in the TAC (Test 1); then we test that with the same control effort, especially at the startup point, what is the control performance achieved by each control scheme (Test 2).

Test 1: The AAC demands much less control effort to get the similar tracking precision obtained by the TAC

The results are shown in Figs. 8.2–8.5, where it is observed that while similar control precision is obtained, as seen in Fig. 8.2, the TAC ($\kappa = 1$) demands much larger

---

[2]According to (8.12) and the Lyapunov function $V(t)$, we have $\frac{1}{2}z^2 \leq V(0) = \frac{1}{2}e^2(0) + \frac{1}{2}\tilde{\theta}^\top(0)\Gamma^{-1}\tilde{\theta}(0)$, then we define z_{max} as $z_{max} = \sqrt{e^2(0) + \tilde{\theta}^\top(0)\Gamma^{-1}\tilde{\theta}(0)}$.

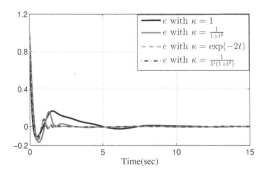

FIGURE 8.2   The evolution of the tracking error with different κ.

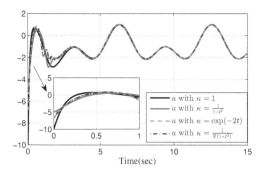

FIGURE 8.3   Test 1: The evolution of the control input with different κ.

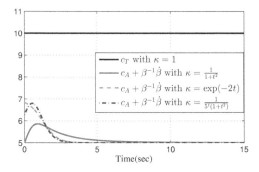

FIGURE 8.4   Test 1: The adaptation control gain with different κ.

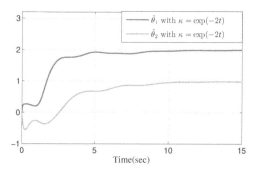

FIGURE 8.5   Test 1: The evolutions of the parameter estimates $\hat{\theta}_1$ and $\hat{\theta}_2$ with $\kappa = \exp(-2t)$.

initial control effort as compared with the proposed control, as clearly seen from Fig.
8.3. It is interesting to notice from Fig. 8.4 that in order to get the tracking precision
as shown in Fig. 8.2, the gain c_T for the TAC maintains a large constant value and
never changes during the entire control process, while the gain $(c_A + \beta^{-1}\dot{\beta})$ for the
proposed control, however, is self-adjusting in that initially it is set as a small value
(i.e., $c_A = 5$) and it keeps increasing for a short period of time, then it keeps reducing
to c_A (since $\beta^{-1}\dot{\beta}$ vanishes as time goes by). Besides, $c_A + \beta^{-1}\dot{\beta}$ is always smaller
than c_T as seen in Fig. 8.4, consequently, for the same control precision, less control
effort is required from the proposed AAC scheme. In addition, the evolution of the
parameter estimate $\hat{\theta} = [\hat{\theta}_1, \hat{\theta}_2]^\top$ is plotted in Fig. 8.5, which shows that all signals
are bounded for $\forall t \geq 0$.

*Test 2: Under the similar amount of control effort, the AAC performs much better
than the TAC*

The results are shown in Figs. 8.6 and 8.7. As seen from Fig. 8.6 and Fig. 8.7, with
the similar initial control effort Fig. 8.7, the proposed AAC gives rise to improved
transient and steady-state performance as shown in Fig. 8.6. It is shown from Fig.
8.2-Fig. 8.7 that the following observations can be made:

(*i*)   the proposed AAC demands much less initial control effort to get similar
        tracking precision obtained by the TAC;
(*ii*) under the similar control effort, especially at the startup point, the proposed
        AAC performs much better than the TAC;
(*iii*) the proposed AAC method allows the main tracking process to be control-
        lable in that the decay rate κ can be pre-specified arbitrarily by the designer.

While the above observations on the AAC method, it becomes apparent that the
proposed design method does exhibit appealing features worthy of further investiga-
tion, which is addressed in what follows.

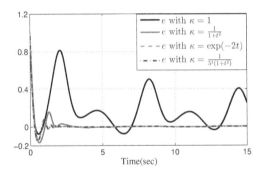

FIGURE 8.6   Test 2: The evolution of the tracking error with different κ.

8.3   ACCELERATED ADAPTIVE CONTROL OF STRICT-FEEDBACK SYSTEMS

8.3.1   PROBLEM FORMULATION

In this chapter, we consider the following parametric strict-feedback nonlinear systems:

$$\begin{cases} \dot{x}_k = x_{k+1} + \theta^\top \varphi_k(\bar{x}_k), \ k = 1, \cdots, n-1, \\ \dot{x}_n = u + \theta^\top \varphi_n(\bar{x}_n), \\ y = x_1, \end{cases} \tag{8.14}$$

where $x_i \in \mathbb{R}$, $i = 1, \cdots, n$, is the system state with $\bar{x}_i = [x_1, \cdots, x_i]^\top \in \mathbb{R}^i$, $u \in \mathbb{R}$ and $y \in \mathbb{R}$ are the control input and system output, respectively, the available function $\varphi_i(\bar{x}_i) \in \mathbb{R}^m$ is $n-i$ times differentiable taking arguments in $\mathbb{R}^i$, $\theta \in \mathbb{R}^m$ represents an unknown parameter vector.

The objective is to design a state-feedback AAC so that:

(i) all signals in the closed-loop systems are bounded;
(ii) the tracking error $e = x_1 - y_d$ converges to zero asymptotically;

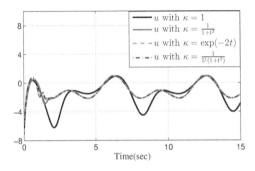

FIGURE 8.7   Test 2: The evolution of the control input with different κ.

(*iii*)   the accelerated decay rate of the tracking error is independent of the system initial conditions and can be explicitly and arbitrarily.

To this end, we impose the following assumptions.

Assumption 8.1 *The systems states are measurable.*

Assumption 8.2 *The reference trajectory $y_d(t)$ and its derivatives up to n-th are known, bounded, and piecewise continuous.*

8.3.2   ACCELERATED ADAPTIVE BACKSTEPPING DESIGN

In this section, we outline the control design based on the developed time-varying scaling function $\beta(t)$ in (8.5) to improve the transient tracking performance. Unlike the coordinate transformation in reference [9], we construct the following transformation:

$$\begin{cases} \zeta = \beta z_1, \\ z_1 = x_1 - y_d, \\ z_k = x_k - \alpha_{k-1}, \ k = 2, \cdots, n, \end{cases} \tag{8.15}$$

where α_{k-1} is the virtual control law which will be designed later.

Now, we begin the controller design by utilizing the adaptive backstepping technique. Compared with the basic backstepping design in reference [9], the major differences lie in Step 1 and Step 2. Thus, the details of Step 1 and Step 2 are elaborated and other steps are omitted.

Step 1. The derivative of ζ as defined in (8.15) w.r.t. time is:

$$\dot{\zeta} = \dot{\beta} z_1 + \beta \dot{z}_1 = \dot{\beta} z_1 + \beta (x_2 + \theta^\top \varphi_1 - \dot{y}_d).$$

As $x_2 = z_2 + \alpha_1$, then we get that:

$$\dot{\zeta} = \dot{\beta} z_1 + \beta (z_2 + \alpha_1 + \theta^\top \varphi_1 - \dot{y}_d). \tag{8.16}$$

The virtual controller α_1 is designed as:

$$\alpha_1 = -c_1 z_1 - \hat{\theta}^\top \varphi_1 + \dot{y}_d - \beta^{-1} \dot{\beta} z_1, \tag{8.17}$$

where c_1 is a positive constant and $\hat{\theta}$ is an estimate of θ.

Substituting the virtual controller α_1 as defined in (8.17) into (8.16), one has:

$$\dot{\zeta} = \beta z_2 - c_1 \zeta + \beta \tilde{\theta}^\top \varphi_1, \tag{8.18}$$

where $\tilde{\theta} = \theta - \hat{\theta}$ denotes the estimated error.

Define the first Lyapunov function candidate V_1 as $V_1 = \frac{1}{2}\zeta^2 + \frac{1}{2}\tilde{\theta}^\top \Gamma^{-1} \tilde{\theta}$, where Γ is a positive definite matrix, then the derivative of V_1 along (8.18) is:

$$\dot{V}_1 = -c_1 \zeta^2 + \zeta \beta z_2 + \tilde{\theta}^\top \Gamma^{-1} \left(\Gamma \tau_1 - \dot{\hat{\theta}} \right),$$

with $\tau_1 = \zeta\beta\varphi_1$ being the first tuning function.

Step 2. As α_1 is the function of variables $x_1, y_d, \dot{y}_d, \beta,$ and $\dot{\beta}$, then the derivative of $z_2 = x_2 - \alpha_1$ is:

$$\dot{z}_2 = \theta^\top \varphi_2 + \alpha_2 + z_3 - \frac{\partial \alpha_1}{\partial x_1}(\theta^\top \varphi_1 + x_2) + \ell_1 - \frac{\partial \alpha_1}{\partial \hat{\theta}}\dot{\hat{\theta}}, \qquad (8.19)$$

with $\ell_1 = -\sum_{k=0}^{1}\left(\frac{\partial \alpha_1}{\partial y_d^{(k)}}y_d^{(k+1)} + \frac{\partial \alpha_1}{\partial \beta^{(k)}}\beta^{(k+1)}\right)$ and $z_3 = x_3 - \alpha_2$.

The second tuning function τ_2 and virtual controller α_2 are designed, respectively, as:

$$\tau_2 = \tau_1 + z_2\omega_2, \quad \omega_2 = \varphi_2 - \frac{\partial \alpha_1}{\partial x_1}\varphi_1,$$

$$\alpha_2 = -\beta\zeta - c_2 z_2 - \hat{\theta}^\top \omega_2 + \frac{\partial \alpha_1}{\partial \hat{\theta}}\Gamma\tau_2 + \frac{\partial \alpha_1}{\partial x_1}x_2 - \ell_1,$$

where $c_2 > 0$ is a constant. Hence, (8.19) can be rewritten as:

$$\dot{z}_2 = z_3 - \beta\zeta - c_2 z_2 + \tilde{\theta}^\top \omega_2 + \frac{\partial \alpha_1}{\partial \hat{\theta}}\left(\Gamma\tau_2 - \dot{\hat{\theta}}\right). \qquad (8.20)$$

Then the derivative of $V_2 = V_1 + \frac{1}{2}z_2^2$ along (8.20) is:

$$\dot{V}_2 = -c_1\zeta^2 - c_2 z_2^2 + z_2 z_3 + \left(\tilde{\theta}^\top \Gamma^{-1} + z_2\frac{\partial \alpha_1}{\partial \hat{\theta}}\right)\left(\Gamma\tau_2 - \dot{\hat{\theta}}\right).$$

Step i $(i = 3, \cdots, n)$. The virtual controller α_i and actual controller u are designed as:

$$\alpha_i = -z_{i-1} - c_i z_i - \hat{\theta}^\top \omega_i + \sum_{k=1}^{i-1}\frac{\partial \alpha_{i-1}}{\partial x_k}x_{k+1} - \ell_{i-1}$$

$$+ \frac{\partial \alpha_{i-1}}{\partial \hat{\theta}}\Gamma\tau_i + \sum_{k=1}^{i-2}z_{k+1}\frac{\partial \alpha_k}{\partial \hat{\theta}}\Gamma\omega_i, \qquad (8.21)$$

$$\ell_{i-1} = -\sum_{k=0}^{i-1}\left(\frac{\partial \alpha_{i-1}}{\partial y_d^{(k)}}y_d^{(k+1)} + \frac{\partial \alpha_{i-1}}{\partial \beta^{(k)}}\beta^{(k+1)}\right), \qquad (8.22)$$

$$\omega_i = \varphi_i - \sum_{k=1}^{i-1}\frac{\partial \alpha_{i-1}}{\partial x_k}\varphi_k, \qquad (8.23)$$

$$\tau_i = \tau_{i-1} + z_i\omega_i, \qquad (8.24)$$

$$u = \alpha_n, \qquad (8.25)$$

$$\dot{\hat{\theta}} = \Gamma\tau_n, \qquad (8.26)$$

with $c_i > 0$.

8.3.3 THEOREM AND STABILITY ANALYSIS

The stability analysis of the closed-loop system under the AAC (8.25) and (8.26) is summarized in the following theorem.

Theorem 8.2 *Consider the closed-loop system consisting of the plant (8.14) and the controller (8.25) with the parameter update law (8.26). Under Assumptions 8.1 and 8.2, the objectives (i)-(iii) are ensured.*

Proof. Upon using the virtual controllers and the actual controller as given in (8.17), (8.21), and (8.25), it is readily proved that:

$$\dot{\zeta} = -c_1\zeta + \beta z_2 + \beta \tilde{\theta}^\top \varphi_1, \tag{8.27}$$

$$\dot{z}_2 = -\beta \zeta - c_2 z_2 + z_3 + \tilde{\theta}^\top \omega_2 - \sum_{k=3}^{n} \frac{\partial \alpha_1}{\partial \hat{\theta}} z_k \Gamma \omega_k, \tag{8.28}$$

$$\dot{z}_j = -z_{j-1} - c_j z_j + z_{j+1} + \tilde{\theta}^\top \omega_j + \sum_{k=1}^{j-2} \frac{\partial \alpha_k}{\partial \hat{\theta}} \Gamma z_{k+1} \omega_i$$

$$- \sum_{k=j+1}^{n} \frac{\partial \alpha_{j-1}}{\partial \hat{\theta}} \Gamma z_k \omega_k, \; j = 3, \cdots, n-1, \tag{8.29}$$

$$\dot{z}_n = -z_{n-1} - c_n z_n + \tilde{\theta}^\top \omega_n + \sum_{k=1}^{n-2} \frac{\partial \alpha_k}{\partial \hat{\theta}} \Gamma z_{k+1} \omega_n. \tag{8.30}$$

Define the Lyapunov function candidate as $V_n = \frac{1}{2}\zeta^2 + \sum_{k=2}^{n} z_k^2 + \frac{1}{2}\tilde{\theta}^\top \Gamma^{-1}\tilde{\theta}$, then the derivative of V_n along (8.27)-(8.30) is:

$$\dot{V}_n = -c_1\zeta^2 - \sum_{k=2}^{n} z_k^2 \leq 0.$$

Using the standard stability analysis in Section 6.2, it is not difficult to prove that $\zeta \in \mathscr{L}_\infty \cap \mathscr{L}_2$, $z_i \in \mathscr{L}_\infty \cap \mathscr{L}_2$, $i = 2, 3, \cdots, n$, and ensure the boundedness of all signals in the closed-loop systems. By following Barbalat's Lemma, it is seen that $\lim_{t \to \infty} \zeta(t) = 0$ and $\lim_{t \to \infty} z_k(t) = 0$, $k = 2, 3, \cdots, n$.

Furthermore, we give a brief explanation of the tracking performance. As ζ is bounded by $|\zeta(t)| \leq \zeta_{max}$, where $\zeta_{max} = \sqrt{\sum_{k=1}^{n} z_k^2(0) + \tilde{\theta}^\top(0)\Gamma^{-1}\tilde{\theta}(0)}$, then according to (8.15), one has:

$$e = \beta^{-1}\zeta = (1 - b_f)\kappa\zeta + b_f\zeta,$$

and

$$|e| = |\beta^{-1}\zeta| \leq (1 - b_f)\kappa\zeta_{max} + b_f\zeta_{max}.$$

Therefore, according to the analysis in Section 8.2, it is shown that not only the tracking error converges to zero at $t \to \infty$, but also the tracking process, before reaching the residual region, is explicitly controllable in that the tracking error converges to the region at an accelerated decay rate determined by $\kappa(t)$. The proof is completed.

8.4 ACCELERATED ADAPTIVE CONTROL OF MIMO SYSTEMS

It is worth mentioning that in Sections 8.2 and 8.3, only the parametric SISO nonlinear system is considered, here we now move beyond the scalar examples and explore the applicability of the above idea to more general MIMO nonlinear systems presented in Chapter 7.

8.4.1 PROBLEM FORMULATION

We consider a class of non-parametric MIMO nonlinear systems:

$$\begin{cases} \dot{x}_k = x_{k+1}, \ k = 1, 2, \cdots, n-1, \\ \dot{x}_n = F(x, p) + G(x, p)u + D(x, p), \\ y = x_1, \end{cases} \tag{8.31}$$

where $x_i = [x_{i1}, \cdots, x_{im}]^\top \in \mathbb{R}^m$, $i = 1, \cdots, n$, and $x = \left[x_1^\top, \cdots, x_n^\top\right]^\top \in \mathbb{R}^{mn}$ is the state vector, $p \in \mathbb{R}^r$ represents the unknown parameter vector inseparable from the system nonlinearities [i.e., $F(x, p)$, $G(x, p)$, and $D(x, p)$ are non-parametric uncertainties], $u = [u_1, \cdots, u_m]^\top \in \mathbb{R}^m$ is the control input, $y \in \mathbb{R}^m$ is the system output, $F(\cdot) = [f_1(\cdot), \cdots, f_m(\cdot)]^\top \in \mathbb{R}^m$ is a smooth but uncertain nonlinear function vector, $G(x, p) \in \mathbb{R}^{m \times m}$ is the control gain matrix, and $D(x, p) = [d_1(\cdot), \cdots, d_m(\cdot)]^\top \in \mathbb{R}^m$ denotes the system modeling uncertainty and external disturbance.

Define the tracking error as:

$$E = x_1 - y_d = [e_1, e_2 \cdots, e_m]^\top, \quad E^{(i)} = \left[e_1^{(i)}, e_2^{(i)}, \cdots, e_m^{(i)}\right]^\top, \tag{8.32}$$

for $i = 1, \cdots, n-1$, where $y_d = [y_{d1}, \cdots, y_{dm}]^\top \in \mathbb{R}^m$ and $y_d^{(i)} = \left[y_{d1}^{(i)}, \cdots, y_{dm}^{(i)}\right]^\top$ are the known reference signal and its i-th derivatives. In this section, we seek for a robust AAC approach capable of achieving the following three control objectives:

(i)   all signals in the closed-loop systems are bounded; the control action is continuously differentiable; and no excessively initial large control effort is involved;
(ii) accelerated tracking is obtained despite non-parametric uncertainties arising from $F(x, p)$, $G(x, p)$, and $D(x, p)$;
(iii) the tracking process is pre-designable in that each component of the tracking error, before reaching the residual set, has its own pre-assigned convergence mode and convergence rate.

To this end, we impose the following conditions.

Assumption 8.3 *The system states are available for control design.*

Assumption 8.4 *The reference trajectory y_d and its up to $(n+1)$-th derivatives are known, bounded, and piecewise continuous.*

Remark 8.1 *Full-state feedback is needed to achieve full-state tracking, and Assumption 8.3 is quite standard and commonly used for tracking control design in practice [9, 39]. Assumption 8.4 is to ensure that the control input to be designed is continuously differentiable.*

Note that both $F(x,p)$ and $D(x,p)$ are non-parametric uncertainties that cannot be expressed as a known regressor multiplying with an unknown parameter vector. However, with some primary information on $F(x,p)$ and $D(x,p)$, it is always possible to establish the inequality as imposed in the following assumptions.

Assumption 8.5 *There exist a non-negative constant a and a smooth function $\phi(x) \geqslant 0$, so that $\|F(x,p) + D(x,p,t)\| \leq a\phi(x)$. In addition, $\phi(x)$, $F(x,p)$, $G(x,p)$, and $D(x,p)$ are bounded if x is bounded.*

Remark 8.2 *Assumption 8.5 is related to the extraction of the core information from the system uncertainties. Note that the unknown constant a, although associated with the physical parameter vector p, bears no physical meaning, thus it is referred to as a virtual parameter. Also note that the scalar function $\phi(x)$ is the computable "core function" for it contains the deep-rooted information of the system, which can be readily obtained with certain crude information on $F(x,p)$ and $D(x,p)$, as confirmed with the robotic systems.*

The unknown control gain matrix $G(x,p)$ represents another major source of challenge and complexity for control design due to the fact that unknown and time-varying $G(x,p)$, through which the control input u enters into the system, actually makes it uncertain how the control input would impact the system. Here, we address such situation and the following assumption is introduced.

Assumption 8.6 *The control gain matrix $G(\cdot) \in \mathbb{R}^{m \times m}$ is square but unnecessarily symmetric yet completely unknown. The only information available for control design is that $G_1 = \frac{G + G^\top}{2}$ is positive definite. Here it is assumed that there exist some unknown bounded constants $\overline{\lambda}_1 > 0$ and $\underline{\lambda}_1 > 0$ such that with the minimum eigenvalue $\lambda_{\min}(t)$ and the maximum eigenvalue $\lambda_{\max}(t)$ of G_1, the following inequality holds: $\underline{\lambda}_1 \leq \lambda_{\min}(t) < \lambda_{\max}(t) \leq \overline{\lambda}_1$.*

8.4.2 ACCELERATED TRACKING CONTROL DESIGN

To achieve satisfactory and adjustable tracking performance for MIMO nonlinear systems with non-parametric uncertainties, based on the rate function developed in Section 8.1.1, we construct the following time-varying scaling function:

$$\beta_i(\kappa_i(t)) = \frac{1}{(1 - b_{if})\kappa_i(t) + b_{if}}, \quad i = 1, 2, \cdots, m, \tag{8.33}$$

where κ_i is the rate function which is selected from the pool of rate functions, b_{if} is a design parameter at the user's disposal and satisfies $0 < b_{if} \ll 1$. From the definition of $\beta_i(\kappa_i)$ as given in (8.33), it is seen that β_i is a monotonically increasing function of time with upper bound and $\beta_i^{(j)}$ $(j = 0, 1, \cdots, \infty)$ is also bounded and continuously differentiable for $t \in [0, \infty)$.

To enable each component of the tracking error vector to have its own mode of convergence, we construct the following matrix function:

$$\beta(t) = \text{diag}\{\beta_1(\kappa_1(t)), \cdots, \beta_m(\kappa_m(t))\}. \tag{8.34}$$

Remark 8.3 *According to the definition of β_i, it is seen that $\beta_i \in \left[1, \frac{1}{b_{if}}\right)$, thus for the matrix function $\beta(t)$, there exist positive constants $\underline{\beta}$ and $\overline{\beta}$ such that $0 < \underline{\beta} \le \text{eig}\{\beta^\top \beta\} \le \overline{\beta} < \infty$.*

To proceed, we conduct the following rate transformation on E to get:

$$\zeta_1 = \beta E, \tag{8.35}$$

and we further define:

$$\zeta_i = \frac{d}{dt}\zeta_{i-1}, \ i = 2, \cdots, n+1. \tag{8.36}$$

Or equivalently,

$$\zeta_1 = \beta E, \tag{8.37}$$

$$\zeta_2 = \beta^{(1)}E + \beta E^{(1)}, \tag{8.38}$$

$$\vdots$$

$$\zeta_i = \sum_{j=0}^{i-1} C_{i-1}^j \beta^{(j)} E^{(i-1-j)}, \ i = 3, \cdots, n-1, \tag{8.39}$$

$$\vdots$$

$$\zeta_n = \sum_{j=0}^{n-1} C_{n-1}^j \beta^{(j)} E^{(n-1-j)}, \tag{8.40}$$

$$\zeta_{n+1} = \sum_{j=0}^{n} C_n^j \beta^{(j)} E^{(n-j)}, \tag{8.41}$$

where $C_n^j = \frac{n!}{j!(n-j)!}$, $0! = 1$, and $\zeta_i = [\zeta_{i1}, \cdots, \zeta_{im}]^\top \in \mathbb{R}^m$ $(i = 1, \cdots, n+1)$, with which we introduce a new filtered variable z as:

$$z = \lambda_1 \zeta_1 + \lambda_2 \zeta_2 + \cdots + \lambda_{n-1} \zeta_{n-1} + \zeta_n, \tag{8.42}$$

where $z = [z_1, \cdots, z_m]^\top \in \mathbb{R}^m$ and $\lambda_1, \cdots, \lambda_{n-1}$ are the design parameters chosen such that $l^{n-1} + \lambda_{n-1} l^{n-2} + \cdots + \lambda_1$ is Hurwitz, so that if z goes to zero (or goes to a small region of the equilibrium point), so does ζ_i $(i = 1, \cdots, n)$.

It is worth noting that the filtered variable z as defined in (8.42) is not directly based on E and $E^{(i)}$ $(i = 1, \cdots, n-1)$, but rather, on the transformed error variables ζ_i $(i = 1, \cdots, n)$, as defined in (8.35) and (8.36), which is essentially different from the commonly used way of defining the filtered variable as [114]:

$$s = \lambda_1 E + \lambda_2 E^{(1)} + \cdots + \lambda_{n-1} E^{(n-2)} + E^{(n-1)}.$$

Such treatment, together with other design skills, allows for the aforementioned three control objectives to be achieved concurrently, as seen shortly.

With (8.42), we then have:

$$\dot{z} = \lambda_1 \zeta_2 + \cdots + \lambda_{n-1} \zeta_n + \sum_{j=0}^{n} C_n^j \beta^{(j)} E^{(n-j)} = \beta \left(\dot{x}_n - y_d^{(n)} \right) + \Psi, \qquad (8.43)$$

where

$$\Psi = \sum_{j=1}^{n-1} \lambda_j \zeta_{j+1} + \sum_{j=1}^{n} C_n^j \beta^{(j)} E^{(n-j)} \qquad (8.44)$$

is a computable function.

Substituting (8.31) into (8.43), we get the following filtered "accelerated error dynamics":

$$\dot{z} = \beta G u + \beta \left(F(x, p) + D(x, p) - y_d^{(n)} + \beta^{-1} \Psi \right) = \beta G u + \Delta, \qquad (8.45)$$

with

$$\Delta = \beta (F(x, p) + D(x, p, t) - y_d^{(n)} + \beta^{-1} \Psi)$$

begin the lumped uncertainties.

Upon utilizing Assumption 8.3, $\Delta(\cdot)$ can be upper bounded by:

$$\|\Delta\| \leq \|\beta\| a \phi(x) + \|\beta\| \left\| y_d^{(n)} \right\| + \|\beta\| \|\beta^{-1} \Psi\| \leq b \|\beta\| \Phi(\cdot),$$

where

$$b = \max\{1, a\} > 0,$$

$$\Phi(\cdot) = \phi(x) + y_d^{(n)\top} y_d^{(n)} + (\beta^{-1} \Psi)^\top (\beta^{-1} \Psi) + \frac{1}{2},$$

with b being a new virtual parameter and Φ being a computable scalar function independent of any uncertainties.

It is then interesting to note that, as (8.45) is converted from (8.31), the previously stated full state tracking control problem for (8.31) boils down to designing a continuous and smooth control u to stabilize z governed by (8.45) in the presence of uncertain G and Δ. By making use of the diagonal rate matrix β as defined in (8.34), we construct the following AAC method:

$$\begin{cases} u = - \left(c_0 + \hat{\theta} \|\beta\|^2 \Phi^2 \right) \beta z, \\ \dot{\hat{\theta}} = \gamma \|\beta\|^2 \|z\|^2 \Phi^2 - \sigma \hat{\theta}, \quad \hat{\theta}(0) \geq 0, \end{cases} \qquad (8.46)$$

where $\hat{\theta}$ is the estimate of the virtual unknown parameter $\theta = b^2$, $\hat{\theta}(0) \geq 0$ is the arbitrarily chosen initial value, $c_0 > 0$, $\gamma > 0$, and $\sigma > 0$ are user-chosen control parameters. It should be noted that as $\gamma\|\beta\|^2\|z\|^2\Phi^2 \geq 0$, it holds that $\hat{\theta}(t) \geq 0$ for $t \in [0,\infty)$.

Note that the proposed control exhibits several appealing structural features: 1) it bears the feedback control form, in which the feedback gain is composed of the constant gain and adaptive gain; 2) it is built upon the feedback of βz, the scaled version of z. These features combined play a vital role in establishing the following results.

8.4.3   THEOREM AND STABILITY ANALYSIS

Theorem 8.3 *Consider the MIMO normal-form nonlinear systems in the presence of non-parametric uncertainties and external disturbance described by (8.31), let Assumptions 8.3–8.6 hold. If the AAC method (8.46) is applied, then the control objectives (i)–(iii) are achieved.*

Proof. Choose the Lyapunov function candidate as:

$$V = \frac{1}{2}z^\top z + \frac{1}{2g\gamma}\tilde{\theta}^2,$$

where $\tilde{\theta} = \theta - g\hat{\theta}$ is the virtual parameter estimation error, $g = \lambda_1\beta > 0$ is an unknown constant with λ_1 and β being the unknown constants as defined in Assumption 8.4 and Remark 8.3, respectively.

Taking the time derivative of V along (8.45) yields:

$$\dot{V} = z^\top\beta Gu + z^\top\Delta - \frac{1}{\gamma}\tilde{\theta}\dot{\hat{\theta}}, \tag{8.47}$$

where Δ is defined as before.

Using Young's inequality, we get:

$$z^\top\Delta \leq b\|z\|\|\beta\|\Phi \leq \frac{1}{4} + \theta\|\beta\|^2\|z\|^2\Phi^2,$$

where $\theta = b^2$ is called the virtual parameter as it bears no physical meaning.

Then we have from (8.47) that:

$$\dot{V} \leq z^\top\beta Gu + \frac{1}{4} + \theta\|\beta\|^2\|z\|^2\Phi^2 - \frac{1}{\gamma}\tilde{\theta}\dot{\hat{\theta}}. \tag{8.48}$$

Substituting the control law as given in (8.46) into $z^\top\beta Gu$, one has:

$$z^\top\beta Gu = -\left(c_0 + \hat{\theta}\|\beta\|^2\Phi^2\right)z^\top\beta G\beta z.$$

To go on, we need to equivalently express G as:

$$G = G_1 + G_2$$

with $G_1 = \frac{G+G^\top}{2}$ being symmetric and $G_2 = \frac{G-G^\top}{2}$ being skew-symmetric, then upon employing Assumption 8.6, it is easily obtained that:

$$-z^\top \beta G \beta z = -(\beta z)^\top G_1(\beta z) - (\beta z)^\top G_2(\beta z)$$
$$\leq -\underline{\lambda}_1(\beta z)^\top(\beta z) = -\underline{\lambda}_1 z^\top \beta^\top \beta z,$$

where the fact that $(\beta z)^\top G_2(\beta z) = 0$ and $\beta = \beta^\top$ have been used.
According to Remark 8.3, one further has:

$$-\underline{\lambda}_1 z^\top \beta^\top \beta z \leq -\underline{\lambda}_1 \underline{\beta} z^\top z = -\underline{g} z^\top z,$$

which leads to:

$$z^\top \beta G u \leq -\underline{g}\left(c_0 + \hat\theta \|\beta\|^2 \Phi^2\right) z^\top z.$$

Therefore, it is seen from (8.48) that:

$$\dot V \leq -\underline{g}(c_0 + \hat\theta\|\beta\|^2\Phi^2)z^\top z + \frac{1}{4} + \theta\|\beta\|^2\|z\|^2\Phi^2 - \frac{1}{\gamma}\tilde\theta\dot{\hat\theta}$$
$$= -\underline{g}c_0 z^\top z + \tilde\theta\|\beta\|^2\|z\|^2\Phi^2 + \frac{1}{4} - \frac{1}{\gamma}\tilde\theta\dot{\hat\theta}. \tag{8.49}$$

Inserting the adaptive law for $\dot{\hat\theta}$ as given in (8.46) into (8.49), then we arrive at:

$$\dot V \leq -\underline{g}c_0\|z\|^2 + \frac{\sigma}{\gamma}\tilde\theta\hat\theta + \frac{1}{4} \leq -\underline{g}c_0\|z\|^2 - \frac{\sigma}{2\underline{g}\gamma}\tilde\theta^2 + \frac{\sigma}{2\underline{g}\gamma}\theta^2 + \frac{1}{4}$$
$$\leq -l_1 V + l_2, \tag{8.50}$$

where $l_1 = \min\{2\underline{g}c_0, \sigma\} > 0$, $l_2 = \frac{\sigma}{2\underline{g}\gamma}\theta^2 + \frac{1}{4} < \infty$. From (8.50) we establish the following important results.

(1) We first prove that the variables z, ζ_i $(i=1,\cdots,n)$, x, and $\hat\theta$ are bounded. From (8.50) it is seen that $V(t)$ is bounded over the interval $[0,\infty)$, then it follows that $z \in \mathcal{L}_\infty$ and $\hat\theta \in \mathcal{L}_\infty$ for $\forall t \in [0,\infty)$, which implies that $z_i \in \mathcal{L}_\infty$ $(i=1,\cdots,m)$. According to the definition of z in (8.42), it is seen that $\zeta_j \in \mathcal{L}_\infty$ $(j=1,\cdots,n)$. Note that $\beta^{(j)}$ is bounded, then from (8.37)-(8.41) it is ensured that $E^{(j)} \in \mathcal{L}_\infty$ $(j=0,1,\cdots,n-1)$, which further implies that $x_j \in \mathcal{L}_\infty$ $(j=1,\cdots,n)$ over the interval $[0,\infty)$ as y_d and its up to $(n+1)$-th derivatives are bounded.

(2) Next we prove that $\dot{\hat\theta}$ is bounded. As x is bounded, then from Assumption 8.3 it follows that $F(x,p)$, $G(x,p)$, and $D(x,p)$ are bounded, and $\phi(x)$ is bounded, which indicates that $\Phi \in \mathcal{L}_\infty$, then from (8.46) it is ensured that $\dot{\hat\theta} \in \mathcal{L}_\infty$.

(3) We further prove that $u \in \mathcal{L}_\infty$, $\dot z \in \mathcal{L}_\infty$, and $\dot u \in \mathcal{L}_\infty$. Since $\hat\theta \in \mathcal{L}_\infty$ and $\Phi \in \mathcal{L}_\infty$, from (8.46) it is ensured that the control law u is bounded (i.e., $\dot x_n \in \mathcal{L}_\infty$), then from (8.43) we have $\dot z \in \mathcal{L}_\infty$. To establish the boundedness of $\dot u$, we compute from (8.46) that

$$\dot u = \frac{\partial u}{\partial\hat\theta}\dot{\hat\theta} + \frac{\partial u}{\partial\beta}\dot\beta + \frac{\partial u}{\partial\|\beta\|}\frac{d}{dt}\|\beta\| + \frac{\partial u}{\partial\Phi}\dot\Phi + \frac{\partial u}{\partial z}\dot z,$$

with

$$\frac{\partial u}{\partial \hat{\theta}} = -\|\beta\|^2 \Phi^2 \beta z, \quad \frac{\partial u}{\partial \|\beta\|} = -2\hat{\theta}\|\beta\|\Phi^2\beta z,$$

$$\frac{\partial u}{\partial \Phi} = -2\hat{\theta}\|\beta\|^2 \Phi\beta z, \quad \frac{\partial u}{\partial \beta} = -(c_0 + \hat{\theta}\|\beta\|^2\Phi^2)z,$$

$$\frac{\partial u}{\partial z} = -(c_0 + \hat{\theta}\|\beta\|^2\Phi^2)\beta, \quad \Phi = \frac{\partial \phi}{\partial x}\dot{x} + 2y_d^{(n)T} y_d^{(n+1)} + 2(\beta^{-1}\Psi)\frac{d}{dt}(\beta^{-1}\Psi).$$

Note that $\|\beta\| = \max\{\beta_i\}$ is bounded and $\dot{\beta}_i$ is bounded, and that all the signals including ζ_i $(i = 1, \cdots, n)$, z, $\dot{z}$, x, $\dot{x}$, $\hat{\theta}$, $\dot{\hat{\theta}}$, β, $\|\beta\|$, and Φ are all bounded and continuous, then it is obvious that $\dot{u}$ is bounded and continuous, i.e., u is continuously differentiable.

(4) We then focus on analyzing the tracking performance of each component of the tracking error e_i. Define $0_{n-2} = [0, \cdots, 0]^\top \in \mathbb{R}^{n-2}$, $\mathbf{e}_{n-1} = [0, \cdots, 0, 1]^\top \in \mathbb{R}^{n-1}$, $I_{n-2} = \text{diag}\{1\} \in \mathbb{R}^{(n-2)\times(n-2)}$, and

$$\Lambda = \begin{bmatrix} 0_{n-2} & I_{n-2} \\ -\lambda_1, & -\lambda_2, & \cdots, & -\lambda_{n-1} \end{bmatrix} \in \mathbb{R}^{(n-1)\times(n-1)}.$$

Denote $\rho_i = [\zeta_{1i}, \cdots, \zeta_{n-1,i}]^\top \in \mathbb{R}^{n-1}$ $(i = 1, \cdots, m)$, and $\dot{\rho}_i = [\zeta_{2i}, \cdots, \zeta_{ni}]^\top \in \mathbb{R}^{n-1}$, then we have from (8.42) that:

$$\dot{\rho}_i = \Lambda\rho_i + \mathbf{e}_{n-1}z_i. \tag{8.51}$$

Solving the differential equation (8.51) yields:

$$\rho_i(t) = \rho_i(0)\exp(\Lambda t) + \int_0^t \exp(\Lambda(t-\tau))\mathbf{e}_{n-1}z_i d\tau.$$

Note that Λ is Hurwitz, there exist some positive constants $a_0 > 0$ and $\eta > 0$ such that:

$$\|\exp(\Lambda t)\| \le a_0\exp(-\eta t),$$

then we have:

$$\|\rho_i(t)\| \le a_0\exp(-\eta t)\|\rho_i(0)\| + a_0\exp(-\eta t)\int_0^t \exp(\eta\tau)|z_i(\tau)|d\tau := B_{1i}(t).$$

Since $|z_i| \le \|z\| \le z_{\max}$ with $z_{\max}$ being a positive constant, then we have:

$$B_{1i}(t) \le a_0\exp(-\eta t)\|\rho_i(0)\| + a_0\exp(-\eta t)z_{\max}\int_0^t \exp(\eta\tau)d\tau$$

$$\le a_0\exp(-\eta t)\|\rho_i(0)\| + \frac{a_0 z_{\max}}{\eta}(1 - \exp(-\eta t))$$

$$= \frac{a_0 z_{\max}}{\eta} + \left(a_0\|\rho_i(0)\| - \frac{a_0 z_{\max}}{\eta}\right)\exp(-\eta t)$$

$$\le \frac{a_0 z_{\max}}{\eta} + a_0\|\rho_i(0)\| := \bar{B}_{1i}.$$

As $|\zeta_{1i}| \leq \|\rho_i\|$, it holds that:

$$|\zeta_{1i}| \leq B_{1i}(t) \leq \bar{B}_{1i}.$$

Noting that $E = [e_1, \cdots, e_m]^\top \in \mathbb{R}^m$ and $e_i = \beta_i^{-1}\zeta_{1i}$, $(i = 1, \cdots, m)$, then we have:

$$e_i = \beta_i^{-1}\zeta_{1i} = b_{if}\zeta_{1i} + (1 - b_{if})\kappa_i\zeta_{1i}, \tag{8.52}$$

$$|e_i| \leq b_{if}\bar{B}_{1i} + (1 - b_{if})\kappa_i\bar{B}_{1i}. \tag{8.53}$$

It is seen from (8.52) and (8.53) that e_i contains two components, both of which are bounded: the first one, $b_{if}\zeta_{1i}$, is strictly confined within the residual region $b_{if}\bar{B}_{1i}$ (adjustable via b_{if}) during the entire control process; while the second one, $(1 - b_{if})\kappa_i\zeta_{1i}$, converges to zero at the rate governed not only by ζ_{1i}, but more importantly, by κ_i as well. It is interesting to note that the rate function κ_i is at the user's disposal and is independent of system initial conditions and any other design parameters. Therefore, we have the freedom to select a proper rate function κ_i to adjust the convergence rate and mode for $(1 - b_{if})\kappa_i\zeta_{1i}$, and the freedom to choose a proper b_{if} to confine $b_{if}\bar{B}_{1i}$ as needed, such that the transient performance of e_i can be significantly improved.

To show the full state well-shaped transient and steady-state tracking performance, note from (8.36) that $\dot{E} = \beta^{-1}(\zeta_2 - \dot{\beta}E)$ and denote that $\varepsilon_1 = \zeta_2 - \dot{\beta}E = [\varepsilon_{11}, \varepsilon_{12}, \cdots, \varepsilon_{1m}]^\top \in \mathbb{R}^m$, we have:

$$\dot{e}_i = (1 - b_{if})\kappa_i\varepsilon_{1i} + b_{if}\varepsilon_{1i}, \tag{8.54}$$

for $i = 1, 2, \cdots, m$. Similarly, since $\ddot{E} = \beta^{-1}(\zeta_3 - \ddot{\beta}E - 2\dot{\beta}\dot{E})$ and denote that $\varepsilon_2 = \zeta_3 - \ddot{\beta}E - 2\dot{\beta}\dot{E} = [\varepsilon_{21}, \varepsilon_{22}, \cdots, \varepsilon_{2m}] \in \mathbb{R}^m$, we have:

$$\ddot{e}_i = (1 - b_{if})\kappa_i\varepsilon_{2i}(t) + b_{if}\varepsilon_{2i}(t), \tag{8.55}$$

Following the same procedure as in deriving (8.54) and (8.55), it is shown that:

$$e_i^{(k)} = (1 - b_{if})\kappa_i\varepsilon_{ki} + b_{if}\varepsilon_{ki},$$

where $\varepsilon_k = \zeta_{k+1} - \sum_{j=1}^{k} C_k^j\beta^{(j)}E^{(k-j)}$, $(k = 2, \cdots, n-1; i = 1, 2, \cdots, m)$. Then based on the analysis similar to (8.52), it is established that the convergence rate for each component of the full state tracking error, $e_i^{(k)}$ $(k = 0, 1, \cdots, n-1; i = 1, 2, \cdots, m)$, can be influenced by κ_i, which can be pre-specified uniformly and arbitrarily.

Note that $\beta(0) = I$ and $\hat{\theta}(0)$ can be set as 0, then the initial control signal $u(0)$ is $u(0) = -c_0z(0)$, according to the above analysis, it is shown that, to achieve better tracking performance, except for increasing the control gain k_0, the designers can adjust the rate function κ_i and b_{if} freely, which implies that the excessively large initial control signal is avoided. The proof is completed.

Remark 8.4 *In developing the control schemes, a number of parameters such as a, b, η, and a_0 are defined and used in the stability analysis, but these parameters are not involved in the control algorithms, thus analytical estimation of them (a nontrivial task) are not needed in the proposed control scheme.*

Remark 8.5 *It is well known that the normal-form nonlinear system is a special case of strict-feedback nonlinear systems, then based on the backstepping technique and the time-varying scaling function, developing an AAC scheme for strict-feedback nonlinear systems in the presence of parametric/non-parametric uncertainties represents an interesting topic, interested readers can perform the corresponding accelerated controller and stability analysis.*

8.5 APPLICATION TO ROBOTIC SYSTEMS

The applicability of the AAC method to robotic systems, a typical class of MIMO nonlinear systems, is examined in this section. Consider a rigid m-joint robot with the following joint space dynamics:

$$H(q,p)\ddot{q} + N_g(q,\dot{q},p)\dot{q} + G_g(q,p) + \tau_d(\dot{q},p) = u, \qquad (8.56)$$

where $q = [q_1, \cdots, q_m]^\top \in \mathbb{R}^m$ denotes the link position, $H(q,p) \in \mathbb{R}^{m \times m}$ denotes the inertia matrix, $p \in \mathbb{R}^l$ denotes the unknown parameter vector, $N_g(q,\dot{q},p) \in \mathbb{R}^{m \times m}$ denotes the centripetal-Coriolis matrix, $G_g(q,p) \in \mathbb{R}^m$ represents the gravitation vector, $\tau_d(\dot{q},p) \in \mathbb{R}^m$ denotes the frictional and disturbing force, and $u \in \mathbb{R}^m$ is the control input.

Define that $q = x_1$ and $\dot{q} = x_2$, then (8.56) can be expressed as:

$$\begin{cases} \dot{x}_1 = x_2, \\ \dot{x}_2 = F(x,p) + G(x,p)u + D(x,p), \end{cases}$$

with $F(x,p_1) = H^{-1}(-N_g x_2 - G_g)$, $D(x,p_2) = -H^{-1}\tau_d$, and $G = H^{-1}$.

The control objective is as follows: given the reference trajectories $y_d(t)$, derive an AAC for the actuator torque and adaptation law for the unknown parameters, such that the manipulator joint position $x_1(t)$ closely tracks the desired position $y_d(t)$ without involving large control effort in the initial period.

To achieve such control objectives, we employ the Assumptions 7.7–7.9.

Now we perform the transformation by using the diagonal rate matrix β as given in (8.34) to get:

$$\zeta_1 = \beta E,$$
$$\zeta_2 = \dot{\beta} E + \beta \dot{E},$$

where $E = x_1 - y_d = [e_1, \cdots, e_m]^\top \in \mathbb{R}^m$, then we introduce a new filtered variable z as:

$$z = \lambda_1 \zeta_1 + \zeta_2,$$

where $\lambda_1 > 0$, and the system dynamics in terms of z is:

$$\dot{z} = \beta(Gu + F + D - \ddot{y}_d) + \lambda_1 \zeta_2 + \ddot{\beta} E + 2\dot{\beta}\dot{E} = \beta Gu + \Delta_1,$$

where $\Delta_1 = \beta(F + D - \ddot{y}_d) + \Psi$ and $\Psi = \lambda_1 \zeta_2 + \ddot{\beta} E + 2\dot{\beta}\dot{E}$.

According to the properties of robotic systems in Chapter 1, we have:

$$\|F\| \le \|G\| (\|N_g\| \|x_2\| + \|G_g\|) \le \overline{\lambda} \left(\gamma_N \|x_2\|^2 + \gamma_G \right),$$
$$\|D\| \le \|G\| \|\tau_d\| \le \overline{\lambda} \gamma_D (1 + \|x_2\|),$$

then the term Δ_1 can be further upper bounded by:

$$\begin{aligned}
\|\Delta_1\| &= \beta \left(F + D - \ddot{y}_d + \beta^{-1} \Psi \right) \\
&\le \|\beta\| \left(\|F\| + \|D\| + \|\ddot{y}_d\| + \|\beta^{-1} \Psi\| \right) \\
&\le \|\beta\| \left[\overline{\lambda} \left(\gamma_N \|x_2\|^2 + \gamma_G \right) + \overline{\lambda} \gamma_D (1 + \|x_2\|) + \|\ddot{y}_d\| + \|\beta^{-1} \Psi\| \right] \\
&\le b \|\beta\| \left[\|x_2\|^2 + 2 + \|x_2\| + \|\ddot{y}_d\| + \|\beta^{-1} \Psi\| \right] \\
&\le b \|\beta\| \Phi,
\end{aligned}$$

with

$$b = \max \left\{ \overline{\lambda} \gamma_N, \overline{\lambda} \gamma_G, \overline{\lambda} \gamma_D, 1 \right\} > 0,$$
$$\Phi(\cdot) = 2 \|x_2\|^2 + \|\ddot{y}_d\|^2 + \left(\beta^{-1} \Psi \right)^{\top} \left(\beta^{-1} \Psi \right) + 3,$$

where b is an unknown virtual parameter and Φ is a computable function.
Now we construct the following AAC scheme for robotic systems:

$$\begin{cases} u = -(c_0 + \hat{\theta} \|\beta\|^2 \Phi^2) \beta z, \\ \dot{\hat{\theta}} = \gamma \|\beta\|^2 \|z\|^2 \Phi^2 - \sigma \hat{\theta}, \ \hat{\theta}(0) \ge 0, \end{cases} \tag{8.57}$$

where $\hat{\theta}$ is the estimate of the virtual unknown parameter $\theta = b^2$, $\hat{\theta}(0) \ge 0$ is the arbitrarily chosen initial value, $c_0 > 0$, $\gamma > 0$, and $\sigma > 0$ are the user-chosen control parameters.

Theorem 8.4 *Consider the nonlinear robotic systems described by (8.56), if the AAC scheme (8.57) is applied, it is ensured that:*

(i) *all signals in the closed-loop systems are bounded, and the control action is continuously differentiable;*
(ii) *accelerated tracking is obtained without involving excessively initial large control effort.*

Proof. The proof is similar to the stability analysis in Theorem 8.3, thus is omitted.

Remark 8.6 *When $\beta = I$, the proposed AAC scheme (8.57) for robotic systems reduces to:*

$$\begin{cases} u = -(c_0 + \hat{\theta} \Phi^2) z, \\ \dot{\hat{\theta}} = \gamma \|z\|^2 \Phi^2 - \sigma \hat{\theta}, \cdot \hat{\theta}(0) \ge 0. \end{cases} \tag{8.58}$$

As a special case of (8.57), here (8.58) is referred to as the TAC method, which, although achieving ultimately uniformly bounded tracking, does not exhibit the appealing features (such as pre-assignable convergence mode) as previously identified and analyzed. In order to achieve better tracking performance under the TAC method (8.58), the design parameters should be chosen appropriately by the method of trial and error, usually the large control effort in the initial period (near the startup point) is inevitable. However, in the proposed AAC method due to the introduced rate transformation as defined in (8.36), a pre-assignable convergence rate of the tracking error is ensured and the large control effort in the initial period can be avoided.

8.6 NUMERICAL SIMULATION

To verify the effectiveness of the proposed AAC method in Section 8.5, we consider the 2-DOF robotic manipulator as defined in (8.56), where the detailed expressions are in Section 7.2.3.

In the simulation, the desired joint trajectory is $y_d(t) = [\sin(t), \sin(t)]^\top$. The rate matrix is:

$$\beta = \text{diag}\{\beta_1(\kappa_1), \beta_2(\kappa_2)\}$$

with κ_i $(i = 1, 2)$ being the rate function in Section 8.1.1. To simplify the simulation and representation, we choose $\kappa_i = \kappa$, $b_{if} = b_f$ for both joints and the rate function κ is chosen as:

(i) traditional adaptive control $(\kappa = 1)$;
(ii) $\kappa = \exp(-1.8t)$;
(iii) $\kappa = \frac{1}{5^t + t + t^4}$.

Furthermore, the initial conditions are given as: $x_1(0) = [0.5, 1]^\top$, $x_2(0) = [1, 1]^\top$, and $\hat{\theta}(0) = 0$.

Note that, for any control scheme, different control efforts (related to its control gain) could lead to different control precision. Hence, to get a fair comparison between the TAC in Remark 8.6 and the proposed AAC in Section 8.5, we first test that with the same or similar control effort, especially at the startup point, what is the control performance achieved by each control scheme (Test 1); then we test what control magnitude (effort) is required from each control scheme in order to get the same or similar tracking precision by increasing the control gain in the TAC (Test 2).

Test 1: To verify that the proposed AAC gives rise to much better control performance as compared with the TAC method under a similar amount of control effort.

In this case, the design parameters in the TAC are given as: $c_0 = 20$, $\gamma = 1$, $\sigma = 0.8$, and $\lambda_1 = 2$; whereas the ones in the proposed AAC are chosen as: $c_0 = 10$, $\gamma = 0.001$, $\sigma = 0.8$, $\lambda_1 = 1$, and $b_f = 0.07$. Using the same initial conditions and the above design parameters, the simulation results are shown in Fig. 8.8–Fig. 8.12, where Fig. 8.8 and Fig. 8.9 are the evolutions of each component of the tracking error, e_1 and e_2, under different rate functions $\left(1, \exp(-1.8t), \text{ and } \frac{1}{5^t + t + t^4}\right)$, showing

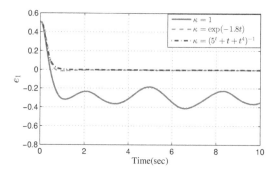

FIGURE 8.8 The evolution of the tracking error e_1 with different κ.

that better tracking performance is obtained with the proposed AAC method, as compared with the TAC method, which confirms the theoretical prediction. The control torques u_1 and u_2 are presented in Fig. 8.10 and Fig. 8.11. It is interesting to observe that the control torque corresponding to different rate functions differs in detail as compared with the TAC scheme. However, the overall magnitude, continuity, and smoothness of the control torques generated by the proposed control schemes are well comparable with those by the TAC method. Furthermore, the evolution of the parameter estimate $\hat{\theta}$ is shown in Fig. 8.12.

Test 2: To verify that the proposed AAC is able to achieve similar tracking performance with much less control effort as compared with the TAC.

As verified in Test 1, the TAC does not produce satisfactory tracking results. Now we change the design parameters for the TAC as $c_0 = 1000$, $\gamma = 1$, $\sigma = 0.8$, and $\lambda_1 = 2$, so that similar satisfactory tracking results are obtained as seen in Figs. 8.13 to 8.16. However, much larger control effort is required from the TAC to achieve such performance as seen in Figs. 8.15 to 8.16. In particular, during the initial period, the TAC involves a large peaking value. Whereas the proposed AAC is able to ensure

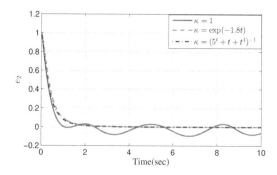

FIGURE 8.9 The evolution of the tracking error e_2 with different κ.

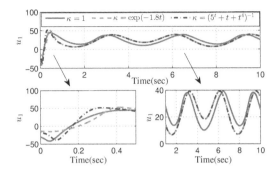

FIGURE 8.10 The evolution of the control input u_1 with different κ.

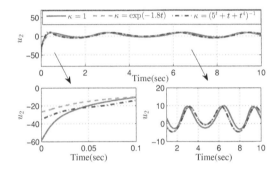

FIGURE 8.11 The evolution of the control input u_2 with different κ.

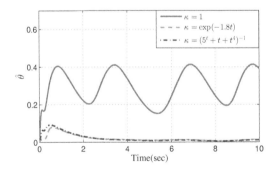

FIGURE 8.12 The evolution of the parameter $\hat{\theta}$ with different rate functions κ.

satisfactory performance without the need for large control magnitude during the initial period.

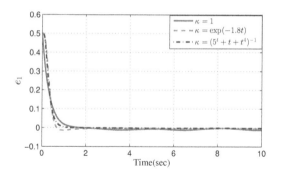

FIGURE 8.13 The evolution of the tracking error e_1 with different κ.

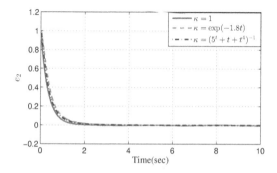

FIGURE 8.14 The evolution of the tracking error e_2 with different κ.

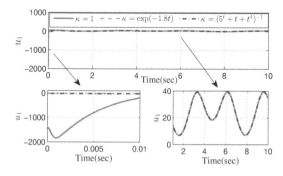

FIGURE 8.15 The evolution of the control input u_1 with different κ.

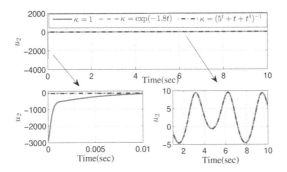

FIGURE 8.16 The evolution of the control input u_2 with different κ.

8.7 NOTES

To close this chapter we offer the following concluding remarks. First, the idea behind using rate transformation to derive accelerated tracking control is simple, and the underlying concept goes with the adaptive (or robust adaptive) control seamlessly, thus is readily apprehensible to the control community.

Second, it is interesting to note that the resultant AAC scheme in Section 8.2 contains a time-varying self-updating gain $\dot{\beta}\beta^{-1}$, in addition to the constant gain k_A, it is such a self-tuning gain that makes the control scheme more energy-efficient as compared with the TAC with a high and constant gain because here the time-varying gain vanishes as time goes by. This is a remarkable feature that contributes to the improvement in the transient and steady-state performance.

Third, for the TAC method, the prevailing intuitive wisdom is that increasing the control gain and speeding up the parameter adaptation rate could result in better control performance. While this is probably true, the more important issue to address is *"how to increase the gain or the updating rate and by how much"*. Prior to this work, the common practice has been to use a high constant gain in the control law and in the updating law, and such gain normally remains unchanged during the entire control process. Now the proposed AAC method suggests that increasing (at a certain point of time) the control gain and adaptation rate does enhance control performance but this has to be done in a quite dedicated way-the control gain and the parameter adaptation rate should be adjusted at varying magnitude and different pace determined by a time-varying and bounded factor related to the rate function, whose analytical expression has been explicitly provided.

Finally, it is worth mentioning that the result presented in Section 8.2 is based on the adaptive control technique that relies on the parametric decomposition of uncertain terms. To deal with the more general case of nonlinear systems with non-parametric uncertainty, a rate transformation-based robust AAC scheme is developed in Section 8.3 such that accelerated tracking with acceptable transient and steady-state tracking performance is achieved without involving large control effort in the initial period. However, although the tracking performance specification in terms of transient and steady-state performance in this chapter can be adjusted by properly choosing the design parameter and rate function, it still depends on some unknown constants and initial conditions. Therefore, how to design a prescribed performance control scheme for a class of nonlinear systems represents an interesting topic in future research.

9 Control with Exponential Convergence and Prescribed Performance

In Chapter 8, an AAC scheme was proposed for a class of nonlinear systems with parametric/non-parametric uncertainties. By introducing a rate function and a time-varying scaling function-based transformation, the transient and steady-state performance in terms of tracking error can be adjusted by properly choosing design parameters without involving large control effort in the initial period. However, the corresponding tracking performance specification still depends on some unknown constants and initial conditions [i.e., the residual term $b_f z(t)$ in (8.13)], which cannot be determined at the designer's disposal. Therefore, how to design a prescribed performance control scheme is an interesting yet a challenging work.

In this chapter, we present two scaling function transformation-based control algorithms for nonlinear systems. In Section 9.1, we first propose a model-dependent traditional backstepping control scheme for strict-feedback nonlinear systems. Although exponential tracking can be guaranteed, the convergence rate relies on control gains. To achieve a uniform exponential tracking result, we propose an exponential transformation, based on which a uniform exponential tracking control algorithm is developed in Section 9.1.3 so that the convergence rate is independent of control gains. However, the system models in the above works must be available for control design. In Section 9.2, we propose a model-independent exponential tracking control scheme for a class of normal-form nonlinear systems. Furthermore, by introducing the concept of prescribed performance function, a novel asymptotic adaptive tracking control algorithm with transient performance for parametric strict-feedback nonlinear systems is developed in Section 9.3. Numerical simulations are conducted to verify the effectiveness and benefits of the proposed methods.

9.1 MODEL-DEPENDENT UNIFORM EXPONENTIAL TRACKING CONTROL

9.1.1 PROBLEM FORMULATION AND PRELIMINARIES

We consider the following strict-feedback nonlinear systems:

$$\begin{cases} \dot{x}_k = x_{k+1} + \phi_k\left(\theta, \bar{x}_k\right), \ k = 1, 2, \cdots, n-1, \\ \dot{x}_n = u + \phi_n\left(\theta, \bar{x}_n\right), \\ y = x_1, \end{cases} \tag{9.1}$$

DOI: 10.1201/9781003474364-9

where $x_i \in \mathbb{R}$, $i = 1, \cdots, n$, is the system state and $\bar{x}_i = [x_1, \cdots, x_i]^\top \in \mathbb{R}^i$, $u \in \mathbb{R}$, and $y \in \mathbb{R}$, are the control input and system output, respectively, $\phi_i \in \mathbb{R}$ is a smooth signal, and $\theta \in \mathbb{R}^r$ represents the system parameter.

Define $e = x_1 - y_d$ as the tracking error with y_d being the reference signal. In this chapter, the control objective is to establish a control strategy such that:

(i)   all signals in the closed-loop systems are bounded;
(ii) the uniform exponential convergence of the tracking error is ensured and the error overshoot can be made as small as desired.

For later technical development, the following assumptions are imposed.

Assumption 9.1 *The system states are available for control design.*

Assumption 9.2 *The reference trajectory $y_d(t)$ and its derivatives up to $(n+1)$-th are known, bounded, and piecewise continuous.*[1]

Assumption 9.3 *The nonlinear function $\phi_i(\theta, \bar{x}_i)$ is available for control design.*

In this section, we consider the strict-feedback nonlinear systems without uncertainties (i.e., $\phi_k(\theta, \bar{x}_k) \in \mathscr{C}^{n+1-k}$, $k = 1, \cdots, n$, is available). Before designing the uniform exponential tracking control for strict-feedback nonlinear systems, we briefly recall the backstepping design procedure of the traditional control scheme.

9.1.2 TRADITIONAL BACKSTEPPING CONTROL DESIGN

Similar to that in reference [9], we develop the following control scheme for (9.1):

$$z_i = x_i - \alpha_{i-1}, \quad i = 1, \cdots, n, \tag{9.2}$$

$$\alpha_1 = -c_1 z_1 - \phi_1 + \dot{y}_d, \tag{9.3}$$

$$\alpha_j = -z_{j-1} - c_j z_j + \dot{\alpha}_{j-1} - \phi_j, \quad j = 2, \cdots, n, \tag{9.4}$$

$$u = \alpha_n, \tag{9.5}$$

$$\dot{\alpha}_{j-1} = \sum_{k=1}^{j-1} \frac{\partial \alpha_{j-1}}{\partial x_k}(x_{k+1} + \phi_k) + \sum_{k=0}^{j-1} \frac{\partial \alpha_{j-1}}{\partial y_d^{(k)}} y_d^{(k+1)}, \quad j = 2, \cdots, n,$$

with $\alpha_0 = y_d$, and $c_1, \cdots, c_n > 0$.

Choosing the Lyapunov function candidate as $V = \frac{1}{2} \sum_{k=1}^{n} z_k^2$, together with the standard procedure in backstepping design, it is readily derived from (9.3) to (9.5) that:

$$\dot{V} = -\sum_{k=1}^{n} c_k z_k^2 \leq -lV, \tag{9.6}$$

---

[1] The boundedness of $y_d^{(n+1)}$ is imposed to establish the boundedness of the control rate $\dot{u}$.

with $l = \min\{2c_1, \cdots, 2c_n\}$. Based on the analysis in reference [9], it is easily ensured that all signals in the closed-loop systems are bounded.

Now we focus on analyzing the convergence rate of the tracking error. By integral operation for (9.6) on $[0,t]$, we have:

$$V(t) \le V(0) \exp(-lt).$$

Note that $\frac{1}{2}z_1^2 \le V(t)$, then it is ensured that:

$$|z_1(t)| \le \sqrt{2V(0)} \exp\left(-\frac{l}{2}t\right) \le \|\mathscr{Z}(0)\| \exp\left(-\frac{l}{2}t\right), \qquad (9.7)$$

where $\mathscr{Z} = [z_1, \cdots, z_n]^\top$. Next we make the following comments.

(1) From (9.7), it is seen that the tracking error converges to zero with the exponential rate no less than $\exp\left(-\frac{l}{2}t\right)$. However, according to the expression of $l = \min\{2c_1, \cdots, 2c_n\}$, it is seen that l depends on the control gain c_k, $k = 1, \cdots, n$.
(2) From (9.7), it is seen that the value of maximum overshoot $\|\mathscr{Z}(0)\|$ relies on $z_1(0), \cdots, z_n(0)$, which are related to $x_1(0), \cdots, x_n(0), y_d(0), \cdots, y_d^{(n)}(0)$, then adjusting the initial values of these variables (especially the initial values of system states) simultaneously to limit the overshoot of the tracking error is required, which is undesirable or even impossible in practice.

9.1.3 UNIFORM EXPONENTIAL TRACKING CONTROL DESIGN

To achieve uniform exponential tracking and at the same time to make the error overshoot as small as desired without the need for re-setting system initial states and/or the reference trajectory, we employ the following error-dependent function transformation:

$$s(t) = \frac{\zeta_1(t)}{(k_{c_1} + \zeta_1(t))(k_{c_2} - \zeta_1(t))} \qquad (9.8)$$

and the time-varying scaling transformation:

$$\zeta_i = \beta(t)z_i, \quad i = 1, \cdots, n, \qquad (9.9)$$
$$\beta(t) = \exp(\lambda t), \quad \lambda > 0, \qquad (9.10)$$
$$z_i = x_i - \alpha_{i-1}, \qquad (9.11)$$

where $\alpha_0 = y_d$ and α_j, $j = 1, \cdots, n-1$, will be given in later, ζ_i is the transformed error, k_{c1} and k_{c2} are positive constants, and the initial condition $\zeta_1(0)$ satisfies $-k_{c_1} < \zeta_1(0) < k_{c_2}$.

To facilitate the analysis, the following lemma is imposed.

Lemma 9.1 *The transformed function $s(t)$ has the following properties:*

(1) *In the interval $\zeta_1 \in (-k_{c_1}, k_{c_2})$, the transformed function s is strictly monotonic w.r.t. ζ_1;*

(2)   *In the interval $\zeta_1(t) \in (-k_{c_1}, k_{c_2})$,*

$$\zeta_1(t) = 0 \Leftrightarrow s(t) = 0.$$

(3)   *If $\zeta_1(0) \in (-k_{c_1}, k_{c_2})$ and $s(t) \in \mathscr{L}_\infty$ holds for $t \geq 0$, then*

$$\zeta_1(t) \in (-k_{c_1}, k_{c_2}), \quad \text{for } t \in [0, \infty)$$

always holds.

Proof. We now prove the properties of Lemma 9.1 one by one.

(1) According to (9.8), the derivation of s w.r.t. ζ_1 can be calculated as:

$$\frac{\partial s}{\partial \zeta_1} = \frac{\zeta_1^2 + k_{c_1} k_{c_2}}{(k_{c_1} + \zeta_1)^2 (k_{c_2} - \zeta_1)^2}.$$

Since $k_{c_1} k_{c_2} > 0$, then $\frac{\partial s}{\partial \zeta_1} > 0$ is established in the internal $\zeta_1(t) \in (-k_{c_1}, k_{c_2})$, implying that the transformed function s is a monotonic function of ζ_1.

(2) In the internal $\zeta_1(t) \in (-k_{c_1}, k_{c_2})$, $(k_{c_1} + \zeta_1)(k_{c_2} - \zeta_1) > 0$ holds for all time, therefore, $\zeta_1(t) = 0$ is ensured if and only if $s(t) = 0$.

(3) Since the transformed function s is strictly monotonic w.r.t. ζ_1, then if and only if

$$\zeta_1 \to -k_{c_1} \quad \text{or} \quad \zeta_1 \to k_{c_2},$$

we have $s \to \pm\infty$. If the initial value of ζ_1 satisfies $\zeta_1(0) \in (-k_{c_1}, k_{c_2})$, and s satisfies $s(t) \in \mathscr{L}_\infty$ for $\forall t > 0$, it implies that $\zeta_1(t)$ is always within the open interval $\zeta_1(t) \in (-k_{c_1}, k_{c_2})$. The proof is completed.

According to the above lemma, it follows from (9.9) and (9.10) that, for $i = 1$, if $-k_{c_1} < \zeta_1(t) < k_{c_2}$, we have:

$$-k_{c_1} \exp(-\lambda t) < z_1(t) < k_{c_2} \exp(-\lambda t), \tag{9.12}$$

which implies that:

(1)   on the one hand, the tracking error converges to zero with the exponential rate $\exp(-\lambda t)$, where λ is at the user's disposal and is independent of the control gain c_k;

(2)   on the other hand, k_{c_2} and $-k_{c_1}$ serve as the upper bound and lower bound of the overshoot, respectively, which are independent of initial conditions. Thus, by choosing the parameters λ, k_{c_1}, and k_{c_2} properly and independently, the transient behavior of the tracking error $z_1(t)$ can be well-shaped.

The above analysis indicates that the key is to design a control scheme to guarantee the boundedness of s.

To this end, we take the derivative of s as defined in (9.8) w.r.t. time:

$$\dot{s} = \mu_1 \dot{\zeta}_1$$

with

$$\mu_1 = \frac{k_{c_1} k_{c_2} + \zeta_1^2}{(k_{c_1} + \zeta_1)^2 (k_{c_2} - \zeta_1)^2}$$

being well defined in the set $\Omega_\zeta := \{\zeta_1 \in \mathbb{R} : -k_{c_1} < \zeta_1(t) < k_{c_2}\}$, which is computable and available for control design.

For ease of description, define $\bar{y}_d^{(i)} = \left[y_d, \cdots, y_d^{(i)}\right]^\top$, $\bar{\zeta}_i = [\zeta_1, \cdots, \zeta_i]^\top$. Similar to the control design in reference [9], the virtual controller α_i and actual controller u are developed as follows:

$$\alpha_1 = -\lambda z_1 - \beta^{-1} \mu_1^{-1} c_1 s - \phi_1 + \dot{y}_d, \tag{9.13}$$

$$\alpha_2 = -\beta^{-1} \mu_1 s - (\lambda + c_2) z_2 - \phi_2 + \dot{\alpha}_1, \tag{9.14}$$

$$\alpha_j = -z_{j-1} - (\lambda + c_j) z_j - \phi_j + \dot{\alpha}_{j-1}, \quad j = 3, \cdots, n, \tag{9.15}$$

$$u = \alpha_n, \tag{9.16}$$

where $\dot{\alpha}_i = \sum_{k=1}^{i} \frac{\partial \alpha_i}{\partial x_k}(x_{k+1} + \phi_k) + \sum_{k=0}^{i} \frac{\partial \alpha_i}{\partial y_d^{(k)}} y_d^{(k+1)} + \sum_{k=1}^{i} \frac{\partial \alpha_i}{\partial \zeta_k} \dot{\zeta}_k + \frac{\partial \alpha_i}{\partial \beta^{-1}}(-\lambda \beta^{-1})$
for $i = 1, \cdots, n-1$, and $c_k \geq 0$ for $k = 1, \cdots, n$.

Now we are ready to present the following theorem and its proof.

Theorem 9.1 *Consider the strict-feedback nonlinear system (9.1) without uncertainty. Let Assumptions 9.1–9.3 hold. If the control algorithms as given in (9.13)–(9.16) are applied, then the objectives (i) and (ii) are achieved, especially the tracking error converges to zero with a prescribed exponential rate, i.e.,*

$$-k_{c_1} \exp(-\lambda t) < z_1(t) < k_{c_2} \exp(-\lambda t). \tag{9.17}$$

By choosing the design parameters λ, k_{c_1}, and k_{c_2} properly and independently, the transient performance including the convergence rate and overshoot can be improved.

Proof. We first prove the convergence rate of the tracking error. Under the control (9.13)–(9.16), we have the following closed-loop system dynamics,

$$\dot{\zeta}_1 = -\mu_1^{-1} c_1 s + \zeta_2, \tag{9.18}$$

$$\dot{s} = -c_1 s + \mu_1 \zeta_2, \tag{9.19}$$

$$\dot{\zeta}_2 = \zeta_3 - c_2 \zeta_2 - \mu_1 s, \tag{9.20}$$

$$\dot{\zeta}_j = \zeta_{j+1} - c_j \zeta_j - \zeta_{j-1}, \tag{9.21}$$

$$\dot{\zeta}_n = -\zeta_{n-1} - c_n \zeta_n, \tag{9.22}$$

for $j = 3, \cdots, n-1$.

Choosing the Lyapunov function candidate as $V = \frac{1}{2} s^2 + \frac{1}{2} \sum_{k=2}^{n} \zeta_k^2$, then the derivative of V along (9.19)–(9.22) is:

$$\dot{V} = -c_1 s^2 - \sum_{k=2}^{n} c_k \zeta_k^2 \leq 0.$$

Hence, for any initial condition $\zeta_1(0) \in \Omega_\zeta$, it is ensured that $s \in L_\infty$ and $\zeta_k \in L_\infty$, $k = 2, \cdots, n$, then one has:

$$-k_{c_1} < \zeta_1(t) < k_{c_2} \quad \text{for } \forall t \geq 0.$$

Note that $z_1 = \exp(-\lambda t)\zeta_1$, then (9.17) is established, it is therefore concluded that, not only the tracking error can converge to zero in the exponential rate $\exp(-\lambda t)$ with λ being chosen freely by the designer, but also the (maximum) overshoot can be improved by adjusting parameters k_{c_1} and k_{c_2} without trajectory re-initialization.

In addition, as $\dot{V} \leq 0$ and $\beta(0) = 1$, it holds that:

$$\frac{1}{2}\zeta_k^2 \leq V(0) = \frac{1}{2}s^2(0) + \frac{1}{2}\sum_{k=2}^{n} z_k^2(0),$$

then there exists a class $\mathscr{K}_\infty$ function $\check{N}(s(0), \mathscr{Z}_1(0))$ with $\mathscr{Z}_1 = [z_2, \cdots, z_n]^\top$ such that:

$$|\zeta_k(t)| \leq \sqrt{s^2(0) + \sum_{k=2}^{n} z_k^2(0)} \leq \check{N}(s(0), \mathscr{Z}_1(0)),$$

where $\check{N}(s(0), \mathscr{Z}_1(0)) = |s(0)| + \|\mathscr{Z}_1(0)\|$.

Furthermore, note that $z_k = \exp(-\lambda t)\zeta_k$, then it is seen that:

$$|z_k(t)| \leq \check{N}\exp(-\lambda t), \quad k = 2, \cdots, n,$$

which shows that the error $z_k(t)$, $k = 2, \cdots, n$, converges to zero exponentially at a uniform rate.

Next we show that all signals in the closed-loop system are bounded. As $z_1 = x_1 - y_d$ and $y_d^{(k)}$, $k = 0, \cdots, n+1$, are bounded, it follows that $x_1 \in \mathscr{L}_\infty$, which implies that $\phi_1(\cdot) \in \mathscr{L}_\infty$. Note that $s \in \mathscr{L}_\infty$, it indicates that ζ_1 remains within the subset of Ω_ζ, which further implies that μ_1 and α_1 are bounded. Since that $z_2 = x_2 - \alpha_1$ is bounded, it holds that $x_2 \in \mathscr{L}_\infty$, it follows that $\phi_2(\cdot) \in \mathscr{L}_\infty$. In addition, from the definition of $\dot{\alpha}_1$, the boundedness of $\dot{\alpha}_1$ and α_2 can be ensured. Using the similar analysis process, it is ensured that α_k, $k = 3, \cdots, n-1$, x_k, ϕ_k, $\dot{\alpha}_{k-1}$, $k = 3, \cdots, n$, and the control input u are bounded, which ensures that $\dot{x}_k \in \mathscr{L}_\infty$, $k = 1, \cdots, n$. Furthermore, it is seen from (9.16) that the control input u is the function of variables $\bar{x}_n$, $\bar{y}_d^{(n)}$, $\bar{\zeta}_n$, and β^{-1}, as all signals in the closed-loop are bounded, then it is easily derived that the rate of the control input, $\dot{u}$, is bounded, i.e., the control input is $\mathscr{C}^1$ smooth. The proof is completed.

To test the effectiveness of the proposed control in Section 9.1.3 and compare with the method in Section 9.1.2, we consider the following strict-feedback nonlinear system:

$$\begin{cases} \dot{x}_1 = x_2 + \phi_1(\theta, x_1), \\ \dot{x}_2 = u, \end{cases}$$

where $\phi_1 = \theta x_1^2$ with $\theta = 1$. In the simulation, the initial conditions are given as $x_1(0) = 0.8$, $x_2(0) = -0.2$, and the reference signal is $y_d = \sin(t)$. Under the proposed control (9.16), the design parameters are chosen as $c_1 = c_2 = 0/0.5$, $\lambda = 1$,

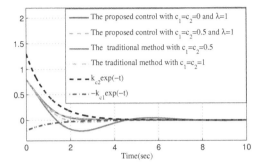

FIGURE 9.1 The evolution of the tracking error under the proposed method (9.16) and the traditional method in Section 9.1.2.

$k_{c_1} = 0.2$, and $k_{c_2} = 1.3$. For the traditional backstepping method in Section 9.1.2, i.e., (9.5), we choose different design parameters, i.e., $c_1 = c_2 = 0.5/1$. The simulation results are shown in Fig. 9.1–Fig. 9.2. From Fig. 9.1, it is seen that, compared with the traditional method, the tracking error converges to zero with exponential rate $\exp(-\lambda t)$ that is independent of control gain c_k, $k = 1, 2$, and the overshoot of the tracking error can be improved under the proposed control method.

9.2 MODEL-INDEPENDENT UNIFORM EXPONENTIAL TRACKING CONTROL

In this section, we introduce a model-independent exponential tracking control scheme for normal-form nonlinear systems.

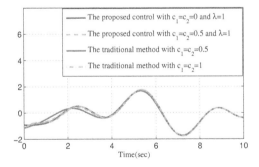

FIGURE 9.2 The evolution of the control input under the proposed method (9.16) and the traditional method in Section 9.1.2.

9.2.1 DYNAMIC MODEL AND PROBLEM FORMULATION

Consider the following MIMO normal-form nonlinear systems:

$$\begin{cases} \dot{x}_k = x_{k+1}, \; k = 1, 2, \cdots, n-1, \\ \dot{x}_n = F(x,p) + G(x)u, \\ y = x_1, \end{cases} \qquad (9.23)$$

where $x_k = [x_{k1}, \cdots, x_{km}]^\top \in \mathbb{R}^m$ $(k = 1, \cdots, n)$ and $x = \left[x_1^\top, \cdots, x_n^\top\right]^\top \in \mathbb{R}^{mn}$ are the state vectors assumed to be available for control design, $F(x,p) \in \mathbb{R}^m$ denotes the non-parametric uncertainty, $G(x) \in \mathbb{R}^{m \times m}$ represents the gain matrix, $p \in \mathbb{R}^r$ is an unknown parameter vector, $y \in \mathbb{R}^m$ is the system output, and $u = [u_1, u_2, \cdots, u_m]^\top \in \mathbb{R}^m$ is the control signal.

Remark 9.1 *Many practical systems can be described by the form (9.23), for example, robotic systems, unmanned aerial vehicles, and inverted pendulums [84, 115]. In addition, non-parametric uncertainty indicates that the uncertain function does not satisfy the parameter decomposition condition, i.e., $F(x,p) \neq \Psi^\top p$ with $\Psi \in \mathbb{R}^{r \times m}$ being a smooth function. Accordingly, the control design of nonlinear systems with non-parametric uncertainty is much more difficult yet challenging compared with the ones with parametric uncertainty.*

Define the tracking error as $E = x_1 - y_d = [e_{11}, \cdots, e_{1m}]^\top$ and $E^{(k)} = x_1^{(k)} - y_d^{(k)} = [e_{k+1,1}, \cdots, e_{k+1,m}]^\top$, $k = 1, \cdots, n-1$, where $y_d = [y_{d1}, \cdots, y_{dm}]^\top$ and $y_d^{(k)} = \left[y_{d1}^{(k)}, \cdots, y_{dm}^{(k)}\right]^\top$ are the known reference signal and its k-th derivative, respectively. Furthermore, it is assumed that y_d and its derivatives up to n-th are known, bounded, and piecewise continuous.

The control objectives in this section are to provide an effective control solution for nonlinear systems (9.23) such that: all signals in the closed-loop systems are bounded; the tracking error converges to zero exponentially regardless of non-parametric uncertainties; and the exponential decay rate is no less than a predefined value that is independent of initial conditions.

To this end, we impose the following assumptions.

Assumption 9.4 *The unknown control gain matrix $G(\cdot) \in \mathbb{R}^{m \times m}$ is symmetric and positive definite, and the Euclidean norm of $G(\cdot)$ is bounded, i.e., there exist unknown positive constants λ_g and λ_G such that, for any $X \in \mathbb{R}^m$, one has $\lambda_g \|X\|^2 \leq X^\top G X \leq \lambda_G \|X\|^2$.*

Assumption 9.5 *There exist an unknown constant a and a non-negative available smooth function $\phi(x)$ such that $\|F(x,p)\| \leq a\phi(x)$, where $\phi(x)$ is bounded if x is bounded.*

Similar to that in Chapter 5, we start our algorithm by designing a filtered error $s(t)$ that converts the relative degree n system to a relative degree one system w.r.t. s.

Let

$$s = \lambda_1 E + \lambda_2 E^{(1)} + \cdots + \lambda_{n-1} E^{(n-2)} + E^{(n-1)}, \tag{9.24}$$

where $\lambda_1, \cdots, \lambda_{n-1}$ are positive constants chosen such that $l^{n-1} + \lambda_{n-1} l^{n-2} + \cdots + \lambda_1$ is Hurwitz.

The derivative of s w.r.t. time is:

$$\dot{s} = \lambda_1 E^{(1)} + \lambda_2 E^{(2)} + \cdots + \lambda_{n-1} E^{(n-1)} + E^{(n)} = \dot{x}_n + L, \tag{9.25}$$

with $L = \lambda_1 E^{(1)} + \lambda_2 E^{(2)} + \cdots + \lambda_{n-1} E^{(n-1)} - y_d^{(n)}$ being available for control design.

Substituting the expression of $\dot{x}_n$ in (9.23) into (9.25), one has:

$$\dot{s} = F + Gu + L. \tag{9.26}$$

Before going further, the following section is developed, which is helpful for stability and performance analysis in Section 9.3.

9.2.2 TRANSFORMED DYNAMICS

Consider a dynamic system described by:

$$\dot{\chi} = f(\chi, p), \quad \chi(0) = \chi_0, \tag{9.27}$$

where $\chi \in \mathbb{R}^m$ is the system state vector, $f \in \mathbb{R}^m$ is a continuous nonlinear vector function of χ and p, and $p \in \mathbb{R}^r$ is the system parameter vector.

Introducing a nonlinear but simple exponential transformation:

$$z(t) = \beta(t)\chi,$$
$$\beta(t) = \exp(\lambda t), \ \lambda > 0,$$

into the system (9.27) leads to the following new nonlinear system:

$$\dot{z} = \lambda z + \beta f(\beta^{-1} z, p) \triangleq h(\lambda, z, p), \tag{9.28}$$

with $z(0) = \chi(0)$ and $h(\cdot)$ being a new nonlinear function.

Note that the transformed system (9.28) differs from the original one as it carries the "transformed" dynamics along with it. It is interesting to observe that if system (9.28) is made stable, i.e., $z(t) \in \mathscr{L}_\infty$ and $\|z(t)\| \leq \bar{z}$ for $t \geq 0$ with $\bar{z}$ being a positive constant, according to $z = \beta \chi$, then we have:

$$\chi(t) = \beta^{-1}(t)z(t) \ \text{and} \ \|\chi(t)\| \leq \exp(-\lambda t)\bar{z},$$

then by properly selecting the parameter λ, the variable $\chi(t)$ can be made converge to zero exponentially. Furthermore, it is worth mentioning that the parameter λ is independent of initial conditions and can be made arbitrarily assigned.

9.2.3   EXPONENTIAL TRANSFORMATION AND CONTROL DESIGN

In order to obtain exponential tracking for MIMO normal-form nonlinear systems in the presence of non-parametric uncertainty, similar to the discussion in Section 9.2.2, we introduce the following exponential transformation depending on the filtered error $s(t)$:

$$\zeta(t) = \beta(t)s. \tag{9.29}$$

The "accelerated" error dynamics $\zeta(t)$ w.r.t. time is:

$$\dot{\zeta} = \beta Gu + \beta(\lambda s + F + L),$$

then the derivative of $V = \frac{1}{2}\zeta^{\top}\zeta$ is:

$$\dot{V} = \zeta^{\top}\beta Gu + \zeta^{\top}\beta(\lambda s + F + L) = \beta\zeta^{\top}u + \check{L}, \tag{9.30}$$

where $\check{L} = \zeta^{\top}\beta(\lambda s + F + L)$ is the lumped uncertainty.

By applying Assumption 9.5, it is easily seen that:

$$\|\check{L}\| \leq \|\zeta\|\beta(\lambda\|s\| + a\phi + \|\ell\|) \leq \beta\|\zeta\|\theta\Phi,$$

where $\theta = \max\{a, 1\}$ denotes an unknown virtual constant, and

$$\Phi = \lambda\|s\| + \phi + \|L\| \tag{9.31}$$

is an available function for control design.

By utilizing Young's inequality, we have:

$$\|\check{L}\| \leq \beta\|\zeta\|\theta\Phi \leq \lambda_g\beta\|\zeta\|^2\Phi^2 + \frac{\theta^2\beta}{4\lambda_g},$$

where λ_g is defined in Assumption 9.4. Hence, (9.30) becomes:

$$\dot{V} \leq \zeta^{\top}\beta Gu + \lambda_g\beta\|\zeta\|^2\Phi^2 + \frac{\theta^2\beta}{4\lambda_g} \tag{9.32}$$

The proposed control is designed as:

$$u = -\left(c_0 + \Phi^2\right)\zeta, \tag{9.33}$$

where c_0 is a positive design parameter.

Before presenting the following theorem, we give some notations and lemmas to facilitate the stability analysis.

Define $0_{n-2} = [0, \cdots, 0]^{\top} \in \mathbb{R}^{n-2}$, $\mathbf{e}_{n-1} = [0, \cdots, 0, 1]^{\top} \in \mathbb{R}^{n-1}$, $I_{n-2} = \text{diag}\{1\} \in \mathbb{R}^{(n-2)\times(n-2)}$, and

$$\Lambda = \begin{bmatrix} 0_{n-2} & & I_{n-2} \\ -\lambda_1, & -\lambda_2, & \cdots, & -\lambda_{n-1} \end{bmatrix} \in \mathbb{R}^{(n-1)\times(n-1)},$$

$\gamma_k = [e_{1k}, e_{2k}, \cdots, e_{n-1,k}]^{\top}$, and $\dot{\gamma}_k = [e_{2k}, e_{3k}, \cdots, e_{nk}]^{\top}$. According to (9.24), it is not difficult to get that:

$$\dot{\gamma}_k = \Lambda\gamma_k + \mathbf{e}_{n-1}s_k, \ k = 1, \cdots, m.$$

Lemma 9.2 *[116] Let* $|s_k|_{[0,t]} = \sup_{\tau \in [0,t]} |s(\tau)|$, *then there exist some finite constants* $a_0 > 0$ *and* $\eta > 0$ *such that:*

$$\|\exp(\Lambda t)\| \leq a_0 \exp(-\eta t), \quad \|\gamma_k(t)\| \leq a_0 \exp(-\eta t)\|\gamma_k(0)\| + \frac{a_0}{\eta}|s_k|_{[0,t]}, \quad (9.34)$$

where η *is determined by the matrix* Λ.

Lemma 9.3 *[117, 118] Consider a first-order differential equation of the form:*

$$\dot{X} = -f_1(t)X + f_2(t),$$

where $X \in \mathbb{R}$, $f_1 \in \mathbb{R}$, *and* $f_2 \in \mathbb{R}$ *are continuous functions, then for any* $t \geq 0$, *the solution* $X(t)$ *satisfies:*

$$X(t) = X(0)\exp\left(-\int_0^t f_1(\tau)d\tau\right) + \int_0^t f_2(\tau)\exp\left(-\int_\tau^t f_1(s)ds\right)d\tau.$$

Now we are ready to present the following theorem and stability analysis.

Theorem 9.2 *Consider the MIMO normal-form nonlinear systems described by (9.23). Let Assumptions 9.4 and 9.5 hold, if the robust control scheme (9.33) is applied, then it is ensured that:*

(i) *the tracking errors,* $e_{11}, \cdots, e_{1m}$, *converge to zero exponentially regardless of non-parametric uncertainties, i.e.,*

$$|e_{1k}| \leq \exp(-\underline{\eta}t)\left[a_0\|\gamma_k(0)\| + \left(1 + \frac{a_0}{\eta}\right)\mathbb{B}\right], \quad k = 1, \cdots, m, \quad (9.35)$$

 where $\mathbb{B} = \|s(0)\| + \sqrt{\frac{2\alpha_2}{\alpha_1}}$, $\alpha_1 = 2\lambda_g c_0$, $\alpha_2 = \frac{\theta^2}{4\lambda_g}$, *and* $\underline{\eta} = \min\{\eta, \lambda\}$ *which is independent of initial conditions and is at the user's disposal;*

(ii) *all signals in the closed-loop systems are bounded.*

Proof. Substituting the robust controller as given in (9.33) into the term $\zeta^\top \beta G u$, it is seen that:

$$\zeta^\top \beta G u = -(c_0 + \Phi^2)\beta\zeta^\top G\zeta \leq -\lambda_g\beta(c_0 + \Phi^2)\|\zeta\|^2,$$

which leads to:

$$\dot{V} \leq -\lambda_g c_0 \beta \|\zeta\|^2 + \frac{\beta\theta^2}{4\lambda_g} = -2\lambda_g c_0 \beta(t)V(t) + \frac{\theta^2}{4\lambda_g}\beta(t). \quad (9.36)$$

By invoking Lemma 9.3, it is shown that $f_1(t) = 2\lambda_g c_0 \beta(t) = \alpha_1 \beta(t)$ and $f_2(t) = \frac{\theta^2}{4\lambda_g}\beta(t) = \alpha_2 \beta(t)$, then we solve the differential inequality (9.36) on $[0, \infty)$ yields:

$$V(t) \leq \exp\left(-\alpha_1\int_0^t \beta(\tau)d\tau\right)V(0) + \alpha_2\int_0^t \beta(\tau)\exp\left(-\alpha_1\int_\tau^t \beta(\varsigma)d\varsigma\right)d\tau. \quad (9.37)$$

For the first term $\exp\left(-\alpha_1 \int_0^t \beta(\tau)d\tau\right) V(0)$, it is deviated that:

$$\exp\left(-\alpha_1 \int_0^t \beta(\tau)d\tau\right) V(0) = \exp\left(-\frac{\alpha_1}{\lambda}\left[\exp(\lambda t) - 1\right]\right) V(0).$$

As λ and α_1 are positive constants, then it is seen that $\exp\left(-\alpha_1 \int_0^t \beta(\tau)d\tau\right) V(0) \leq V(0)$ for $\forall t \geq 0$.

For the second term in the right-hand side of (9.37), one has:

$$\alpha_2 \int_0^t \beta(\tau) \exp\left(-\alpha_1 \int_\tau^t \beta(\varsigma)d\varsigma\right) d\tau$$

$$= \alpha_2 \int_0^t \beta(\tau) \exp\left(\alpha_1 \left(-\int_0^t \beta(\varsigma) + \int_0^\tau \beta(\varsigma)\right) d\varsigma\right) d\tau$$

$$= \alpha_2 \exp\left(-\alpha_1 \int_0^t \beta(\tau)d\tau\right) \int_0^t \exp\left(\alpha_1 \int_0^\tau \beta(\varsigma)d\varsigma\right) d\left(\int_0^\tau \beta(\varsigma)d\varsigma\right)$$

$$= \frac{\alpha_2}{\alpha_1} \exp\left(-\alpha_1 \int_0^t \beta(\tau)d\tau\right) \exp\left(\alpha_1 \int_0^\tau \beta(\varsigma)d\varsigma\right) \Big|_0^t$$

$$= \frac{\alpha_2}{\alpha_1} \exp\left(-\alpha_1 \int_0^t \beta(\tau)d\tau\right) \left(\exp\left(\alpha_1 \int_0^t \beta(\varsigma)d\varsigma\right) - 1\right)$$

$$= \frac{\alpha_2}{\alpha_1} \left(1 - \exp\left(-\alpha_1 \int_0^t \beta(\tau)d\tau\right)\right)$$

$$\leq \frac{\alpha_2}{\alpha_1}.$$

Therefore, it is shown from (9.37) that $V(t) \leq V(0) + \frac{\alpha_2}{\alpha_1}$, which indicates that $V(t) \in \mathscr{L}_\infty$ and $\zeta(t) \in \mathscr{L}_\infty$.

According to the definition of $V(t)$, it is seen that $\|\zeta(t)\|^2 \leq 2V(t) \leq 2V(0) + \frac{2\alpha_2}{\alpha_1}$. Notice that $2V(0) = \|\zeta(0)\|^2 = \|s(0)\|^2$, then it follows that:

$$\|\zeta(t)\|^2 \leq \|s(0)\|^2 + \frac{2\alpha_2}{\alpha_1},$$

namely, $\|\zeta(t)\| \leq \|s(0)\| + \sqrt{\frac{2\alpha_2}{\alpha_1}}$. As $s(t) = \beta^{-1}\zeta(t)$, then it is seen that:

$$\|s(t)\| \leq \exp(-\lambda t)\left(\|s(0)\| + \sqrt{\frac{2\alpha_2}{\alpha_1}}\right) := \exp(-\lambda t)\mathbb{B}. \qquad (9.38)$$

To derive the bound for e_{1k}, $k = 1, \cdots, m$, we introduce a new variable as:

$$\bar{\gamma}_k = \left[\gamma_k^\top, s_k\right]^\top. \qquad (9.39)$$

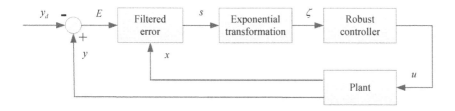

FIGURE 9.3 The diagram of the control design procedure.

Since $|s_k| \leq \|s(t)\|$, then it is seen from (9.34) and (9.39) that:

$$\|\overline{\gamma}_k\| \leq \|\gamma_k\| + |s_k| \leq a_0 \exp(-\eta t)\|\gamma_k(0)\| + \left(1 + \frac{a_0}{\eta}\right)|s_k|$$

$$\leq a_0 \exp(-\eta t)\|\gamma_k(0)\| + \left(1 + \frac{a_0}{\eta}\right)\exp(-\lambda t)\mathbb{B}$$

$$\leq \exp(-\underline{\eta} t)\left[a_0\|\gamma_k(0)\| + \left(1 + \frac{a_0}{\eta}\right)\mathbb{B}\right], \tag{9.40}$$

where $\underline{\eta} = \min\{\eta, \lambda\}$.

As $\overline{\gamma}_k = \left[e_{1k}, \cdots, e_{n-1,k}, s_k\right]^{\top}$ and $|e_{1k}| \leq \|\overline{\gamma}_k\|$, then we arrive at (9.35). Since ζ and s are bounded over $[0, \infty)$, then it follows that E, $E^{(1)}$, $\cdots$, and $E^{(n-1)}$ are bounded. As $y_d^{(k)}$ ($k = 0, 1, \cdots, n$) is bounded, then from the definition of $E^{(k)}$ it is seen that the boundedness of $x_1, x_2, \cdots, x_n$ is ensured, which implies that the available functions L and $\phi(x)$ are bounded, then it further shows that Φ as given in (9.31) is bounded, according to the definition of u as shown in (9.33), we have $u \in \mathcal{L}_\infty$ for $\forall t \geq 0$. The proof is completed.

To recap the main design procedure, a control diagram that clearly illustrates the structure of the control loop is presented in Fig. 9.3.

Remark 9.2 *It should be noted that, under the exponential tracking control scheme, in spite of the introduced exponential function $\beta(t)$ that would grow to infinity as $t \to \infty$, the boundedness of signals in the original closed-loop system can still be ensured. If the measurement noise is considered, the exponential transformation-based control will cause robustness issues and may lead to the instability of the closed-loop system. To avoid the possibly perceived "danger", a novel prescribed performance tracking control method will be introduced in Section 9.3.*

9.2.4 NUMERICAL SIMULATION

In this section, in order to validate the effectiveness of the proposed control in Section 9.2.3, a 2-DOF robotic system, borrowed from Chapter 7, is considered:

$$H(x_1)\dot{x}_2 + N_g(x_1, x_2)x_2 + G_g(x_1) + \tau_d(x_2) = u.$$

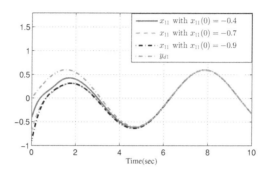

FIGURE 9.4 The evolutions of x_{11} and y_{d1} under the proposed method (9.33) with different initial conditions.

In the simulation, the design parameters are chosen as: $c_0 = 10$, $\lambda_1 = 2$, and $\lambda = 1$. It should be noted that, for the robotic systems under the proposed control, as $\Lambda = -\lambda_1$ and $\eta = \lambda_1$, it follows from the definition of η that $\underline{\eta} = \min\{\eta, \lambda\} = 1$. The reference signal is $y_d = [y_{d1}, y_{d2}]^\top = [0.6\sin(t), 0.3\sin(t)]^\top$. To verify that the exponential decay rate is independent of initial conditions, we use the following initial values of states: $x_1 = [x_{11}(0), x_{12}(0)]^\top = [-0.4, 0.5]^\top, [-0.7, 0.8]^\top, [-0.9, 1]^\top$, and $x_2 = [x_{21}(0), x_{22}(0)]^\top = [0.5, 0.5]^\top$.

Under the developed control scheme (9.33), the simulation results are shown in Fig. 9.4–Fig. 9.10. The evolutions of x_{1k} and y_{dk} $(k = 1, 2)$ are shown in Fig. 9.4 and Fig. 9.5, respectively, verifying that the fairly good tracking performance is obtained. The evolutions of tracking errors e_{11} and e_{12} are plotted in Fig. 9.6 and Fig. 9.7, which not only confirm that the tracking error converges to zero exponentially, but also indicate that the exponential decay rate is independent of initial conditions and other parameters, satisfying the theoretical analysis. Furthermore, the evolutions of the control inputs u_1 and u_2 are shown in Fig. 9.8 and Fig. 9.9, respectively, from which we observe that the control inputs are continuous and uniformly bounded. The evolution of $\|\zeta(t)\|$ is plotted in Fig. 9.10. From all the simulation results, it is shown that excellent tracking performance under the proposed control can be achieved, even in the presence of non-parametric uncertainties.

9.3   ADAPTIVE CONTROL WITH PRESCRIBED PERFORMANCE GUARANTEES

9.3.1   SYSTEM MODEL WITH PRESCRIBED PERFORMANCE

In this section, we consider the parametric strict-feedback nonlinear systems:

$$\begin{cases} \dot{x}_k = x_{k+1} + \theta^\top \varphi_k(\bar{x}_k), \ k = 1, \cdots, n-1, \\ \dot{x}_n = u + \theta^\top \varphi_n(\bar{x}_n), \\ y = x_1, \end{cases} \qquad (9.41)$$

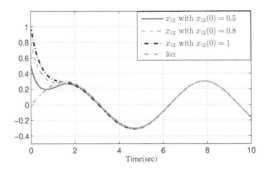

FIGURE 9.5 The evolutions of x_{12} and y_{d2} under the proposed method (9.33) with different initial conditions.

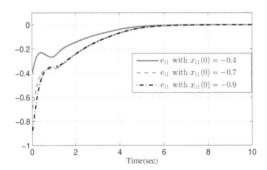

FIGURE 9.6 The evolution of the tracking error e_{11} under the proposed method (9.33) with different initial conditions.

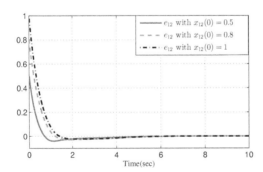

FIGURE 9.7 The evolution of the tracking error e_{12} under the proposed method (9.33) with different initial conditions.

where $x_i \in \mathbb{R}$, $i = 1, \cdots, n$, is the system state with $\bar{x}_i = [x_1, \cdots, x_i]^\top \in \mathbb{R}^i$, $u \in \mathbb{R}$ and $y \in \mathbb{R}$ are the control input and system output, respectively, the available func-

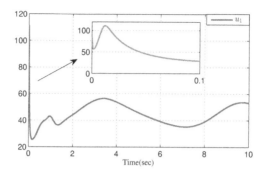

FIGURE 9.8   The evolution of the control input u_1 with $x_1(0) = [-0.4, 0.5]^\top$.

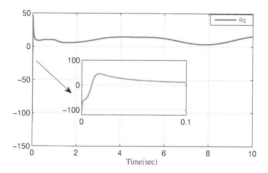

FIGURE 9.9   The evolution of the control input u_2 with $x_1(0) = [-0.4, 0.5]^\top$.

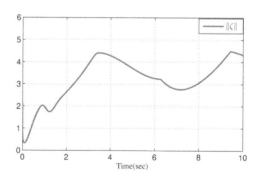

FIGURE 9.10   The evolution of $\|\zeta\|$ with $x_1(0) = [-0.4, 0.5]^\top$.

tion $\varphi_i(\bar{x}_i) \in \mathbb{R}^m$ is $n - i$ times differentiable, and $\theta \in \mathbb{R}^m$ represents an unknown parameter vector.

To show the prescribed performance, define the tracking error as $e = x_1 - y_d$, where $y_d(t)$ is the reference signal. The prescribed performance on the tracking error can be stated mathematically as:

$$-\beta(t) < e(t) < \beta(t), \tag{9.42}$$

or

$$-k_{c_1}\beta(t) < e(t) < k_{c_2}\beta(t), \tag{9.43}$$

with

$$\beta(t) = (\beta_0 - b_f)\exp(-\lambda t) + b_f > 0,$$

where $k_{c_1} > 0$, $k_{c_2} > 0$, $\beta_0 > b_f > 0$, and $\lambda > 0$ are design parameters, and $\beta(t)$ is the prescribed performance function.

If $k_{c_1} = k_{c_2} = 1$, (9.43) is simplified to the symmetric form (9.42). In this section, we only consider the case of symmetric prescribed performance.

In this section, the objective is to design an adaptive prescribed performance control scheme for parametric strict-feedback nonlinear systems (9.41) such that:

(*i*)   all signals in the closed-loop system are bounded;
(*ii*) the tracking error remains in the prescribed region for all time;
(*iii*) the tracking error asymptotically approaches zero.

9.3.2   PRESCRIBED PERFORMANCE-BASED TRANSFORMATION

The problem of prescribed performance is essentially a constraint problem. It is seen from Chapter 6 that the control design of strict-feedback nonlinear systems is quite complicated, thus, the control of such systems with prescribed performance is more difficult and more challenging. In this section, we develop the following tracking error-based nonlinear function, driving the problem of prescribed performance constraint on tracking error converted into the boundedness of a new variable:

$$\zeta(t) = \frac{e(t)}{(\beta(t) - e(t))(\beta(t) + e(t))}, \tag{9.44}$$

with the initial error $e(0)$ satisfying $-\beta(0) < e(0) < \beta(0)$.

According to Lemma 9.1, the problem of prescribed performance is converted into the boundedness of the transformed function $\zeta(t)$. The derivative of $\zeta(t)$ w.r.t. time is:

$$\dot{\zeta} = \zeta_1 \dot{e} + \zeta_2, \tag{9.45}$$

where

$$\zeta_1 = \frac{\beta^2 + e^2}{(\beta^2 - e^2)^2}, \quad \zeta_2 = \frac{-2e\beta\dot{\beta}}{(\beta^2 - e^2)^2} \tag{9.46}$$

are computable in the interval $e(t) \in (-\beta(t), \beta(t))$. Furthermore, $\zeta_1 > 0$ holds for $e(t) \in (-\beta(t), \beta(t))$.

Together with the model of nonlinear system (9.41), the problem of prescribed performance on tracking error is transformed into the stabilization of the following nonlinear system:

$$\begin{cases} \dot{\zeta} = \zeta_1 \left(\theta^\top \varphi_1(x_1) + x_2 - \dot{y}_d \right) + \zeta_2, \\ \dot{x}_k = \theta^\top \varphi_k(\bar{x}_k) + x_{k+1}, \quad k = 2, 3, \cdots, n-1, \\ \dot{x}_n = \theta^\top \varphi_n(\bar{x}_n) + u, \\ y = x_1. \end{cases} \tag{9.47}$$

9.3.3   CONTROLLER DESIGN AND STABILITY ANALYSIS

For the strict-feedback nonlinear system (9.47), we define the following coordinate transformation:

$$\begin{cases} z_1 = \zeta, \\ z_k = x_k - \alpha_{k-1}, \quad k = 2, \cdots, n, \end{cases} \tag{9.48}$$

where α_{k-1} denotes the virtual control law. As we have given a detailed description for the tuning-function-based backstepping design in the previous chapters, we only give the first two key steps in the design procedure.

Step 1. Using $x_2 = z_2 + \alpha_1$, the derivative of z_1 can be expressed as:

$$\dot{z}_1 = \zeta_1 \left(\theta^\top \varphi_1 + z_2 + \alpha_1 - \dot{y}_d \right) + \zeta_2. \tag{9.49}$$

Designing the virtual controller α_1 as:

$$\alpha_1 = \frac{1}{\zeta_1} \left(-c_1 z_1 - \zeta_2 \right) - \hat{\theta}^\top \varphi_1 + \dot{y}_d, \tag{9.50}$$

where $c_1 > 0$ is a design parameter and $\hat{\theta}$ is the estimate of the unknown parameter θ.

Substituting (9.50) into (9.49), one has $\dot{z}_1 = -c_1 z_1 + \zeta_1 \left(z_2 + \tilde{\theta}^\top \varphi_1 \right)$, where $\tilde{\theta} = \theta - \hat{\theta}$ denotes the estimation error.

Choosing the quadratic function as $V_1 = \frac{1}{2} z_1^2 + \frac{1}{2} \tilde{\theta}^\top \Gamma^{-1} \tilde{\theta}$, where $\Gamma \in \mathbb{R}^{r \times r}$ is a symmetric and positive definite parameter matrix, so the derivative of V_1 is:

$$\dot{V}_1 = -c_1 z_1^2 + \zeta_1 z_1 z_2 + \tilde{\theta}^\top \Gamma^{-1} \left(\Gamma \tau_1 - \dot{\hat{\theta}} \right), \tag{9.51}$$

where $\tau_1 = \zeta_1 z_1 \varphi_1$, the term $\zeta_1 z_1 z_2$ will be handled in the following step.

Step 2. Since α_1 is the function of $x_1, y_d, \dot{y}_d, \beta, \dot{\beta}$, and $\hat{\theta}$, then according to (9.48), the derivative of virtual error z_2 w.r.t. time can be expressed as:

$$\dot{z}_2 = z_3 + \alpha_2 + \hat{\theta}^\top \omega_2 + \tilde{\theta}^\top \omega_2 - \frac{\partial \alpha_1}{\partial x_1} x_2 + \ell_1 - \frac{\partial \alpha_1}{\partial \hat{\theta}} \dot{\hat{\theta}} \tag{9.52}$$

where $\omega_2 = \varphi_2 - \frac{\partial \alpha_1}{\partial x_1} \varphi_1$ and $\ell_1 = -\sum_{k=0}^{1} \left(\frac{\partial \alpha_1}{\partial y_d^{(k)}} y_d^{(k+1)} + \frac{\partial \alpha_1}{\partial \beta^{(k)}} \beta^{(k+1)} \right)$.

Choosing the virtual controller α_2 as:

$$\alpha_2 = -c_2 z_2 - \zeta_1 z_1 - \hat{\theta}^\top \omega_2 + \frac{\partial \alpha_1}{\partial x_1} x_2 - \ell_1 + \frac{\partial \alpha_1}{\partial \hat{\theta}} \Gamma \tau_2, \tag{9.53}$$

where $\tau_2 = \tau_1 + \omega_2 z_2$ and $c_2 > 0$.

Substituting the expression of the virtual controller (9.53) into (9.52), one has:

$$\dot{z}_2 = z_3 - c_2 z_2 - \zeta_1 z_1 + \tilde{\theta}^\top \omega_2 + \frac{\partial \alpha_1}{\partial \hat{\theta}} \left(\Gamma \tau_2 - \dot{\hat{\theta}} \right).$$

Selecting the quadratic function V_2 as $V_2 = V_1 + \frac{1}{2} z_2^2$, then its derivative is:

$$\dot{V}_2 = -\sum_{k=1}^{2} c_k z_k^2 + z_2 z_3 + z_2 \tilde{\theta}^\top \omega_2 + \tilde{\theta}^\top \Gamma^{-1} \left(\Gamma \tau_1 - \dot{\hat{\theta}} \right) + z_2 \frac{\partial \alpha_1}{\partial \hat{\theta}} \left(\Gamma \tau_2 - \dot{\hat{\theta}} \right). \tag{9.54}$$

Note that $\Gamma \tau_1 - \dot{\hat{\theta}}$ can be decomposed as $\Gamma \tau_1 - \dot{\hat{\theta}} = \Gamma \tau_1 - \Gamma \tau_2 + \Gamma \tau_2 - \dot{\hat{\theta}} = -\Gamma \omega_2 z_2 + \Gamma \tau_2 - \dot{\hat{\theta}}$, thus, (9.54) can be rewritten as:

$$\dot{V}_2 = -\sum_{k=1}^{2} c_k z_k^2 + z_2 z_3 + \left(z_2 \frac{\partial \alpha_1}{\partial \hat{\theta}} + \tilde{\theta}^\top \Gamma^{-1} \right) \left(\Gamma \tau_2 - \dot{\hat{\theta}} \right).$$

Step j $(j = 3, \cdots, n)$. The design procedures in the following steps are identical with those in Chapter 6, therefore, the detailed step is omitted. By using $x_{j+1} = z_{j+1} + \alpha_j$, we have:

$$\dot{z}_j = z_{j+1} + \alpha_j + \theta^\top \varphi_j - \sum_{k=1}^{j-1} \frac{\partial \alpha_{j-1}}{\partial x_k} \left(x_{k+1} + \theta^\top \varphi_k \right) + \ell_{j-1} - \frac{\partial \alpha_{j-1}}{\partial \hat{\theta}} \dot{\hat{\theta}},$$

where $\ell_{j-1} = -\sum_{k=0}^{j-1} \frac{\partial \alpha_{j-1}}{\partial y_d^{(k)}} y_d^{(k+1)} - \sum_{k=0}^{j-1} \frac{\partial \alpha_{j-1}}{\partial \beta^{(k)}} \beta^{(k+1)}$.

Designing the virtual controller and actual controller as:

$$\alpha_j = -c_j z_j - z_{j-1} - \hat{\theta}^\top \omega_j + \sum_{k=1}^{j-1} \frac{\partial \alpha_{j-1}}{\partial x_k} x_{k+1} - \ell_{j-1}$$

$$+ \frac{\partial \alpha_{j-1}}{\partial \hat{\theta}} \Gamma \tau_j + \sum_{k=2}^{j-1} \frac{\partial \alpha_{k-1}}{\partial \hat{\theta}} \Gamma \omega_j z_k, \tag{9.55}$$

$$\tau_j = \tau_{j-1} + \omega_j z_j, \quad \omega_j = \varphi_j - \sum_{k=1}^{j-1} \frac{\partial \alpha_{j-1}}{\partial x_k} \varphi_k, \tag{9.56}$$

$$u = \alpha_n, \tag{9.57}$$

$$\dot{\hat{\theta}} = \Gamma \tau_n, \tag{9.58}$$

with $c_j > 0$.

We now give the system stability analysis under the prescribed performance adaptive controller (9.57) in the following theorem.

Theorem 9.3 *For the strict-feedback nonlinear system (9.41), if the adaptive controller (9.57) with the adaptive law (9.58) is applied, then the objectives (i)–(iii) are achieved.*

Proof. According to the virtual controller (9.55) and actual controller (9.57), the following closed-loop system is obtained:

$$\dot{z}_1 = -c_1 z_1 + \zeta_1 (z_2 + \tilde{\theta}^\top \varphi_1),$$

$$\dot{z}_2 = z_3 - c_2 z_2 - \zeta_1 z_1 + \tilde{\theta}^\top \omega_2 - \sum_{k=3}^{n} \frac{\partial \alpha_1}{\partial \hat{\theta}} \Gamma z_k \omega_k,$$

$$\dot{z}_j = z_{j+1} - c_j z_j - z_{j-1} + \tilde{\theta}^\top \omega_j + \sum_{k=1}^{j-2} \frac{\partial \alpha_k}{\partial \hat{\theta}} \Gamma z_{k+1} \omega_j - \sum_{k=j+1}^{n} \frac{\partial \alpha_{j-1}}{\partial \hat{\theta}} \Gamma z_k \omega_k,$$

$$\dot{z}_n = -c_n z_n - z_{n-1} + \tilde{\theta}^\top \omega_n + \sum_{k=1}^{n-2} \frac{\partial \alpha_k}{\partial \hat{\theta}} \Gamma z_{k+1} \omega_n,$$

for $j = 3, 4, \cdots, n-1$.

Constructing the Lyapunov function candidate as $V = \frac{1}{2} \sum_{k=1}^{n} z_k^2 + \frac{1}{2} \tilde{\theta}^\top \Gamma^{-1} \tilde{\theta}$, the derivative of the Lyapunov function candidate is:

$$\dot{V} = -\sum_{k=1}^{n} c_k z_k^2 \le 0. \tag{9.59}$$

Integrating (9.59) on $[0,t]$, one has:

$$V(t) + c_1 \int_0^t z_1^2(\tau)d\tau + \cdots + c_n \int_0^t z_n^2(\tau)d\tau = V(0). \tag{9.60}$$

Firstly, we prove the boundedness of signals in the closed-loop system and the prescribed tracking performance. According to (9.59), $V(t)$ is bounded, thus z_k, $(k = 1, 2, \cdots, n)$, and the parameter estimate error $\hat{\theta}$ are all bounded. Note that $z_1 = \zeta$, then, if the initial tracking error satisfies $-\beta(0) < e(0) < \beta(0)$, by using the analysis in Lemma 9.1,

$$-\beta(t) < e(t) < \beta(t)$$

holds for $\forall t \ge 0$, i.e., the tracking error always remains in the predefined region. Furthermore, there exist some functions $\beta_1(t)$ and $\beta_2(t)$ and some constants υ_1 and υ_2 such that:

$$-\beta(t) < -\beta_1(t) \le e(t) \le \beta_2(t) < \beta(t),$$
$$\beta(t) - \beta_2(t) \ge \upsilon_2, \quad \beta(t) - \beta_1(t) \ge \upsilon_1.$$

Therefore, $\zeta_1(t)$ as given in (9.46) is bounded and there exist constants $\overline{\zeta}_1$ and $\underline{\zeta}_1$ such that:

$$0 < \underline{\zeta}_1 \le \zeta(t) \le \overline{\zeta}_1 < \infty.$$

Since $e = x_1 - y_d$ and $y_d \in \mathscr{L}_\infty$, then the system state x_1 is bounded, then it is not difficult to obtain the boundedness of the virtual control α_1 and $\dot{z}_1$. As $k = 2$, since z_2 and α_1 are bounded, the system state x_2 is bounded, then it follows that $\varphi_2(\bar{x}_2) \in \mathscr{L}_\infty$. In addition, due to the smoothness and boundedness of the virtual control α_1, the boundedness of $\frac{\partial \alpha_1}{\partial x_1}$, $\frac{\partial \alpha_1}{\partial y_d}$, $\frac{\partial \alpha_1}{\partial \dot{y}_d}$, $\frac{\partial \alpha_1}{\partial \hat{\theta}}$, $\frac{\partial \alpha_1}{\partial \beta}$, and $\frac{\partial \alpha_1}{\partial \dot{\beta}}$ are ensured, then $\omega_2 \in \mathscr{L}_\infty$, $\tau_2 \in \mathscr{L}_\infty$, $\alpha_2 \in \mathscr{L}_\infty$. Using the similar analysis process, it follows that the system state x_i, $(i = 3, 4, \cdots, n)$, is bounded, the nonlinear function φ_i is bounded, each partial derivative of the virtual controller is bounded, the virtual control law is bounded, the actual controller u is bounded, the adaptive law $\dot{\hat{\theta}}$ is bounded, and the derivative of the virtual errors, $\dot{z}_i$, is bounded. Therefore, all signals in the closed-loop system are bounded.

Secondly, we prove that the tracking error converges to zero asymptotically. With the aid of (9.60), it is shown that $z_k \in \mathscr{L}_2$ $(k = 1, \cdots, n)$. Note that $z_k \in \mathscr{L}_\infty$ and $\dot{z}_k \in \mathscr{L}_\infty$, by utilizing Barbarlat's Lemma, one has $\lim_{t \to \infty} z_k(t) \to 0$. Therefore, it follows that $\lim_{t \to \infty} \zeta(t) \to 0$. By using the second property of Lemma 9.1, it is not difficult to obtain that $\lim_{t \to \infty} e(t) \to 0$. The proof is completed.

Remark 9.3 *Significant efforts in funnel control have also been made to improve the prescribed performance [119]. Funnel control can be viewed as an extension of the adaptive high-gain control methodology with the advancement of replacing the monotonically increasing control gain by a time-varying scaling function which admits high values only when the distance between the output and the funnel boundary becomes small, resulting in a nonlinear and time-varying proportional control scheme. Inspired by this idea, funnel control has been used for various kinds of systems and practical applications to ensure that the tracking error evolves within the performance funnel [119, 120, 121, 122, 123, 124, 125, 126]. Recently, by utilizing the backstepping technique or filtered error, funnel control for a class of nonlinear systems was reported in references [127] and [128] with relative degree of two and in references [129, 130, 131] with relative degree r $(r > 2)$. However, it should be emphasized that, when the relative degrees r is equal to or greater than two, although the tracking error evolves within the prescribed funnel, only the UUB tracking (rather than asymptotic tracking) can be ensured. Moreover, by utilizing the first $r - 1$ derivatives of the tracking error, funnel controllers for uncertain nonlinear systems in normal form with arbitrary relative degree were developed in references [127] and [129], whereas such condition is difficult (or even impossible) to satisfy for strict-feedback nonlinear systems.*

9.3.4 NUMERICAL SIMULATION

To demonstrate the effectiveness of the proposed adaptive control scheme in Section 9.3.3, the following second-order strict-feedback nonlinear system is considered:

$$
\begin{cases}
\dot{x}_1 = x_2 + \theta \varphi_1(x_1), \\
\dot{x}_2 = u,
\end{cases}
$$

where $\theta = 1$ and $\varphi_1(x_1) = x_1^2$.

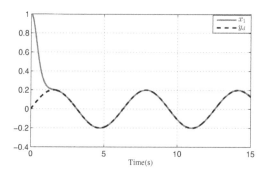

FIGURE 9.11 The evolutions of x_1 and y_d.

In the simulation, the reference signal is given as $y_d = 0.2\sin(t)$. The design parameters are given as: $c_1 = 3$, $c_2 = 6$, $\Gamma = 1$, $\beta_0 = 1.5$, $b_f = 0.06$, and $\lambda = 0.8$. The initial values of system states and parameter estimate are: $x_1(0) = 1$, $x_2(0) = -1$, and $\hat{\theta}(0) = 0$. Under the developed control scheme in Section 9.3.3, the simulation results are shown in Fig. 9.11–Fig. 9.14. The evolutions of the system output x_1 and the reference signal y_d are plotted in Fig. 9.11. The evolution of the tracking error e under the prescribed performance constraint is shown in Fig. 9.12, which indicates that the tracking error remains in the predefined region for $\forall t \geq 0$ and at the same time asymptotically converges to zero, verifying the effectiveness of the corresponding control scheme in Section 9.3.3. In addition, the evolutions of the control input and parameter estimate are plotted in Fig. 9.13 and Fig. 9.14, respectively, which are bounded for all time.

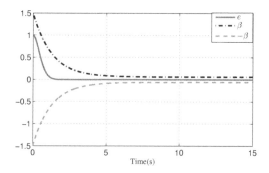

FIGURE 9.12 The evolution of the tracking error e with prescribed performance.

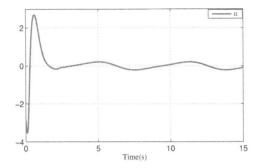

FIGURE 9.13   The evolution of the control input u.

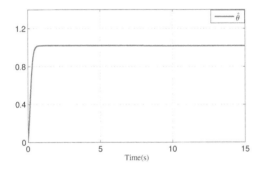

FIGURE 9.14   The evolution of the parameter estimate $\hat{\theta}$.

9.4   NOTES

In this chapter, the tracking control of strict-feedback nonlinear systems with exponential/prescribed tracking performance is investigated. For the system without uncertainty, by introducing an exponential transformation, the proposed control ensures that the tracking error converges to zero exponentially in a uniform manner, which is independent of control gains and initial conditions. Subsequently, we extend the exponential transformation method to a class of normal-form nonlinear systems.

It should be noted that, under the proposed exponential tracking control scheme, in spite of the introduced exponential function $\beta(t)$ that would grow infinity as $t \to \infty$, the boundedness of signals in the original closed-loop system (i.e., the system state, tracking error, and control input u) can still be ensured. If the measurement noise is considered, the exponential transformation-based control will cause robustness issues and may lead to the instability of the closed-loop system. To avoid the possibly perceived "danger", a novel prescribed performance tracking control method is introduced in Section 9.3. We construct a time-varying but bounded scaling function, based on which the proposed control ensures that the tracking error is always within the pre-given region.

Noting that, although the above control methods are able to improve the transient tracking performance, especially the convergence rate, tracking precision, and overshoot, a key index, convergence time, is not considered in this chapter. Therefore, we will introduce the finite-time and prescribed-time control schemes for nonlinear systems in the following chapters.

10 Finite-time Control

Convergence time is a key performance index to evaluate control algorithms. However, in most of the existing control schemes, the convergence rate is at asymptotic (or is at best exponential) with *infinite settling time* [9, 84, 85, 132, 133]. That is, it needs infinite time to achieve stabilization, regulation, or tracking. The finite-time control means that the system state (or tracking error) converges to zero within a finite time. Compared with the nonfinite-time control, the finite-time control not only ensures that the closed-loop system has a faster convergence rate and higher convergence accuracy but also guarantees that the corresponding control has stronger disturbance rejection and better robustness when there are external disturbances. Therefore, the study of the finite-time control has obtained more and more attention in the control community [43, 134, 135, 136, 137, 138, 139, 140, 141].

Before introducing the finite-time controllers, we first introduce several commonly used finite-/fixed-time stability criteria for nonlinear systems.

10.1 CRITERIA FOR FINITE-TIME CONTROL

In order to facilitate the description of the following contents, two nonlinear system models are given:

$$\dot{x}(t) = f(x), \tag{10.1}$$
$$\dot{x}(t) = f(x) + d(t), \tag{10.2}$$

where $x(t) \in \mathbb{R}$ is the system state, $f(\cdot) \in \mathbb{R}$ represents the nonlinear function, and $d(t) \in \mathbb{R}$ denotes an external bounded disturbance.

Criterion 10.1 *Consider the nonlinear system (10.1), if there exists a continuously differentiable function $V(x) > 0$ such that:*

$$\dot{V}(x) \leq -lV^{\alpha}(x),$$

where $l > 0$, $0 < \alpha < 1$, then the system is finite-time stable and the settling time is

$$T = \frac{1}{l(1-\alpha)}V^{1-\alpha}(x_0).$$

Criterion 10.2 *Consider the nonlinear system (10.1), if there exists a continuously differentiable function $V(x) > 0$ such that:*

$$\dot{V}(x) \leq -l_1 V^{\alpha}(x) - l_2 V(x),$$

where $l_1 > 0$, $l_2 > 0$, $0 < \alpha < 1$, then the system is finite-time stable and the corresponding settling time is

$$T = \frac{\ln\left(1 - \frac{l_1}{l_2}V^{1-\alpha}(x_0)\right)}{l_2(1-\alpha)}.$$

DOI: 10.1201/9781003474364-10

Definition 10.1 *[142] (**Practical Finite-time Stability**) The practical finite-time stability of the nonlinear system (10.1) [or (10.2)] is said to be reached if for any initial
condition $x(0) = x_0$, there exist constants $\vartheta > 0$ and finite time $T(x_0, \omega) < \infty$ (in
which ω is a parameter vector), such that $|x(t)| < \vartheta$ for all $t \geq T(x_0, \omega)$.*

Criterion 10.3 *Consider the nonlinear system (10.2), if there exists a continuously
differentiable function $V(x) > 0$ such that:*

$$\dot{V}(x) \leq -lV^\alpha(x) + \vartheta, \tag{10.3}$$

*where $l > 0$, $\vartheta > 0$, $0 < \alpha < 1$, then the system is practical finite-time stable and the
settling time is*

$$T = \frac{V^{1-\alpha}(x_0)}{l\theta_0(1-\alpha)}, \quad 0 < \theta_0 < 1.$$

Furthermore, the trajectory of the nonlinear system (10.2) is within

$$\Omega = \left\{ x \middle| V(x) \leq \left(\frac{\vartheta}{l(1-\theta_0)} \right)^{\frac{1}{\alpha}} \right\}.$$

*Here the practical finite-time stable means that the system states converge to the
neighborhood of the equilibrium point, see Definition 10.1.*

Criterion 10.4 *Consider the nonlinear system (10.2), if there exists a continuously
differentiable function $V(x) > 0$ such that:*

$$\dot{V}(x) \leq -l_1 V^\alpha(x) - l_2 V(x) + \vartheta,$$

*where $l_1 > 0$, $l_2 > 0$, $\vartheta > 0$, and $0 < \alpha < 1$, then the system is fast practical finite-
time stable and the settling time is*

$$\begin{cases} T = \max\{T_1; T_2\}, \\ T_1 = \frac{1}{l_1\theta_0(1-\alpha)} \ln\left(\frac{l_1\theta_0 V^{1-\alpha}(x_0) + l_1}{l_1} \right), \\ T_2 = \frac{1}{l_1(1-\alpha)} \ln\left(\frac{l_1 V^{1-\alpha}(x_0) + \theta_0 l_1}{l_1} \right), \\ 0 < \theta_0 < 1. \end{cases}$$

*Here fast practical finite-time stable means that the system states converge to the
neighborhood of the equilibrium point faster than the traditional finite-time stable.*

Criterion 10.5 *Consider the nonlinear system (10.1), if there exists a continuously
differentiable function $V(x) > 0$ such that:*

$$\dot{V}(x) \leq -(l_1 V^p(x) + l_2 V^q(x))^\alpha,$$

*where $l_1 > 0$, $l_2 > 0$, $q > 0$, $\alpha > 0$, $p\alpha < 1$, $q\alpha > 1$, then the system is fixed-time
stable (fixed-time stable means that the system states can converge to the equilibrium*

point within a finite time, and the upper bound of the convergence time is independent of the initial states), and the settling time is

$$T \leq T_{\max} := \frac{1}{l_1^\alpha (1 - p\alpha)} + \frac{1}{l_2^\alpha (q\alpha - 1)}.$$

Criterion 10.6 *Consider the nonlinear system (10.1), if there exists a continuously differentiable function $V(x) > 0$ such that:*

$$\dot{V}(x) \leq -l_1 V^p(x) - l_2 V^q(x),$$

where $l_1 > 0$, $l_2 > 0$, $p = 1 - 1/(2\gamma)$, $q = 1 + 1/(2\gamma)$, and $\gamma > 1$, then the system is fixed-time stable, and the settling time is

$$T \leq T_{\max} := \frac{\pi\gamma}{\sqrt{l_1 l_2}}.$$

Criterion 10.7 *Consider the nonlinear system (10.1), if there exists a continuously differentiable function $V(x) > 0$ such that:*

$$\dot{V}(x) \leq -l_1 V^{2 - \frac{p}{q}}(x) - l_2 V^{\frac{p}{q}}(x),$$

where $l_1 > 0$, $l_2 > 0$, $q > p > 0$, p and q are odd integers, then the system is fixed-time stable, and the settling time is

$$T \leq T_{\max} := \frac{q\pi}{2\sqrt{l_1 l_2}(q - p)}.$$

Criterion 10.8 *Consider the nonlinear system (10.1), if there exists a continuously differentiable function $V(x) > 0$ such that:*

$$\dot{V}(x) \leq -l_1 V^{\frac{m}{n}}(x) - l_2 V^{\frac{p}{q}}(x),$$

where $l_1 > 0$, $l_2 > 0$, $q > p > 0$, $m > n > 0$, p, q, m, and n are odd integers, then the system is fixed-time stable, and the settling time is

$$T \leq T_{\max} := \frac{1}{l_1} \frac{n}{m - n} + \frac{1}{l_2} \frac{q}{q - p}.$$

10.2 FINITE-TIME CONTROL OF FIRST-ORDER SYSTEMS

To show the advantage of the prescribed-time control in Chapter 11, we briefly introduce the signum function feedback-based finite-time control and the fractional power error feedback-based finite-time control in Section 10.2 and Section 10.3, respectively.

Here we consider the following first-order nonlinear system:

$$\dot{x} = f(x) + g(x)u, \tag{10.4}$$

where $x \in \mathbb{R}$ is the system state, $f \in \mathbb{R}$ is a smooth nonlinear function, and $g \in \mathbb{R}$ denotes the control coefficient.

10.2.1 MODEL-DEPENDENT FINITE-TIME CONTROL

In this subsection, we assume that the nonlinear function $f(x)$ and the control gain $g(x)$ are available for control design. The objective is to design a model-dependent finite-time tracking control scheme such that the tracking error $e = x - y_d$ converges to zero within a finite time, where y_d is the reference signal.

To this end, we impose the following assumptions.

Assumption 10.1 *The reference signal and its first derivative are known, bounded, and piecewise continuous.*

Assumption 10.2 *The control gain is available and bounded for control design, and there exists a positive constant $\underline{g}$ such that $0 < \underline{g} \le g(x) < \infty$.*

Now we carry out the control design.

According to the definition of the tracking error e, one has:

$$\dot{e} = \dot{x} - \dot{y}_d = f + gu - \dot{y}_d. \tag{10.5}$$

Since the nonlinear function f and control gain g are available, then the signum function-based finite-time tracking controller is designed as:

$$u = \frac{1}{g}\left[-f + \dot{y}_d - c|e|^{\frac{1}{3}}\mathrm{sgn}(e)\right], \tag{10.6}$$

with $c > 0$ being a design parameter.

Theorem 10.1 *Consider the first-order nonlinear system (10.4). Under the Assumptions 10.1 and 10.2, if the model-dependent controller (10.6) is applied, then all signals in the closed-loop systems are bounded and the tracking error converges to zero in a finite time.*

Proof. Substituting the proposed finite-time tracking controller as given in (10.6) into (10.5), one has:

$$\dot{e} = -c|e|^{\frac{1}{3}}\mathrm{sgn}(e). \tag{10.7}$$

Choosing the Lyapunov function candidate as $V = \frac{1}{2}e^2$, then the derivative of V along (10.7) is:

$$\dot{V} = e\dot{e} = -c|e|^{\frac{4}{3}} = -2^{\frac{2}{3}}cV^{\frac{2}{3}}. \tag{10.8}$$

According to Criterion 10.1, it is shown that V is bounded and the tracking error converges to zero within a finite time and the corresponding settling time is:

$$T = \frac{1}{2^{\frac{2}{3}}c(1-\frac{2}{3})}V^{1-\frac{2}{3}}(0) = \frac{3}{2^{\frac{5}{3}}c}e^{\frac{2}{3}}(0).$$

As $V \in \mathscr{L}_\infty$, it follows that $e \in \mathscr{L}_\infty$. From the definition of $e = x - y_d$, it is shown from Assumption 10.1 that x is bounded, which implies that $f(x)$ and $g(x)$ are bounded, with the aid of Assumption 10.2, the developed finite-time controller is bounded for all time. The proof is completed.

Remark 10.1 *It should be emphasized that motivated by the proposed finite-time tracking controller (10.6), we develop the following general finite-time tracking controller:*

$$u = \frac{1}{g}\left[-f + \dot{y}_d - c|e|^{\frac{q}{p}} sgn(e)\right],$$

where q and q are some odd integers and $0 < \frac{q}{p} < 1$. Under such control, the derivative of $V = \frac{1}{2}e^2$ is $\dot{V} = -2^{\frac{p+q}{2p}} cV^{\frac{p+q}{2p}}$ and the settling time is

$$T = \frac{2p}{2^{\frac{p+q}{2p}} c(p-q)} V^{\frac{p-q}{2p}}(0)$$

The proof is omitted.

10.2.2   NUMERICAL SIMULATION

To test the effectiveness of the proposed finite-time control, we consider the first-order nonlinear system (10.4) with $f(x) = x^2 \cos(x)$ and $g(x) = 1 + 0.2 \sin(x)$ under the initial condition $x(0) = 1$. The reference signal is given as $y_d(t) = \sin(t)$. The control parameter is chosen as $c = 1$. The simulation results are shown in Fig. 10.1 and Fig. 10.2, from which it is seen that the tracking error converges to zero in a finite time ($T \leq 1.5s$). Furthermore, it should be emphasized that due to the employment of the signum function in the controller design, although finite-time tracking can be achieved, the phenomenon of chattering may happen, which causes an adverse effect on the actuator.

In addition, to verify that the settling time T relies on the initial conditions and design parameters, we choose that $c = 1$ or 2 and $x(0) = 1$ or 2. Under the proposed finite-time controller (10.6), the simulation result is shown in Fig. 10.3, from which it is seen that different initial values and different design parameters may lead to different settling time T.

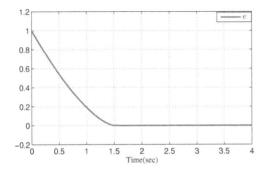

FIGURE 10.1   The evolution of the tracking error e.

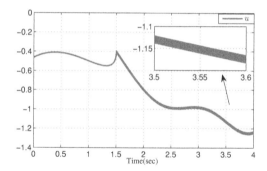

FIGURE 10.2   The evolution of the control input u.

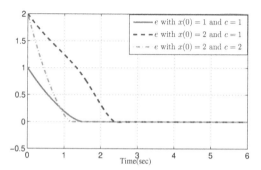

FIGURE 10.3   The evolution of the tracking error e under different initial values and design parameters.

10.2.3   MODEL-INDEPENDENT FINITE-TIME CONTROL

It must be emphasized that, in Section 10.2.1, the nonlinear function f and control gain g must be available for control design, which is conservative in practice. Therefore, here we propose a new signum function feedback-based finite-time control scheme to relax the conditions on the nonlinear function f and control gain g.

Assumption 10.3 *The control gain $g(x)$ is unknown and bounded, but there exists a known positive constant $\underline{g}$ such that $0 < \underline{g} \leq g(x) < \infty$.*

Assumption 10.4 *The nonlinear function $f(x)$ satisfies the following condition, i.e., $|f(x)| \leq a\phi(x)$, where $a \geq 0$ is a known constant and $\phi(x) \geq 0$ is available for control design.*

The signum function feedback-based finite-time controller is designed as:

$$u = \frac{1}{\underline{g}}\left[-c|e|^{\frac{1}{3}} - a\phi(x) - |\dot{y}_d|\right]\mathrm{sgn}(e), \qquad (10.9)$$

with $c > 0$.

Theorem 10.2 *Consider the first-order nonlinear system (10.4). Under the Assumptions 10.1, 10.3, and 10.4, if the developed controller (10.9) is applied, all signals in the closed-loop systems are bounded and the tracking error converges to zero in a finite time.*

Proof. With the aid of (10.9), the derivative of $V = \frac{1}{2}e^2$ along (10.5) is:

$$\dot{V} = e\dot{e} = e(f + gu - \dot{y}_d) = \frac{g}{g}e\left(-c|e|^{\frac{1}{3}} - a\phi - |\dot{y}_d|\right)\operatorname{sgn}(e) + ef - e\dot{y}_d$$

$$\leq -\frac{g}{g}c|e|^{\frac{4}{3}} - \frac{g}{g}a|e|\phi - \frac{g}{g}|e||\dot{y}_d| + a|e|\phi + |e||\dot{y}_d|.$$

It is seen from Assumption 10.3 that $\frac{g}{g} \geq 1$, then it follows that:

$$\dot{V} \leq -\frac{g}{g}c|e|^{\frac{4}{3}} \leq -c|e|^{\frac{4}{3}} = -2^{\frac{2}{3}}cV^{\frac{2}{3}}.$$

By using the same process as that in Theorem 10.1, it is ensured that all signals in the closed-loop systems are bounded and the tracking error converges to zero in a finite time. The proof is completed.

Remark 10.2 *The proposed finite-time tracking controller (10.9) can also be extended into the following form:*

$$u = \frac{1}{g}\left(-k|e|^{\frac{q}{p}} - a\phi(x) - |\dot{y}_d|\right)\operatorname{sgn}(e),$$

where q and q are some odd integers satisfying $0 < \frac{q}{p} < 1$.

To test the effectiveness of the proposed finite-time control, we consider the system (10.4) with $f(x) = x^2\cos(x)$ and $g(x) = 1 + 0.2\sin(x)$ under the initial condition $x(0) = 1$. The reference signal is $y_d(t) = \sin(t)$. The control parameter is chosen as $c = 1$. The simulation results are shown in Fig. 10.4 and Fig. 10.5, which confirm that the tracking error converges to zero in a finite time ($T \leq 0.5s$).

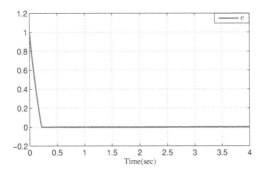

FIGURE 10.4   The evolution of the tracking error e.

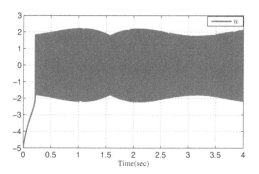

FIGURE 10.5 The evolution of the control input u.

10.3 PRACTICAL FINITE-TIME CONTROL OF FIRST-ORDER SYSTEMS

It is obvious from Theorem 10.2 that the constant a must be available for control design. If a is unknown, the signum function feedback-based finite-time control (10.9) may fail. Furthermore, it should be emphasized that due to the employment of the signum function in Section 10.2, although finite-time tracking can be achieved, the phenomenon of chattering may happen. To solve the above issues, with the aid of adaptive technology, we introduce the fractional power error feedback-based practical finite-time control for first-order nonlinear systems.

Assumption 10.5 *The control gain $g(x)$ is unknown and bounded, but there exist some positive unknown constants $\underline{g}$ and $\overline{g}$ such that $0 < \underline{g} \leq g(x) \leq \overline{g} < \infty$.*

Assumption 10.6 *For the nonlinear function $f(x)$, there exist an unknown parameter $a \geq 0$ and an available smooth function $\phi(x) \geq 0$ such that $|f(x)| \leq a\phi(x)$.*

10.3.1 CONTROL DESIGN

Now we carry out the control design.

Choosing the first part of Lyapunov function candidate as $V_1 = \frac{1}{1+h}e^{1+h}$, where $h = \frac{h_1}{h_2}$ with h_1 and h_2 being some odd integers and $0 < h < 1$, then the derivative of V_1 along (10.5) yields:

$$\dot{V}_1 = e^h \dot{e} = e^h gu + e^h(f - \dot{y}_d). \tag{10.10}$$

Noting that $e^h(f - \dot{y}_d)$ can be upper bounded by:

$$e^h(f - \dot{y}_d) \leq |e^h|(a\phi + |\dot{y}_d|) \leq |e^h|b\Phi,$$

with $b = \max\{a, 1\}$ being a virtual parameter and $\Phi = \phi + \dot{y}_d^2 + \frac{1}{4}$ being a computable function. With the aid of Young's inequality, (10.10) becomes:

$$\dot{V}_1 \leq e^h gu + |e^h|b\Phi \leq e^h gu + be^{2h}\Phi^2 + \frac{b}{4}. \tag{10.11}$$

The fractional power error feeback-based practical finite-time tracking controller is designed as:

$$u = -\left(c + \hat{b}\Phi^2\right)e^h, \quad c > 0, \tag{10.12}$$

where $\hat{b}$ is the parameter estimate of the unknown parameter b, which is updated by:

$$\dot{\hat{b}} = \gamma e^{2h}\Phi^2 - \sigma\hat{b}, \quad \hat{b}(0) \geq 0, \tag{10.13}$$

with $\hat{b}(0)$ being the initial value of $\hat{b}(t)$, $\gamma > 0$, and $\sigma > 0$ being design parameters. Furthermore, as $\gamma e^{2h}\Phi \geq 0$, then $\hat{b}(t) \geq 0$ holds for any $\hat{b}(0) \geq 0$.

Theorem 10.3 *Consider the first-order nonlinear system (10.4). Under the Assumptions 10.1, 10.5, and 10.6, if the developed controller (10.12) with the adaptive law (10.13) is applied, then all signals in the closed-loop systems are bounded and there exists a finite time satisfying*

$$T \leq \frac{V^{1-\alpha}(0)}{l\theta_0(1-\alpha)}, \quad 0 < \theta_0 < 1, \tag{10.14}$$

such that the tracking error converges to a residual region around zero in a finite time T, where $V(0) = \frac{1}{2\gamma g}\tilde{b}^2(0) + \frac{1}{1+h}e^{1+h}(0)$ with $\tilde{b} = b - g\hat{b}$ and $\alpha = \frac{2h}{1+h} < 1$, $l = \min\{gc(1+h)^\alpha, \sigma\} > 0$.

Proof. Substituting the controller as shown in (10.12) into the term $e^h gu$, we have:

$$e^h gu = -(c + \hat{b}\Phi^2)ge^{2h}.$$

As $\hat{b}(t) \geq 0$ for $\forall t \geq 0$ and $0 < \underline{g} \leq g$, then one has:

$$e^h gu = -(c + \hat{b}\Phi^2)ge^{2h} \leq -\underline{g}(c + \hat{b}\Phi^2)e^{2h},$$

which leads to:

$$\dot{V}_1 \leq -\underline{g}(c + \hat{b}\Phi^2)e^{2h} + be^{2h}\Phi^2 + \frac{b}{4} = -\underline{g}ce^{2h} + \tilde{b}e^{2h}\Phi^2 + \frac{b}{4}, \tag{10.15}$$

with $\tilde{b} = b - g\hat{b}$ being the estimate error of the virtual parameter b.

Choosing the second part of Lyapunov function candidate as $V_2 = \frac{1}{2g\gamma}\tilde{b}^2$, where $\gamma > 0$ is a design parameter, then its derivative is $\dot{V}_2 = -\frac{1}{\gamma}\tilde{b}\dot{\hat{b}}$. Define the whole Lyapunov function candidate as $V = V_1 + V_2$, together with (10.15), one has:

$$\dot{V} \leq -\underline{g}ce^{2h} + \tilde{b}e^{2h}\Phi^2 + \frac{b}{4} - \frac{1}{\gamma}\tilde{b}\dot{\hat{b}}. \tag{10.16}$$

Substituting the adaptive law for $\dot{\hat{b}}$ as shown in (10.13) into (10.16), it is checked that:

$$\dot{V} \leq -\underline{g}ce^{2h} + \frac{\sigma}{\gamma}\tilde{b}\hat{b} + \frac{b}{4} \leq -\underline{g}ce^{2h} - \frac{\sigma}{2g\gamma}\tilde{b}^2 + \frac{\sigma}{2g\gamma}b^2 + \frac{b}{4}, \tag{10.17}$$

where the fact that $\frac{\sigma}{\gamma}\tilde{b}\hat{b} \leq -\frac{\sigma}{2g\gamma}\tilde{b}^2 + \frac{\sigma}{2g\gamma}b^2$ is used.

Denote $\alpha = \frac{2h}{1+h}$ and $0 < \alpha < 1$, then we have:

$$e^{2h} = (1+h)^\alpha \left(\frac{1}{1+h} e^{1+h} \right)^\alpha = (1+h)^\alpha V_1^\alpha.$$

Choosing $\mathbf{a} = \alpha$, $\mathbf{b} = 1 - \alpha$, $\varsigma_1 = V_2$, $\varsigma_2 = 1$, and $\vartheta = \frac{1}{\alpha}$, then by applying Theorem 2.13, it is established that:

$$V_2^\alpha = \left(\frac{1}{2g\gamma}\tilde{b}^2 \right)^\alpha \leq \frac{1}{2g\gamma}\tilde{b}^2 + (1-\alpha)\alpha^{\frac{\alpha}{1-\alpha}},$$

namely,

$$-\frac{1}{2g\gamma}\tilde{b}^2 \leq -V_2^\alpha + (1-\alpha)\alpha^{\frac{\alpha}{1-\alpha}},$$

therefore, (10.17) can be written as:

$$\dot{V} \leq -gc(1+h)^\alpha V_1^\alpha - \sigma V_2^\alpha + \sigma(1-\alpha)\alpha^{\frac{\alpha}{1-\alpha}} + \frac{\sigma}{2g\gamma}b^2 + \frac{b}{4} \qquad (10.18)$$

$$\leq -l(V_1^\alpha + V_2^\alpha) + \vartheta, \qquad (10.19)$$

where $\vartheta = \sigma(1-\alpha)\alpha^{\frac{\alpha}{1-\alpha}} + \frac{\sigma}{2g\gamma}b^2 + \frac{b}{4} < \infty$, and $l = \min\{gc(1+h)^\alpha, \sigma\} > 0$.

According to Theorem 2.14 and the definition of V, we have $V^\alpha = (V_1 + V_2)^\alpha \leq V_1^\alpha + V_2^\alpha$, which leads to:

$$\dot{V} \leq -lV^\alpha + \vartheta.$$

Then by using Criterion 10.3, it is ensured that V is bounded and there exists a finite time T satisfying (10.14) such that:

$$V(t) \leq \left(\frac{\vartheta}{l(1-\theta_0)} \right)^{\frac{1}{\alpha}}, \quad \forall t \geq T,$$

where $0 < \theta_0 < 1$, which implies that the tracking error $e \in \mathscr{L}_\infty$ and $\tilde{b} \in \mathscr{L}_\infty$. As $\tilde{b} = b - g\hat{b}$ and b and g are constants, it follows that $\hat{b}$ is bounded. Noting that the reference signal y_d is bounded, then the system state x is bounded, thus according to Assumption 10.6 we have $\phi(x) \in \mathscr{L}_\infty$. From the definition of Φ, it follows that $\Phi \in \mathscr{L}_\infty$. According to (10.12) and (10.13), it is ensured that $u \in \mathscr{L}_\infty$ and $\dot{\hat{b}} \in \mathscr{L}_\infty$. Therefore, all signals in the closed-loop systems are bounded.

Furthermore, (10.18) can be further expressed as $\dot{V} \leq -gce^{2h} + \vartheta$, which implies that $\dot{V} < 0$ if e is outside of the compact region

$$\Omega_e = \left\{ e \in \mathbb{R} : |e(t)| \leq \left(\frac{\vartheta + v}{gc} \right)^{\frac{1}{2h}} \right\} \qquad (10.20)$$

with v being a small constant, then the tracking error is confined in the set Ω_e. Hence, from the definition of Ω_e in (10.20), there are two ways to adjust the size of region Ω_e: one way is to adjust the design parameters (i.e., c, γ, σ); and another way is to adjust the fractional power h. The proof is completed.

Remark 10.3 *Actually, we can also design the practical finite-time tracking control scheme by utilizing the robust method, namely, the estimate of unknown parameter is not required. Compared with the adaptive practical finite-time control, the main difference is shown in (10.11). If the robust method is employed, the term $|e^h|b\Phi$ can be handled by using the following form: $|e^h|b\Phi \leq \underline{g}e^{2h}\Phi^2 + \frac{b^2}{4\underline{g}}$. Then (10.11) becomes:*

$$\dot{V}_1 \leq e^h gu + |e^h|b\Phi \leq e^h gu + \underline{g}e^{2h}\Phi^2 + \frac{b^2}{4\underline{g}}.$$

The practical robust finite-time tracking controller is designed as:

$$u = -\left(c+\Phi^2\right)e^h, \quad c > 0.$$

By choosing the Lyapunov function candidate as $V = V_1$ and employing the same analysis procedure in Theorem 10.3, it is not difficult to ensure that all signals in the closed-loop systems are bounded, and the tracking error converges to a compact set around zero within a finite time. The detailed proof is omitted.

10.3.2 NUMERICAL SIMULATION

To verify the effectiveness of the proposed practical finite-time tracking control algorithm, we use the same example in Section 10.2.3 and the initial conditions are given as $x(0) = 1$ and $\hat{b}(0) = 0$. The design parameters are chosen as $c = 1$, $\gamma = 1$, $\sigma = 0.1$, and $h = \frac{1}{3}$. The simulation results are depicted in Fig. 10.6-Fig. 10.9, from which it is seen that all signals in the closed-loop systems are bounded, and the tracking error converges to a compact set around zero within a finite time.

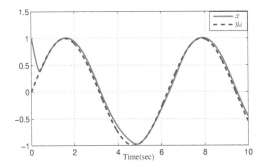

FIGURE 10.6 The evolutions of the system state x and reference signal y_d.

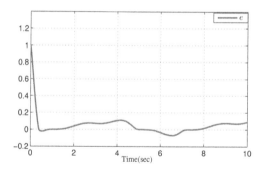

FIGURE 10.7 The evolution of the tracking error e.

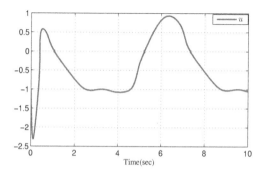

FIGURE 10.8 The evolution of the control input u.

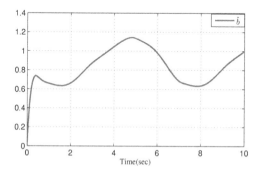

FIGURE 10.9 The evolution of the parameter estimate $\hat{b}$.

10.4 PRACTICAL FINITE-TIME CONTROL OF NORMAL-FORM SYSTEMS

10.4.1 PROBLEM FORMULATION AND PRELIMINARIES

In this section, we consider the following class of normal-form nonlinear systems:

$$\begin{cases} \dot{x}_k = x_{k+1}, \quad k = 1,\cdots,n-1, \\ \dot{x}_n = f(x,p) + g(x)u, \end{cases} \tag{10.21}$$

where $x = [x_1,\cdots,x_n]^\top \in \mathbb{R}^n$ is the system state, $u \in \mathbb{R}$, and $y = x_1 \in \mathbb{R}$ are the control input and system output, respectively, $f(x,p) \in \mathbb{R}$ is an unknown smooth nonlinear function, $p \in \mathbb{R}^l$ is an unknown parameter vector, and $g(x) \in \mathbb{R}$ denotes the control coefficient.

The objective is to develop a fractional power error feedback-based practical finite-time tracking control solution for the normal-form nonlinear system (10.21) such that:

(*i*)   the tracking error $e = x_1 - y_d$ is regulated to an adjustable compact set around zero within a finite time, where y_d is the reference signal;

(*ii*)   all signals in the closed-loop systems are bounded.

To this end, the following assumptions are imposed.

Assumption 10.7 *For the unknown control gain $g(x)$, there exist some positive unknown constants $\underline{g}$ and $\overline{g}$ such that $0 < \underline{g} \leq g(x) \leq \overline{g} < \infty$.*

Assumption 10.8 *The reference trajectory $y_d(t)$ and its derivative up to n-th are assumed to be known, bounded, and piecewise continuous. Furthermore, the system states are available for control design.*

Assumption 10.9 *For the uncertain function $f(x,p)$, there exist an unknown positive constant a and a known smooth scalar function $\phi(x) \geq 0$ such that $|f(x,p)| \leq a\phi(x)$. In addition, $f(x,p)$ is bounded if x is bounded.*

10.4.2 PRACTICAL FINITE-TIME CONTROL DESIGN

To proceed, we introduce a filtered error $s(t)$ as:

$$s = \lambda_1 e + \lambda_2 \dot{e} + \cdots + \lambda_{n-1} e^{(n-2)} + e^{(n-1)}, \tag{10.22}$$

where λ_k, $k = 1,2,\cdots,n-1$, denotes a constant selected by the designer such that the polynomial $l^{n-1} + \lambda_{n-1} l^{n-2} + \cdots + \lambda_1$ is Hurwitz.

Taking the derivative of $s(t)$ w.r.t. time yields:

$$\dot{s} = \dot{x}_n - y_d^{(n)} + \lambda_1 \dot{e} + \lambda_2 \ddot{e} + \cdots + \lambda_{n-1} e^{(n-1)}. \tag{10.23}$$

Substituting the expression of $\dot{x}_n$ in (10.21) into (10.23), we have:

$$\dot{s} = f + gu + L \tag{10.24}$$

where $L = -y_d^{(n)} + \lambda_1 \dot{e} + \lambda_2 \ddot{e} + \cdots + \lambda_{n-1} e^{(n-1)}$ is available for control design.

By using the fractional power feedback of the filtered error s, we construct the following practical finite-time tracking controller:

$$\begin{cases} u = -\left(c + \hat{b}\Phi^2\right) s^h, \\ \dot{\hat{b}} = \gamma \Phi^2 s^{2h} - \sigma \hat{b}, \ \hat{b}(0) \geq 0, \end{cases} \tag{10.25}$$

where $h = \frac{h_1}{h_2}$ with h_1 and h_2 being some odd integers and $0 < h < 1$, $\hat{b}$ is the estimate of the virtual unknown parameter $b = \max\{1, a\}$, $\hat{b}(0) \geq 0$ is the arbitrarily chosen initial estimate, $c > 0$, $\gamma > 0$, and $\sigma > 0$ are user-chosen design parameters, and $\Phi = L^2 + \phi + \frac{1}{4}$ is a computable function.

Theorem 10.4 *Under Assumptions 10.7–10.9, if the control strategy (10.25) is applied for the normal-form nonlinear system (10.21), it is ensured that: (1) all signals in the closed-loop systems are bounded; and (2) there exists a finite time T satisfying*

$$T \leq \frac{V^{1-\alpha}(0)}{l\theta_0(1-\alpha)}, \quad 0 < \theta_0 < 1, \tag{10.26}$$

where $V(0) = \frac{1}{2\gamma g}\tilde{b}^2(0) + \frac{1}{1+h}\varepsilon^{1+h}(0)$ *with* $\tilde{b} = b - g\hat{b}$ *and* $\alpha = \frac{2h}{1+h} < 1$, $l = \min\{\underline{g}c(1+h)^\alpha, \sigma\} > 0$, *such that the tracking error e converges to a compact set within a finite time T.*

Proof. The proof of Theorem 10.4 consists of two steps.

Step 1. We show the stability of the closed-loop system. Choosing the Lyapunov function candidate as $V = V_{11} + V_{12}$, where $V_{11} = \frac{1}{1+h}\varepsilon^{1+h}$ and $V_{12} = \frac{1}{2g\gamma}\tilde{b}^2$, in which $\tilde{b} = b - g\hat{b}$ is the virtual parameter estimation error, with $g > 0$ being the unknown constant as defined in Assumption 10.7.

Taking the derivative of V w.r.t. time yields:

$$\dot{V} = s^h \dot{s} - \frac{1}{\gamma}\tilde{b}\dot{\hat{b}} = s^h(f + gu + L) - \frac{1}{\gamma}\tilde{b}\dot{\hat{b}}. \tag{10.27}$$

Note that

$$s^h(f + L) \leq \left|s^h\right|(|L| + a\phi) \leq \left|s^h\right| b\Phi,$$

where a is an unknown virtual parameter and

$$\Phi = L^2 + \phi + \frac{1}{4}$$

is a computable function available for control design.

By using Young's inequality, we have:

$$s^h(f+L) \le |s^h|b\Phi \le b\Phi^2 s^{2h} + \frac{b}{4},$$

thus,

$$\dot{V} \le s^h gu + b\Phi^2 s^{2h} + \frac{b}{4} - \frac{1}{\gamma}\tilde{b}\dot{\hat{b}}. \tag{10.28}$$

Inserting the finite-time controller as given in (10.25) into (10.28), we get:

$$\dot{V} \le -g\left(c+\hat{b}\Phi^2\right)s^{2h} + b\Phi^2 s^{2h} + \frac{b}{4} - \frac{1}{\gamma}\tilde{b}\dot{\hat{b}}$$

$$\le -\underline{g}cs^{2h} + \tilde{b}\Phi^2 s^{2h} + \frac{b}{4} - \frac{1}{\gamma}\tilde{b}\dot{\hat{b}}, \tag{10.29}$$

where the facts that $0 < \underline{g} \le g$ and $\hat{b}(t) \ge 0$ are used.

By applying the adaptive law for $\hat{b}$ as given in (10.25), we can rewrite (10.29) as:

$$\dot{V} \le -\underline{g}cs^{2h} - \frac{\sigma}{2\underline{g}\gamma}\tilde{b}^2 + \frac{\sigma}{2\underline{g}\gamma}b^2 + \frac{b}{4}, \tag{10.30}$$

where the fact that $\frac{\sigma}{\gamma}\tilde{b}\hat{b} \le -\frac{\sigma}{2\underline{g}\gamma}\tilde{b}^2 + \frac{\sigma}{2\underline{g}\gamma}b^2$ is used.

Denote $\alpha = \frac{2h}{1+h}$ and $0 < \alpha < 1$, then we have:

$$s^{2h} = (1+h)^\alpha \left(\frac{1}{1+h}s^{1+h}\right)^\alpha = (1+h)^\alpha V_{11}^\alpha, \tag{10.31}$$

by applying Theorem 2.13, it is established that:

$$V_{12}^\alpha = \left(\frac{1}{2\underline{g}\gamma}\tilde{a}^2\right)^\alpha \le \frac{1}{2\underline{g}\gamma}\tilde{a}^2 + (1-\alpha)\alpha^{\frac{\alpha}{1-\alpha}}. \tag{10.32}$$

Therefore, together with (10.31) and (10.32), (10.30) can be written as:

$$\dot{V} \le -\underline{g}c(1+h)^\alpha V_{11}^\alpha - \sigma V_{12}^\alpha + \sigma(1-\alpha)\alpha^{\frac{\alpha}{1-\alpha}} + \frac{\sigma}{2\underline{g}\gamma}b^2 + \frac{b}{4} \tag{10.33}$$

$$\le -l\left(V_{11}^\alpha + V_{12}^\alpha\right) + \vartheta, \tag{10.34}$$

where $\vartheta = \sigma(1-\alpha)\alpha^{\frac{\alpha}{1-\alpha}} + \frac{\sigma}{2\underline{g}\gamma}b^2 + \frac{b}{4} < \infty$, and $l = \min\{\underline{g}c(1+h)^\alpha, \sigma\} > 0$.

According to Theorem 2.14, we have $V^\alpha = (V_{11}+V_{12})^\alpha \le V_{11}^\alpha + V_{12}^\alpha$, which leads to $\dot{V} \le -lV^\alpha + \vartheta$. Then by using Criterion 10.3, there exists a finite time T satisfying (10.26) such that:

$$V(t) \le \left(\frac{\vartheta}{l(1-\theta_0)}\right)^{\frac{1}{\alpha}}, \forall t \ge T, \tag{10.35}$$

where $0 < \theta_0 < 1$. From (10.35), it is seen that $V \in \mathscr{L}_\infty$, which implies that $\varepsilon \in \mathscr{L}_\infty$ and $\hat{b} \in \mathscr{L}_\infty$, from (10.22) it indicates that $e, \cdots, e^{(n-1)}$ are bounded, then it follows that $x_1, \cdots, x_n$ are bounded as y_d and its derivatives up to $(n+1)$-th are bounded, thus according to Assumption 10.9, $\phi(x)$ is bounded. From the definition of Φ, it follows that $\Phi \in \mathscr{L}_\infty$. According to (10.25), it is ensured that $u \in \mathscr{L}_\infty$ and $\dot{\hat{b}} \in \mathscr{L}_\infty$. In addition, from (10.24) one gets that $\dot{s} \in \mathscr{L}_\infty$. Therefore, all the signals in the closed-loop systems are bounded.

Furthermore, (10.33) can be further expressed as $\dot{V} \leq -\underline{g}cs^{2h} + \vartheta$, which implies that $\dot{V} < 0$ if $|s|$ is outside of the compact region

$$\Omega_s = \left\{ s \in \mathbb{R} : |s(t)| \leq \left(\frac{\vartheta + v}{\underline{g}c} \right)^{\frac{1}{2h}} \right\} \tag{10.36}$$

with v being a small constant; then the filtered error is confined in the set Ω_s. Hence from the definition of Ω_s as defined in (10.36), there are two ways to adjust the size of region Ω_s: one way is to adjust the design parameters (i.e., c, γ, σ); and another way is to adjust fractional power h.

Step 2. We show that the tracking error $e(t)$ converges to a compact set around zero within a finite time T with the rate of convergence $\exp(-\lambda t)$.

For easy analysis, we express (10.22) equivalently as:

$$\begin{cases} s = \dot{s}_{n-1} + \alpha_{n-1} s_{n-1}, \\ s_{n-1} = \dot{s}_{n-2} + \alpha_{n-2} s_{n-2}, \\ \vdots \\ s_2 = \dot{s}_1 + \alpha_1 s_1 = \dot{e} + \alpha_1 e, \\ s_1 = e, \end{cases} \tag{10.37}$$

in which the coefficient $\alpha_i > 0$, $i = 1, 2, \cdots, n-1$, is determined by the constant λ_k.

Solving the first differentiate equation (10.37) over the interval $[0,t]$ yields:

$$s_{n-1}(t) = s_{n-1}(0) \exp(-\alpha_{n-1}t) + \exp(-\alpha_{n-1}t) \int_0^t \exp(\alpha_{n-1}\tau) s(\tau) d\tau. \tag{10.38}$$

From the aforementioned stability analysis, it is ensured that $s(t)$ is bounded, then:

$$\int_0^t \exp(\alpha_{n-1}\tau) s(\tau) d\tau \leq \int_0^t \exp(\alpha_{n-1}\tau) |s(\tau)| d\tau \leq \frac{\bar{s}}{\alpha_{n-1}} (\exp(\alpha_{n-1}t) - 1),$$

where $\bar{s} = \sup_{\tau \in [0,t]} |s(\tau)|$, then (10.38) is upper bounded by:

$$|s_{n-1}(t)| \leq |s_{n-1}(0)| \exp(-\alpha_{n-1}t) + \frac{\bar{s}}{\alpha_{n-1}}, \tag{10.39}$$

which implies that s_{n-1} is bounded, and when $t = T$ we have:

$$|s_{n-1}(T)| \leq |s_{n-1}(0)| \exp(-\alpha_{n-1}T) + \frac{\bar{s}_n}{\alpha_{n-1}} := \Omega_{n-1}, \tag{10.40}$$

where $\bar{s}_n = \sup_{\tau \in [0,T]} |s(\tau)|$, which implies that $s_{n-1}(t)$ converges to the residual set Ω_{n-1} with the decay rate $\exp(-\alpha_{n-1}t)$ in a finite time T.

From (10.40), it is clearly seen that increasing c and α_{n-1} makes $s_{n-1}(t)$ converge to a smaller compact set Ω_{n-1} in a finite time T. By using the same analysis procedure, we have:

$$|s_{n-i}(T)| \leq |s_{n-i}(0)| \exp(-\alpha_{n-i}T) + \frac{\bar{s}_{n-i+1}}{\alpha_{n-i}},$$

$$|e(T)| = |s_1(T)| \leq |e(0)| \exp(-\alpha_1 T) + \frac{\bar{s}_2}{\alpha_1},$$

where $\bar{s}_{n-i+1} = \sup_{\tau \in [0,T]} |s_{n-i+1}(\tau)|, (i = 2, \cdots, n-1)$, which implies that the tracking error $e(t)$ converges to a residual set

$$\Omega_e = \left\{ e \in \mathbb{R} : |e(t)| \leq |e(0)| \exp(-\alpha_1 T) + \frac{\bar{s}_2}{\alpha_1} \right\},$$

with the decay rate $\exp(-\alpha_1 t)$ in a finite time T as defined in (10.26). Employing the similar analysis process, it is concluded that increasing c and α_1 leads to smaller Ω. The proof is completed.

10.5   NOTES

It is worth noting that, although the signum function feedback-based finite-time control in Section 10.2 and the fractional power error feedback-based finite-time control in Section 10.3 and Section 10.4 are able to ensure the tracking error $e(t)$ to converge to zero or a compact set Ω_e within a finite time T, this method suffers from the following shortcomings.

(1) The control effort is non-smooth.
(2) The settling time T, as indicated in (10.26), depends on some parameters and the initial value of the Lyapunov function $V(0)$, as a result, such a finite time, although existing, cannot be specified by the designer directly. Moreover, if the initial values are not available, the settling time is difficult to determine.
(3) For normal-form nonlinear systems, in order to make the compact set Ω_e small, one has to increase α_1 and decrease $\bar{s}_2$ when h is fixed, which means to increase the design parameters c and α_1. Note that although larger c and α_1 leads to higher tracking precision, larger control effort is demanded, especially at the initial moment. In addition, while the fractional power h can be properly adjusted to enhance the disturbance rejection and improve the control precision without the need for large c and α_1, the fractional power h might lead to the chattering of the control action (see simulation section for detail).

(4) Such a control method relies on the fractional power Lyapunov differential inequality for stability analysis and controller design, making it rather complicated for n-order normal-formal nonlinear systems.

Therefore, in the next chapter we will introduce a novel regular error feedback-based practical prescribed-time control method to avoid the above shortcomings.

11 Practical Prescribed-time Control

In Chapter 10, we described the control design and stability analysis for finite-time control of nonlinear systems. In this chapter, we introduce the practical prescribed-time control scheme for nonlinear systems.

11.1 PRESCRIBED-TIME RATE FUNCTION AND SCALING FUNCTION

To gain insight into the idea of the practical prescribed-time control algorithm, we first give the concept of practical prescribed-time stability and the prescribed-time scaling function.

Definition 11.1 *(**Practical Prescribed-time Stability**) The practical prescribed-time stability of nonlinear systems, $\dot{x} = f(x)$ with $x \in \mathbb{R}$ being the system state, is said to be reached if for any initial condition $x(0) = x_0$, there exist a constant $\vartheta > 0$ and a finite time $t_f < \infty$ (in which t_f is independent of initial conditions and other design parameters), such that $|x(t)| < \vartheta$ for all $t \geq t_f$.*

Definition 11.2 *A function $\kappa(t)$ is called a prescribed-time rate function if it has the following properties:*

(i) *the initial value of $\kappa(t)$ is 1, i.e., $\kappa(0) = 1$;*
(ii) *$\kappa(t)$ is strictly monotonically decreasing on the interval $[0, t_f)$ and then keeps zero for $\forall t \geq t_f$ with $t_f > 0$ being the settling time;*
(iii) *$\kappa(t)$ is at least $\mathscr{C}^{n+1}$ with n being the considered system order;*
(iv) *$\kappa(t)$ and its derivatives (at least) up to $(n+1)$-th are known, bounded, and piecewise continuous.*

Immediate examples of such $\kappa(t)$ include (but are not limited to) the following functions with $t \geq 0$:

$$\kappa(t) = \begin{cases} \left(\frac{t_f - t}{t_f}\right)^{n+2}, & 0 \leq t < t_f, \\ 0, & t \geq t_f, \end{cases}$$

$$\kappa(t) = \begin{cases} \left(\frac{t_f - t}{t_f}\right)^{n+2} \exp(-t), & 0 \leq t < t_f, \\ 0, & t \geq t_f, \end{cases}$$

$$\kappa(t) = \begin{cases} \left(\frac{t_f - t}{t_f}\right)^{n+2} \frac{1}{1+t}, & 0 \leq t < t_f, \\ 0, & t \geq t_f, \end{cases}$$

$$\vdots$$

DOI: 10.1201/9781003474364-11

In this chapter, we give a detailed form of the rate function as follows:

$$\kappa(t) = \begin{cases} \left(\frac{t_f - t}{t_f}\right)^{n+2} \varpi(t), & 0 \le t < t_f, \\ 0, & t \ge t_f, \end{cases} \tag{11.1}$$

where $\varpi(t)$ is any non-increasing function of time satisfying:

(i) $\varpi(t)$ is (at least) $\mathscr{C}^{n+1}$;
(ii) $\varpi^{(i)}(t) \in \mathscr{L}_\infty$, $i = 0, 1, \cdots, n+1$, for $\forall t \in [0, \infty)$;
(iii) $\varpi(0) = 1$ and $\varpi(t) > 0$ for $\forall t > 0$.

Immediate examples of such $\varpi(t)$ include (but are not limited to) the following functions with $t \ge 0$: $\varpi(t) = 1, \exp(-t), \frac{1}{1+t^2}$.

Based on the proposed rate function, we develop a prescribed-time scaling function as follows:

$$\beta(t) = \frac{1}{(1 - b_f)\kappa(t) + b_f}, \quad 0 < b_f < 1, \tag{11.2}$$

which has the following important properties for any κ as defined in Definition 11.2:

(i) β is at least $\mathscr{C}^{n+1}$;
(ii) $\beta^{(i)}$ ($i = 0, 1, \cdots, n+1$) is bounded for $\forall t \ge 0$;
(iii) $\beta(t)$ is strictly increasing with time in the interval $[0, t_f)$ with $\beta(0) = 1$ and then keeps a constant $\frac{1}{b_f}$ for $t \ge t_f$.

11.2   PRACTICAL PRESCRIBED-TIME CONTROL OF FIRST-ORDER SYSTEMS

Here, based on the proposed time-varying scaling function $\beta(t)$, we develop a practical prescribed-time tracking control algorithm for the following first-order nonlinear system:

$$\dot{x} = f(x) + g(x)u, \tag{11.3}$$

such that:

(i)   all signals in the closed-loop systems are bounded;
(ii) the tracking error $e = x - y_d$ converges to an adjustable region around zero within a finite time, where y_d is the reference signal;
(iii) the settling time is at the user's disposal and is independent of design parameters and initial conditions.

In addition to Assumptions 10.1, 10.5, and 10.6, we impose the following assumption.

Assumption 11.1 *Let $T_c > 0$ be the small time interval necessary for signal processing/computing and transmission. For the prescribed time t_f, we have $t_f \ge T_c$.*

11.2.1 CONTROL DESIGN

Now, we carry out the control design based on the time-varying scaling function (11.2).

Note that the derivative of the tracking error e is $\dot{e} = f + gu - \dot{y}_d$, by using the following transformation

$$z = \beta e, \tag{11.4}$$

we have:

$$\dot{z} = \dot{\beta}e + \beta\dot{e} = \dot{\beta}e + \beta(f + gu - \dot{y}_d). \tag{11.5}$$

Choosing the quadratic function as $V_1 = \frac{1}{2}z^2$, then the derivative of V_1 along (11.5) is:

$$\dot{V}_1 = z\dot{z} = \beta gzu + \Xi, \tag{11.6}$$

with

$$\Xi = \beta zf - \beta z\dot{y}_d + \dot{\beta}ze$$

being the lumped uncertainty.

Upon using Assumptions 10.5 and 10.6 and employing Young's inequality, one has:

$$\beta zf \leq \beta|z|a\phi \leq a^2\beta^2z^2\phi^2 + \frac{1}{4},$$

$$-\beta z\dot{y}_d \leq \beta|z||\dot{y}_d| \leq \beta^2z^2\dot{y}_d^2 + \frac{1}{4},$$

$$\dot{\beta}ze \leq \beta|\beta^{-1}\dot{\beta}||z||e| \leq \beta^2(\beta^{-1}\dot{\beta})^2z^2e^2 + \frac{1}{4},$$

then Ξ can be upper bounded by:

$$\Xi \leq a^2\beta^2z^2\phi^2 + \beta^2z^2\dot{y}_d^2 + \beta^2(\beta^{-1}\dot{\beta})^2z^2e^2 + \frac{3}{4} \leq b\beta^2z^2\Phi + \frac{3}{4},$$

where $b = \max\{a^2, 1\}$ is a virtual parameter and $\Phi = \phi^2 + \dot{y}_d^2 + (\beta^{-1}\dot{\beta})^2e^2$ is an available function.

Therefore, (11.6) can be further expressed as:

$$\dot{V}_1 \leq \beta gzu + b\beta^2z^2\Phi + \frac{3}{4}. \tag{11.7}$$

The regular error feedback-based practical prescribed-time controller is designed as:

$$\begin{cases} u = -ce - \hat{b}\beta z\Phi, \\ \dot{\hat{b}} = \gamma\beta^2z^2\Phi - \sigma\hat{b}, \ \hat{b}(0) \geq 0, \end{cases} \tag{11.8}$$

where c, γ, and σ are positive design parameters, $\hat{b}$ denotes the estimate of the unknown parameter b, and $\hat{b}(0) \geq 0$ is the arbitrarily chosen initial estimate. Noting that $\gamma\beta^2z^2\Phi \geq 0$, then $\hat{b}(t) \geq 0$ holds for $\forall t \geq 0$.

Theorem 11.1 *Consider the first-order nonlinear system (11.3). Under Assumptions 10.1, 10.5, 10.6, and 11.1. If the proposed practical prescribed-time controller (11.8) is applied, the objectives (i)–(iii) can be ensured.*

Proof. Substituting the control law as shown in (11.8) into the term βgzu, one has:

$$\beta gzu = -gcz^2 - g\hat{b}\beta^2 z^2 \Phi.$$

As $g \geq \underline{g} > 0$ and $\hat{b}(t) \geq 0$ for $\forall t \geq 0$, we further have:

$$\beta gzu \leq -\underline{g}cz^2 - \underline{g}\hat{b}\beta^2 z^2 \Phi,$$

which leads to:

$$\dot{V}_1 \leq -\underline{g}cz^2 + \tilde{b}\beta^2 z^2 \Phi + \frac{3}{4}, \tag{11.9}$$

where $\tilde{b} = b - \underline{g}\hat{b}$ denotes the estimate error.

Choosing the Lyapunov function candidate as $V = V_1 + \frac{1}{2\gamma}\tilde{b}^2$, then the derivative of V along (11.9) yields:

$$\dot{V} \leq -\underline{g}cz^2 + \tilde{b}\beta^2 z^2 \Phi + \frac{3}{4} - \frac{1}{\gamma}\tilde{b}\dot{\hat{b}}. \tag{11.10}$$

Inserting the adaptive law in (11.8) into (11.10), one has:

$$\dot{V} \leq -\underline{g}cz^2 + \frac{3}{4} + \frac{\sigma}{\gamma}\tilde{b}\hat{b} \leq -\underline{g}cz^2 - \frac{\sigma}{2\underline{g}\gamma}\tilde{b}^2 + \frac{3}{4} + \frac{\sigma}{2\underline{g}\gamma}b^2 \leq -l_1 V + l_2, \tag{11.11}$$

where $l_1 = \min\{2\underline{g}c, \sigma\} > 0$ and $l_2 = \frac{\sigma}{2\underline{g}\gamma}b^2 + \frac{3}{4} > 0$.

We first prove that all signals in the closed-loop systems are bounded. It is seen from (11.11) that $V(t) \in \mathcal{L}_\infty$, which implies that $z \in \mathcal{L}_\infty$ and $\hat{b} \in \mathcal{L}_\infty$. Note that $e = \beta^{-1}z$, then e is bounded, as $\beta^{-1} \in (b_f, 1]$ is bounded, thus x is bounded due to the boundedness of the reference signals y_d and $\dot{y}_d$, then ϕ is bounded. As β and $\beta^{-1}\dot{\beta}$ are bounded for all $t \geq t_0$, it is seen from (11.8) that the control input u and the adaptive law are bounded and implementable.

Next, we prove that the tracking error converges to an adjustable region in a prescribed time at the prescribed decay rate. Note that $e = \beta^{-1}z$, by the definition of β as in (11.2), we have:

$$e(t) = \begin{cases} (1-b_f)\left(\frac{t_f-t}{t_f}\right)^{n+2}\varpi z + b_f z, & 0 \leq t < t_f, \\ b_f z, & t \geq t_f. \end{cases} \tag{11.12}$$

Since $V(t) \leq \left(V(0) - \frac{l_2}{l_1}\right)\exp(-l_1 t) + \frac{l_2}{l_1} \leq V(0) + \frac{l_2}{l_1}$, then according to the definition of $V(t)$, we have $z^2(t) \leq 2V(t) \leq 2\Theta$ with $\Theta = V(0) + \frac{l_2}{l_1}$, namely, $|z(t)| \leq \sqrt{2\Theta}$, then (11.12) becomes:

$$|e(t)| \leq \frac{(1-b_f)(t_f-t)^{n+2}\varpi}{t_f^{n+2}}\sqrt{2\Theta} + b_f\sqrt{2\Theta}, \quad 0 \leq t < t_f, \tag{11.13}$$

$$|e(t)| \leq b_f\sqrt{2\Theta}, \quad t \geq t_f. \tag{11.14}$$

Noting that $\left(\frac{t_f-t}{t_f}\right)^{n+2}\varpi$ is a strictly decreasing function in the interval $[0,t_f)$ and $\left(\frac{t_f-t}{t_f}\right)^{n+2}\varpi = 0$ for $\forall t \geq t_f$, then it shows from (11.13) that the tracking error $e(t)$ converges to a residual region $\Omega := \{e \in \mathbb{R} : |e(t)| \leq b_f\sqrt{2\Theta}\}$ in a prescribed time t_f at the decay rate no less than $\left((t_f-t)/t_f\right)^{n+2}\varpi$. Note that b_f and t_f are free design parameters and $\varpi(t)$ is at the user's disposal, then they can be selected explicitly and arbitrarily so that the tracking error converges to an adjustable compact set Ω with an assignable decay rate. The proof is completed.

11.2.2 NUMERICAL SIMULATION

To test the effectiveness of the proposed practical prescribed-time control scheme, we consider the nonlinear system (11.3) with $f(x) = x^2\cos(x)$ and $g(x) = 1+0.2\sin(x)$ under the initial condition $x(0) = 1$. The reference signal is given as $y_d(t) = \sin(t)$. The time-varying scaling function is selected as $\beta(t) = \frac{1}{(1-b_f)\kappa(t)+b_f}$ with $\kappa(t) = \left(\frac{t_f-t}{t_f}\right)^3$ and $b_f = 0.05$. The design parameters are chosen as: $c = 2$, $\gamma = 0.1$, and $\sigma = 1$. The prescribed time t_f is given as 3 seconds. The simulation results are shown in Fig. 11.1–Fig. 11.4. It is seen from Fig. 11.1 and Fig. 11.2 that the tracking error converges to a residual set around zero within the finite time, which is independent of initial conditions and other design parameters. The evolution of the control input u is plotted in Fig. 11.3, which shows that the control input is smooth everywhere. Furthermore, the evolution of the parameter estimate $\hat{b}$ is plotted in Fig. 11.4, which is bounded for all time.

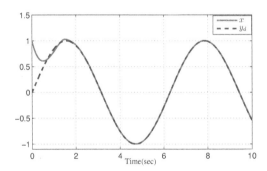

FIGURE 11.1   The evolutions of the system state x and reference signal y_d.

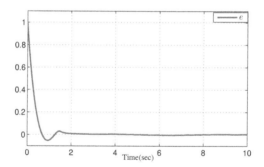

FIGURE 11.2   The evolution of the tracking error e.

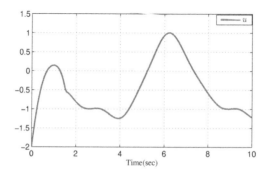

FIGURE 11.3   The evolution of the control input u.

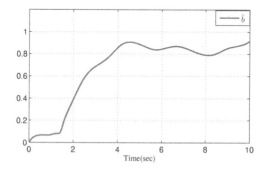

FIGURE 11.4   The evolution of the parameter estimate $\hat{b}$.

11.3 PRACTICAL PRESCRIBED-TIME CONTROL OF NORMAL-FORM SYSTEMS

In this section, the objective is to develop a practical prescribed-time tracking control solution with $\mathscr{C}^1$ smooth control action for the normal-form nonlinear system (10.21) such that the tracking error $e = x_1 - y_d$ is regulated to an adjustable compact set around zero within a prescribed time t_f, where y_d is the reference signal. The assumptions are same as those in Section 10.4.

11.3.1 FILTERED ERROR-BASED PRACTICAL PRESCRIBED-TIME CONTROL

In this section, we first make use of the transferred filtered error version to develop a regular feedback-based practical prescribed-time control to reach an adjustable tracking precision within a prescribed time.

We make use of β as defined in (11.2) to carry out the transformation:

$$\zeta = \beta s, \tag{11.15}$$

where $s = \lambda_1 e + \lambda_2 \dot{e} + \cdots + \lambda_{n-1} e^{(n-2)} + e^{(n-1)}$ denotes the filtered error, λ_k, $k = 1, 2, \cdots, n-1$, denotes a constant selected by the designer such that the polynomial $l^{n-1} + \lambda_{n-1} l^{n-2} + \cdots + \lambda_1$ is Hurwitz.

Then the derivative of ζ along (11.15) is:

$$\dot{\zeta} = \dot{\beta} s + \beta \dot{s} = \beta(\delta s + gu + f + L), \tag{11.16}$$

where $\delta = \beta^{-1} \dot{\beta}$ and $L = -y_d^{(n)} + \lambda_1 \dot{e} + \lambda_2 \ddot{e} + \cdots + \lambda_{n-1} e^{(n-1)}$.

Choosing the Lyapunov function candidate as $V = \frac{1}{2}\zeta^2 + \frac{1}{2g\gamma}\tilde{b}^2$, where $\tilde{b} = b - g\hat{b}$ is the virtual parameter estimate error of the unknown parameter $b = \max\{1, a^2\}$ and $\gamma > 0$ is a user-defined constant. Then the derivative of V is:

$$\dot{V} = \beta \zeta gu + \beta \zeta (\delta s + f + L) - \frac{1}{\gamma}\tilde{b}\dot{\hat{b}}. \tag{11.17}$$

Upon using Young's inequality, one has:

$$\beta \zeta \delta s \leq \beta^2 \zeta^2 \delta^2 s^2 + \frac{1}{4},$$

$$\beta \zeta f \leq |\beta \zeta| a\phi \leq \beta^2 \zeta^2 a^2 \phi^2 + \frac{1}{4},$$

$$\beta \zeta L \leq \beta^2 \zeta^2 L^2 + \frac{1}{4}.$$

Therefore, we further have:

$$\beta \zeta (\delta \varepsilon + f + L) \leq b\beta^2 \zeta^2 \Phi + 1,$$

where $b = \max\{1, a^2\}$ is an unknown constant, and

$$\Phi = \delta^2 s^2 + L^2 + \phi^2$$

is an available function for control design.

Hence, (11.17) becomes:

$$\dot{V} \leq \beta \zeta g u + b \beta^2 \zeta^2 \Phi + 1 - \frac{1}{\gamma} \tilde{b} \dot{\hat{b}}. \tag{11.18}$$

By using the time-varying transformation (11.15), we construct the following regular filtered error feedback-based practical prescribed-time tracking control scheme:

$$\begin{cases} u = -cs - \hat{b} \beta^2 s \Phi, \\ \dot{\hat{b}} = \gamma \beta^4 s^2 \Phi - \sigma \hat{b}, \ \hat{b}(0) \geq 0, \end{cases} \tag{11.19}$$

where $\hat{b}$ is the estimate of the virtual unknown parameter b, $\hat{b}(0) \geq 0$ is the arbitrarily chosen initial estimate, $c > 0$, $\gamma > 0$, and $\sigma > 0$ are user-chosen design parameters, and Φ is the computable function. It should be noted that as $\gamma \beta^4 s^2 \Phi \geq 0$, then for any $\hat{b}(0) \geq 0$, it holds that $\hat{b}(t) \geq 0$ for $\forall t \geq 0$.

Now we are ready to present the following theorem.

Theorem 11.2 *Consider the normal-form nonlinear system (10.21). Let Assumptions 10.7–10.9, and 11.1 hold. If the control strategy (11.19) is applied, then the following objectives are achieved:*

(*i*)   *all signals in the closed-loop systems are globally bounded;*
(*ii*)   *the filtered error and tracking error converge to the adjustable compact sets around zero within a prescribed time t_f;*
(*iii*) *the settling time t_f is user-assignable and is independent of system initial values and any other design parameter.*

Proof. Substituting the controller as defined in (11.19) into (11.18), we have:

$$\dot{V} \leq -\beta \zeta g \left(cs + \hat{b} \beta^2 s \Phi \right) + b \beta^2 \zeta^2 \Phi + 1 - \frac{1}{\gamma} \tilde{b} \dot{\hat{b}}$$

$$\leq -\underline{g} c \zeta^2 + \tilde{b} \beta^2 \zeta^2 \Phi + 1 - \frac{1}{\gamma} \tilde{b} \dot{\hat{b}}, \tag{11.20}$$

where the fact that $0 < \underline{g} \leq g$ is used.

Inserting the adaptive law $\dot{\hat{b}}$ as given in (11.19) into (11.20), it follows that:

$$\dot{V} \leq -\underline{g} c \zeta^2 + \tilde{b} \beta^2 \zeta^2 \Phi + 1 - \frac{1}{\gamma} \tilde{b} \left(\gamma \beta^4 s^2 \Phi - \sigma \hat{b} \right)$$

$$\leq -\underline{g} c \zeta^2 + \frac{\sigma}{\gamma} \tilde{b} \hat{b} + 1$$

$$\leq -\underline{g} c \zeta^2 - \frac{\sigma}{2 \underline{g} \gamma} \tilde{b}^2 + C$$

$$\leq -l_1 V + l_2,$$

where $l_1 = \min\{2\underline{g}c, \sigma\}$, $l_2 = \frac{\sigma}{2g\gamma}b^2 + 1$, and the fact that $\frac{1}{\gamma}\tilde{b}\hat{b} \leq -\frac{\sigma}{2g\gamma}\tilde{b}^2 + \frac{\sigma}{2g\gamma}b^2$ is utilized. Similar to the analysis in Chapter 10, it is ensured that all signals in the closed-loop systems are bounded.

Note that $\frac{1}{2}\zeta^2 \leq V(t) \leq V(0) + \frac{l_2}{l_1}$ and $s = \beta^{-1}\zeta$, we have $|\zeta(t)| \leq \sqrt{2\left(V(0) + \frac{l_2}{l_1}\right)} =: \mathbb{B}_\zeta$, by the definition of β as in (11.2), we further have:

$$|s(t)| \leq (1 - b_f)\kappa\mathbb{B}_\zeta + b_f\mathbb{B}_\zeta, \quad 0 \leq t < t_f,$$
$$|s(t)| \leq b_f\mathbb{B}_\zeta, \quad t \geq t_f,$$

from which it is shown that the filtered error converges to the compact set

$$\Omega_s := \{s(t) \in \mathbb{R} : |s(t)| \leq b_f\mathbb{B}_\zeta\}$$

at the decay rate no less than κ within the finite time t_f. It should be emphasized that the finite time t_f is completely at the designer's disposal, regardless of the other design parameters and the initial value of the Lyapunov function. Furthermore, similar to the process in dealing with the relationship between the filtered error $s(t)$ and the tracking error $e(t)$ in Section 10.4, it is shown that the tracking error $e(t)$ converges to a compact set with the rate of convergence $\exp(-\alpha_1 t)$ within a prescribed time t_f. The proof is completed.

Compared with the traditional fractional power error feedback-based control method in Section 10.4, the settling time t_f of the filtered error transformation-based control algorithm (11.19) in this section can be pre-specified and the control effort is $\mathscr{C}^1$ smooth. However, under such a control scheme we can only directly achieve the tracking performance for the filtered error s [rather than the tracking error e]. To clearly reveal the nature of how to improve the tracking performance of the tracking error $e(t)$ within a prescribed time t_f, some complicated analysis should be carried out and some design parameters should be selected properly (see Lemma 5.1). To avoid this issue, we develop a regular tracking error feedback-based practical prescribed-time control in the following section to directly analyze the feature of the tracking error within a prescribed time.

11.3.2 TRACKING ERROR-BASED PRACTICAL PRESCRIBED-TIME CONTROL

We now develop a regular tracking error feedback-based practical prescribed-time control scheme, capable of achieving adjustable tracking precision within a user-assignable finite time at the prescribed convergence rate.

Define the full state tracking error vector as follows:

$$E = \left[e, e^{(1)}, \cdots, e^{(n-1)}\right]^\top \in \mathbb{R}^n, \tag{11.21}$$

where $e^{(i)} = x_{i+1} - y_d^{(i)}$, $i = 0, \cdots, n-1$. Let $I_{n-1} = \mathrm{diag}\{1\} \in \mathbb{R}^{(n-1)\times(n-1)}$, $0_{n-1} = [0, \cdots, 0]^\top \in \mathbb{R}^{n-1}$, and $B = \left[0_{n-1}^\top, 1\right]^\top \in \mathbb{R}^n$.

Taking the derivative of E along (10.21) yields:

$$\dot{E} = AE + B\left(f(x,p) + gu - y_d^{(n)}\right), \tag{11.22}$$

where $A = \begin{bmatrix} 0_{n-1} & I_{n-1} \\ 0 & 0_{n-1}^\top \end{bmatrix}$.

We make use of β as defined in (11.2) to carry out the transformation:

$$Z = \beta E, \tag{11.23}$$

which converts (11.22) into:

$$\dot{Z} = (\delta I_n + A)Z + \beta B\left(f(x,p) + gu - y_d^{(n)}\right), \tag{11.24}$$

where $\delta = \beta^{-1}\dot{\beta}$ and I_n is the unit matrix.

By adding and subtracting $BK(\delta)Z$ in the right-hand side of (11.24), we arrive at:

$$\dot{Z} = (\delta I_n + A + BK(\delta))Z + \beta Bgu + \beta B\Delta_z(\cdot), \tag{11.25}$$

where $K(\delta) = [k_1(\delta), \cdots, k_n(\delta)] \in \mathbb{R}^{1 \times n}$ is chosen by the designer and $\Delta_z(\cdot) = f(x,p) - y_d^{(n)} - K(\delta)E$. In addition, let $\bar{A}(\delta) = \delta I_n + A + BK(\delta)$.

To facilitate the control design, the following assumptions are imposed.

Assumption 11.2 *There exists a triple of vector/matrix-valued functions $(K(\delta)$, $P(\delta)$, $Q(\delta))$ such that $K(\delta) \in \mathscr{C}^1$, $P(\delta) \in \mathscr{C}^1$, and $Q(\delta) \in \mathscr{C}^1$ obey the following Lyapunov equation:*

$$\bar{A}(\delta)^\top P(\delta) + P(\delta)\bar{A}(\delta) = -Q(\delta), \tag{11.26}$$

where $P(\delta)$ and $Q(\delta)$ are symmetric and positive definite.

Assumption 11.3 *The quantities*

$$\underline{\lambda}_Q = \inf \lambda_{\min}(Q(\delta)) > 0, \ \bar{\lambda}_P = \sup \lambda_{\max}(P(\delta)) > 0, \ \underline{\lambda}_P = \inf \lambda_{\min}(P(\delta)) > 0$$

exist, where $\lambda_{\min}(\cdot)$ and $\lambda_{\max}(\cdot)$ denote the minimum and maximum eigenvalues, respectively, of the symmetric positive definite matrix.

Remark 11.1 *Using pole placement to place the eigenvalues of $\bar{A}$ at $-\rho$ ($\rho > 0$), it is easily obtained the detailed expression of $K(\delta)$. Note that $Q(\delta)$ are selected and computed by the designer, then a corresponding expression of $P(\delta)$ satisfying (11.26) can be achieved, which implies that Assumption 11.2 is reasonable, and similar suppose is also used in reference [39]. Furthermore, as $\delta \in \mathscr{L}_\infty$, then Assumption 11.3 can be ensured. It should be stressed that constants $\underline{\lambda}_Q$, $\underline{\lambda}_P$, and $\bar{\lambda}_P$ are used for stability analysis, not for control design, thus there is no need for analytical estimation or computation of such parameters.*

The regular tracking error feedback-based practical prescribed-time controller is designed as:

$$\begin{cases} u = -\hat{b}\Phi\beta^2 E^\top P(\delta)B, \\ \dot{\hat{b}} = \gamma\Phi\beta^4 \left(E^\top P(\delta)B\right)^2 - \sigma\hat{b}, \ \hat{b}(0) \geq 0, \end{cases} \tag{11.27}$$

where γ and σ are positive parameters chosen by the designer, $\hat{b}$ is the estimate of the virtual parameter $b = \max\left\{1, a^2\right\}$, $\hat{b}(0) \geq 0$ is the arbitrarily chosen initial estimate, and

$$\Phi = \phi^2 + \left(y_d^{(n)}\right)^2 + (K(\delta)E)^2$$

is a scalar and readily computable function.

Theorem 11.3 *Consider the nonlinear system (10.21). Let Assumptions 10.7–10.9, 11.1–11.3 hold. If the practical prescribed-time control scheme (11.27) is applied, then the following objectives are achieved:*

(i) *all signals in the closed-loop systems are bounded and the control action is $\mathscr{C}^1$ smooth;*
(ii) *the tracking error converges to an adjustable compact set around zero within a prescribed time;*
(iii) *the settling time t_f is user-assignable, which is independent of system initial values and other design parameters.*

Proof. We concentrate on proving the system stability.

Constructing the Lyapunov function candidate as $V = \frac{1}{2}Z^\top P(\delta)Z + \frac{1}{2g\gamma}\tilde{b}^2$, with $\tilde{b} = b - g\hat{b}$ being the virtual parameter estimate error. Taking the derivative of V w.r.t. time along the dynamical model (11.25) yields:

$$\begin{aligned} \dot{V} &= \frac{1}{2}\dot{Z}^\top P(\delta)Z + \frac{1}{2}Z^\top P(\delta)\dot{Z} + \frac{1}{2}Z^\top \frac{\partial P(\delta)}{\partial \delta}\dot{\delta}Z - \frac{1}{\gamma}\tilde{b}\dot{\hat{b}} \\ &= \frac{1}{2}Z^\top \left(\delta I_n + A + BK(\delta)\right)^\top P(\delta)Z + \beta^2 E^\top P(\delta)B\Delta_z \\ &\quad + \frac{1}{2}Z^\top P(\delta)\left(\delta I_n + A + BK(\delta)\right)Z + \beta^2 E^\top P(\delta)Bgu \\ &\quad + \frac{1}{2}Z^\top \frac{\partial P(\delta)}{\partial \delta}\dot{\delta}Z - \frac{1}{\gamma}\tilde{b}\dot{\hat{b}}. \end{aligned} \tag{11.28}$$

Upon using Assumption 11.2, we get:

$$\begin{aligned} \dot{V} &= -\frac{1}{2}Z^\top Q(\delta)Z + \beta^2 E^\top P(\delta)B\Delta_z + \beta^2 E^\top P(\delta)Bgu \\ &\quad + \frac{1}{2}Z^\top \frac{\partial P(\delta)}{\partial \delta}\dot{\delta}Z - \frac{1}{\gamma}\tilde{b}\dot{\hat{b}}. \end{aligned} \tag{11.29}$$

According to Young's inequality, we have:

$$\beta^2 E^\top P(\delta) Bf(x,p) \leq a^2 \beta^4 (E^\top P(\delta)B)^2 \phi^2 + \frac{1}{4},$$

$$-\beta^2 E^\top P(\delta) B y_d^{(n)} \leq \beta^4 (E^\top P(\delta)B)^2 \left(y_d^{(n)}\right)^2 + \frac{1}{4},$$

$$-\beta^2 E^\top P(\delta) BKE \leq \beta^4 (E^\top P(\delta)B)^2 (KE)^2 + \frac{1}{4}.$$

Then it holds that:

$$\beta^2 E^\top P(\delta) B\Delta_z \leq b\Phi\beta^4 (E^\top P(\delta)B)^2 + 1$$

where $b = \max\{1, a^2\}$ is an unknown parameter and $\Phi = \phi^2(x) + \left(y_d^{(n)}\right)^2 + (K(\delta)E)^2$ is a computable function that can be used for control design. Then it further follows that:

$$\dot{V} \leq -\frac{1}{2} Z^\top Q(\delta)Z + b\Phi\beta^4 (E^\top P(\delta)B)^2 + 1 + \beta^2 E^\top P(\delta) Bgu$$
$$+ \frac{1}{2} Z^\top \frac{\partial P(\delta)}{\partial \delta} \dot{\delta}Z - \frac{1}{\gamma}\tilde{b}\dot{\hat{b}}. \tag{11.30}$$

Inserting the control law and adaptive law as given in (11.27) into (11.30) yields:

$$\dot{V} \leq -\frac{1}{2} Z^\top Q(\delta)Z + b\Phi\beta^4 (E^\top P(\delta)B)^2 + 1 - g\hat{b}\Phi\beta^4 \left(E^\top P(\delta)B\right)^2$$
$$+ \frac{1}{2} Z^\top \frac{\partial P(\delta)}{\partial \delta} \dot{\delta}Z - \frac{1}{\gamma}\tilde{b}\left(\gamma\Phi\beta^4 \left(E^\top P(\delta)B\right)^2 - \sigma\hat{b}\right)$$
$$\leq -\frac{1}{2} Z^\top Q(\delta)Z + \frac{1}{2} Z^\top \frac{\partial P(\delta)}{\partial \delta} \dot{\delta}Z + \frac{\sigma}{\gamma}\tilde{b}\hat{b} + 1$$
$$\leq -\frac{1}{2}\underline{\lambda}_Q\|Z\|^2 + \frac{1}{2} Z^\top \frac{\partial P(\delta)}{\partial \delta} \dot{\delta}Z - \frac{\sigma}{2g\gamma}\tilde{b}^2 + l_2, \tag{11.31}$$

where the facts that $0 < \underline{g} \leq g$, $\frac{\sigma}{\gamma}\tilde{b}\hat{b} \leq -\frac{\sigma}{2g\gamma}\tilde{b}^2 + \frac{\sigma}{2g\gamma}b^2$, and $\underline{\lambda}_Q\|Z\|^2 \leq Z^\top Q(\delta)Z$ are used and $l_2 = \frac{\sigma}{2g\gamma}b^2 + 1 < \infty$.

Note that $\frac{\partial P(\delta)}{\partial \delta}$ depends on δ, as δ and $\dot{\delta}$ are bounded and continuous, then $\frac{\partial P(\delta)}{\partial \delta}$ is bounded and admits a finite upper bound. We denote $M_{P\delta} = \sup\left\|\frac{\partial P(\delta)}{\partial \delta}\right\|$, $M_\delta = \sup\left|\dot{\delta}\right|$. Then we have:

$$Z^\top \frac{\partial P(\delta)}{\partial \delta} \dot{\delta}Z \leq \overline{M}\|Z\|^2,$$

where $\overline{M} = M_{P\delta} M_\delta$, then from (11.31) we get:

$$\dot{V} \leq -\frac{1}{2}\left(\underline{\lambda}_Q - \overline{M}\right)\|Z\|^2 - \frac{\sigma}{2g\gamma}\tilde{b}^2 + l_2.$$

Consequently, by choosing the symmetric and positive definite matrix $Q(\delta)$ appropriately such that:

$$\lambda_Q - \overline{M} > 0,$$

and denoting $l_q = \lambda_Q - \overline{M}$, and also noting that $Z^\top P(\delta)Z \le \overline{\lambda}_P \|Z\|^2$, we have:

$$\dot{V} \le -\frac{l_q}{2\overline{\lambda}_P} Z^\top P(\delta)Z - \frac{\sigma}{2\underline{g}\gamma}\tilde{b}^2 + l_2 \le -l_1 V + l_2, \qquad (11.32)$$

where $l_1 = \min\left\{ l_q/\overline{\lambda}_P, \sigma \right\}$. From (11.32), we establish the following important results:

(1) We first prove that Z, $\hat{b}$, E, and x are bounded. From (11.32) it is ensured that $V \in \mathcal{L}_\infty$, which indicates that $Z \in \mathcal{L}_\infty$ and $\hat{b} \in \mathcal{L}_\infty$. As $E = \beta^{-1}Z$ and β is bounded, then $E \in \mathcal{L}_\infty$, namely, $e^{(i)} \in \mathcal{L}_\infty$ $(i = 0,1,\cdots,n-1)$, which further implies that $x_i \in \mathcal{L}_\infty$ $(i = 1,2,\cdots,n)$.

(2) Next we prove that $\dot{E}$, $\dot{Z}$, $\dot{\hat{b}}$, $\dot{x}$, and the control signal u are bounded. By applying Assumption 10.9, it follows that $\phi \in \mathcal{L}_\infty$. Note that $K(\delta)$ and $P(\delta)$ are bounded as $\delta \in \mathcal{L}_\infty$, it holds that $\Phi \in \mathcal{L}_\infty$, then it is ensured that $u \in \mathcal{L}_\infty$ and $\dot{\hat{b}} \in \mathcal{L}_\infty$, therefore, it is seen that $\dot{x} = [\dot{x}_1, \cdots, \dot{x}_n]^\top$ is bounded, which is readily concluded from (11.22) and (11.25) that $\dot{E} \in \mathcal{L}_\infty$ and $\dot{Z} \in \mathcal{L}_\infty$.

(3) We further prove that $\dot{u}$ is bounded, namely, u is continuously differentiable for $t \ge 0$. To this end, we compute from (11.27) that:

$$\dot{u} = \frac{\partial u}{\partial \hat{b}}\dot{\hat{b}} + \frac{\partial u}{\partial \beta}\dot{\beta} + \frac{\partial u}{\partial \Phi}\dot{\Phi} + \frac{\partial u}{\partial E^\top P(\delta)B}\frac{d\left(E^\top P(\delta)B\right)}{dt}.$$

Note that all the signals in the closed-loop systems including E, $\dot{E}$, x, $\dot{x}$, $\hat{b}$, $\dot{\hat{b}}$, ϕ, β, δ, $\dot{\delta}$, Φ, $P(\delta)$, $K(\delta)$, $\frac{\partial K(\delta)}{\partial \delta}$, and $\frac{\partial P(\delta)}{\partial \delta}$ are continuously bounded, then it is readily established that $\dot{u} \in \mathcal{L}_\infty$, i.e., u is $\mathscr{C}^1$ smooth.

(4) We examine the tracking performance of the developed control strategy within a prescribed time. From (11.32), we have $V(t) \le \frac{l_2}{l_1} + V(0)$. Note that $\frac{1}{2}\underline{\lambda}_P \|Z\|^2 \le \frac{1}{2}Z^\top P(\delta)Z \le V$, then it is derived that $\|Z\| \le \sqrt{\frac{\frac{2l_2}{l_1} + 2V(0)}{\underline{\lambda}_P}} := \mathbb{B}_z$. Since $E = \beta^{-1}Z$ and $\left|e^{(i)}(t)\right| \le \|E\|$ $(i = 0,1,\cdots,n-1)$, by the definition of β as in (11.2), we further have:

$$\left|e^{(i)}(t)\right| \le (1 - b_f)\kappa \mathbb{B}_z + b_f \mathbb{B}_z, \ 0 \le t < t_f,$$

$$\left|e^{(i)}(t)\right| \le b_f \mathbb{B}_z, \ t \ge t_f,$$

for $i = 0,1,\cdots,n-1$, from which it is seen that the tracking error converges to the compact set

$$\Omega_{1i} := \left\{ e^{(i)} \in \mathbb{R} : \left|e^{(i)}\right| \le b_f \mathbb{B}_z \right\}, \ i = 0,1,\cdots,n-1,$$

at the decay rate no less than κ within the prescribed time t_f. Since b_f is a free design parameter, it can be selected arbitrarily small to render the compact set Ω_{1i} as small as desired. Meanwhile, it should be emphasized that the finite time t_f is completely at the designer's disposal, regardless of the other design parameters and the initial value of the Lyapunov function.

Furthermore, note that $\hat{b}(0)$ can be set as 0, then from (11.27) the initial control signal $u(0)$ is $u(0) = 0$, thus the excessively large initial control signal is avoided. The proof is completed.

Remark 11.2 *In the proposed control method, as matrix $Q(\delta)$ is chosen by the designer and $\frac{\partial P(\delta)}{\partial \delta}$ and $\dot{\delta}$ are computable, then a rough value of $\overline{M}$ can be done easily by using MATLAB Software for example. Thus, we can appropriately adjust the symmetric and positive definite matrix $Q(\delta)$ such that the inequality $\underline{\lambda}_Q - \overline{M} > 0$ holds.*

Remark 11.3 *Upon comparing the fractional power error feedback-based practical finite-time control in Section 10.4 with the proposed regular error feedback-based practical prescribed-time control (11.27), we have the following comments.*

(1) *In the fractional power error feedback-based practical finite-time control, except for adjusting the design parameters c, γ, and σ, the designers can choose the fractional power h $(0 < h < 1)$ properly to obtain better transient performance and to avoid the large control effort in the initial moment, however, the control scheme (10.25) is non-smooth and the analysis is complicated due to the introduction of the fractional power h. Whereas the proposed practical prescribed-time control (11.27) in Section 11.3.2 is built upon the regular error transformation, rather than the fractional power error feedback, rendering the control action not only continuous but also $\mathscr{C}^1$ smooth and making the analysis process simple.*

(2) *As clearly seen from the expression of T in (10.26), the actual finite time T depends on l, θ_0, α, as well as the initial value of the Lyapunov function V(0), as a result, such a finite time, although existing, cannot be specified by the designer directly, if the initial values are not available, the settling time will be difficult to obtain.*

(3) *It is worth noting that the proposed practical prescribed-time control (11.27) features with a special structure as it involves β^2 in its controller and β^4 in its adaptive law, respectively. Both β^2 and β^4 function as the soft accelerator to control gain and adaptation gain because they strictly but gently increase from 1 at t = 0 to the upper bound at t = t_f, it is such a feature that gives rise to the desirable control performance.*

Remark 11.4 *It should be emphasized that, although the fixed-time control also has the appealing features of a faster convergence rate, better disturbance rejection, and stronger robustness against uncertainties [24, 75, 143], the control input is still non-smooth and the settling time T in fixed-time control still cannot be pre-assigned arbitrarily within any physically possible range because the upper bound of T is* *subject to certain restrictions. From this aspect, the proposed practical prescribed-time control essentially differs from the fixed-time control.*

11.3.3 NUMERICAL SIMULATION

To verify the effectiveness of the proposed regular tracking error feedback-based practical prescribed-time control method, the one-link manipulator is considered:

$$\mathscr{D}\dot{x}_2 + \mathscr{B}x_2 + \mathscr{N}\sin(x_1) = u,$$

where $x_1 \in \mathbb{R}$ and $x_2 \in \mathbb{R}$ denote the link angular position and velocity, respectively, and $u \in \mathbb{R}$ is the control input. The parameter values with appropriate units are given by $\mathscr{D} = 1$, $\mathscr{B} = 1$, and $\mathscr{N} = 10$. Noting that $\dot{x}_1 = x_2$, then we have:

$$\begin{cases} \dot{x}_1 = x_2, \\ \dot{x}_2 = f(x,p) + g(x)u, \end{cases} \tag{11.33}$$

where $f(\cdot) = -\frac{\mathscr{B}}{\mathscr{D}}x_2 - \frac{\mathscr{N}}{\mathscr{D}}\sin(x_1)$ and $g(x) = \frac{1}{\mathscr{D}}$. According to (11.33), we have $A = \begin{bmatrix} 0 & 1 \\ 0 & 0 \end{bmatrix}$, $B = [0,1]^\top$, thus $\bar{A} = \begin{bmatrix} \delta & 1 \\ k_1(\delta) & k_2(\delta)+\delta \end{bmatrix}$.

Using pole placement to place the eigenvalues of $\bar{A}$ at $-\rho$ and $-\rho$ ($\rho > 0$), we solve the equation $\det(\bar{A}+\rho I) = 0$ to get $K(\delta) = [k_1(\delta), k_2(\delta)]$ with $k_1(\delta) = -(\rho + \delta)^2$ and $k_2(\delta) = -2(\rho+\delta)$. By selecting $Q(\delta) = \begin{bmatrix} \eta & 0 \\ 0 & \eta \end{bmatrix}$ with η being any positive number, then from (11.26) it is obtained that:

$$P(\delta) = \begin{bmatrix} p_1(\delta) & p_2(\delta) \\ p_2(\delta) & p_3(\delta) \end{bmatrix},$$

with

$$p_1(\delta) = \frac{\eta(2\rho(2\rho+\delta)+(\rho+\delta)^2+(\rho+\delta)^4)}{4\rho^3},$$

$$p_2(\delta) = \frac{\eta(2\rho+\delta+\delta(\rho+\delta)^2)}{4\rho^3},$$

$$p_3(\delta) = \frac{\eta(2\rho^3+2\rho+\delta+\delta(\rho+\delta)^2)}{4\rho^3(2+\delta)}.$$

Now we test the proposed regular tracking error feedback-based practical prescribed-time control (11.27) on the system (11.33), and make a comparative study with the fractional power error feedback-based finite-time control (10.25) in Section 10.4. In the simulation, the reference signal is $y_d = \sin(t)$ and the initial conditions are $x_1(0) = 0.3$, $x_2(0) = 0.5$, and $\hat{b}(0) = 0$. To verify the effectiveness of the proposed control, we carry out the following two tests.

Test 1. Comparison between the fractional power error feedback-based finite-time control in Section 10.4 and the regular tracking error feedback-based practical prescribed-time control in Section 11.3.2.

Under the proposed practical prescribed-time control method in Section 11.3.2, we choose $\varpi = \exp(-t)$, $\gamma = 0.05$, $\sigma = 0.5$, $b_f = 0.06$, $\eta = 4$, $\rho = 40$, $\varpi = \exp(-t)$, and $t_f = 3$. For the traditional fractional power error feedback-based finite-time control in Section 10.4, the design parameters are $c = 5$ or 10, $\lambda_1 = 1$ or 5, $\gamma = 0.001$, $\sigma = 0.5$, and $h = \frac{1}{3}$ or $\frac{3}{11}$. By using the same initial conditions, the simulation results are shown in Fig. 11.5–Fig. 11.8. It is clearly seen from Fig. 11.5 that, by using appropriate design parameters c, λ_1, and h, the tracking precision achieved by the fractional power error feedback-based finite-time control can be made comparable with that obtained by the proposed practical prescribed-time control, however, observed from Fig. 11.7 and Fig. 11.8, the control effort based on the fractional power finite-time control method is non-smooth (even chattering involves), whereas the proposed practical prescribed-time control method leads to smooth control action as seen in Fig. 11.6. In addition, the evolutions of parameter estimate $\hat{b}$ under different control methods are plotted in Fig. 11.9.

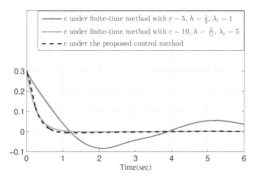

FIGURE 11.5 The tracking error $e(t)$ under the fractional power error feedback-based finite-time control (10.25) and proposed practical prescribed-time control (11.27).

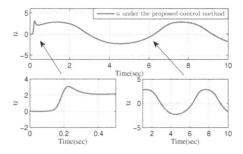

FIGURE 11.6 The control input $u(t)$ under the proposed practical prescribed-time control (11.27).

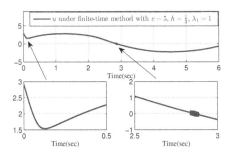

FIGURE 11.7 The control input u under the fractional power error feedback-based finite-time control (10.25) with $c = 5, h = \frac{1}{3}$, and $\lambda_1 = 1$.

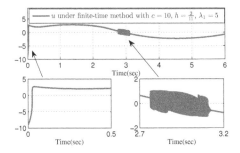

FIGURE 11.8 The control input u under the fractional power error feedback-based finite-time control (10.25) with $c = 10, h = \frac{3}{11}$, and $\lambda_1 = 5$.

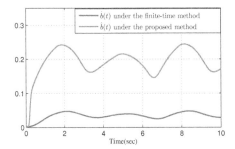

FIGURE 11.9 The parameter estimate $\hat{b}(t)$ under the fractional power error feedback-based finite-time control (10.25) with $c = 5, \lambda_1 = 1, \gamma = 0.001, \sigma = 0.5$, and $h = \frac{1}{3}$ and the proposed practical prescribed-time control (11.27).

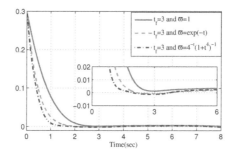

FIGURE 11.10   The evolution of the tracking error with the practical prescribed-time control (11.27) under different $\varpi(t)$.

Test 2. To verify that, under the practical prescribed-time control in Section 11.3.2, t_f and $\varpi(t)$ influence the transient performance of the tracking error.

We focus on testing that the selection of different settling time t_f and different $\varpi(t)$ can influence the convergence rate of the tracking error, we now use an example to verify this point. All design parameters and the initial states are the same, such as $\gamma = 0.05$, $\sigma = 0.5$, $b_f = 0.015$, $\eta = 10$, $\rho = 50$, except for t_f and $\varpi(t)$. We choose $t_f = 1.5, 3, 5$ seconds and $\varpi(t) = 1$, $\exp(-t)$, $4^{-t}(1+t^4)^{-1}$. The evaluation index is selected as $\Omega_{10} := \{e(t) \in \mathbb{R} : -0.01 < e(t) < 0.01\}$ and the simulation results are shown in Fig. 11.10 and Fig. 11.11, from which it is confirmed that different t_f and different $\varpi(t)$ can produce different tracking performance, and the tracking error can converge to the same compact set Ω_{10} within the prescribed time under different convergence rates, which confirms our theoretical prediction.

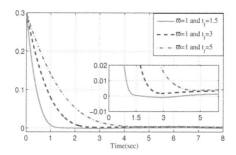

FIGURE 11.11   The evolution of the tracking error with the practical prescribed-time control (11.27) under different t_f.

11.4   PRACTICAL PRESCRIBED-TIME CONTROL OF STRICT-FEEDBACK SYSTEMS

11.4.1   PROBLEM FORMULATION AND PRELIMINARIES

In this section, we consider the following strict-feedback nonlinear systems in the presence of uncertainties:

$$\begin{cases} \dot{x}_k = g_k(\bar{x}_k)x_{k+1} + f_k(\theta_k, \bar{x}_k), \ k = 1, \cdots, n-1 \\ \dot{x}_n = g_n(\bar{x}_n)u + f_n(\theta_n, \bar{x}_n), \\ y = x_1, \end{cases} \tag{11.34}$$

where $x_1 \in \mathbb{R}, x_2 \in \mathbb{R}, \cdots, x_n \in \mathbb{R}$ are the system states and $\bar{x}_i = [x_1, \cdots, x_i]^\top \in \mathbb{R}^i, i = 1, 2, \cdots, n, u \in \mathbb{R}$ and $y \in \mathbb{R}$ are the control input and system output, respectively, $f_i \in \mathbb{R}$ is a smooth nonlinear function, $\theta_i \in \mathbb{R}^{r_i}$ represents an unknown constant parameter, and $g_i(\bar{x}_i) \in \mathbb{R}$ denotes the time-varying control coefficient.

Define $z_1 = x_1 - y_d$ as the tracking error with y_d being the reference signal. The control objective is to establish a practical prescribed-time tracking control strategy such that:

(i)   all signals in the closed-loop system are bounded for all time;
(ii) the tracking error converges to an adjustable residual region around zero in a prescribed time with an assignable decaying rate.

Assumption 11.4 *The system states are available for control design.*

Assumption 11.5 *The reference trajectory $y_d(t)$ and its derivatives up to $(n+1)$-th are known, bounded, and piecewise continuous.*

Assumption 11.6 *Certain crude structural information on $f_i(\theta_i, \bar{x}_i)$ is available to allow an unknown constant a_i and a known smooth function $\phi_i(\bar{x}_i)$ to be extracted, such that $|f_i(\theta_i, \bar{x}_i)| \le a_i \phi_i(\bar{x}_i)$ for $\forall t \ge 0$. In addition, $f_i(\theta_i, \bar{x}_i)$ is bounded if $\bar{x}_i$ is bounded.*

Assumption 11.7 *There exist positive constants $\underline{g}_i$ and $\bar{g}_i$ such that $0 < \underline{g}_i \le g_i(\bar{x}_i) \le \bar{g}_i < \infty$.*

11.4.2   CONTROL DESIGN

We employ the time-varying scaling function $\beta(t)$ as defined in (11.2) and construct the following coordinate transformation:

$$\begin{cases} \zeta = \beta(t)z_1, \\ z_1 = x_1 - y_d, \\ z_k = x_k - \alpha_{k-1}, \ k = 2, 3, \cdots, n, \end{cases} \tag{11.35}$$

with α_{k-1} being the virtual controller which will be given later.

Now we carry out the control design by using the backstepping technique.

Step 1. The derivative of $\dot{\zeta}$ is given by:

$$\dot{\zeta} = \dot{\beta}z_1 + \beta(f_1 + g_1z_2 + g_1\alpha_1 - \dot{y}_d), \tag{11.36}$$

where the fact that $x_2 = z_2 + \alpha_1$ is used, then the derivative of the quadratic function $\frac{1}{2}\zeta^2$ w.r.t time along (11.36) yields:

$$\zeta\dot{\zeta} = \zeta\dot{\beta}z_1 + \zeta\beta(g_1z_2 + g_1\alpha_1 + f_1 - \dot{y}_d) = \zeta\beta g_1\alpha_1 + \Xi_1 + \zeta\beta g_1z_2, \tag{11.37}$$

where $\Xi_1 = \zeta\dot{\beta}z_1 + \zeta\beta(f_1 - \dot{y}_d)$.

From Assumptions 11.6 and 11.7, with the aid of Young's inequality, it follows that:

$$\zeta\dot{\beta}z_1 \leq \underline{g}_1\zeta^2\dot{\beta}^2z_1^2 + \frac{1}{4\underline{g}_1}, \tag{11.38}$$

$$\zeta\beta f_1 \leq \underline{g}_1 a_1^2\zeta^2\beta^2\phi_1^2 + \frac{1}{4\underline{g}_1}, \tag{11.39}$$

$$-\zeta\beta\dot{y}_d \leq \underline{g}_1\zeta^2\beta^2\dot{y}_d^2 + \frac{1}{4\underline{g}_1}. \tag{11.40}$$

Therefore, we have:

$$\Xi_1 \leq \underline{g}_1 b_1\zeta^2\Phi_1 + \frac{3}{4\underline{g}_1},$$

where $b_1 = \max\{1, a_1^2\} > 0$ is an unknown constant parameter and $\Phi_1 = \dot{\beta}^2z_1^2 + \beta^2\phi_1^2 + \beta^2\dot{y}_d^2 \geq 0$ is a computable function. Then (11.37) becomes:

$$\zeta\dot{\zeta} \leq \zeta\beta g_1\alpha_1 + \underline{g}_1 b_1\zeta^2\Phi_1 + \frac{3}{4\underline{g}_1} + \zeta\beta g_1z_2. \tag{11.41}$$

We define the virtual controller α_1 with the adaptive law as:

$$\begin{cases} \alpha_1 = -c_1z_1 - \hat{b}_1z_1\Phi_1, \\ \dot{\hat{b}}_1 = \gamma_1\zeta^2\Phi_1 - \gamma_1\hat{b}_1, \quad \hat{b}_1(0) \geq 0, \end{cases} \tag{11.42}$$

where c_1, γ_1, and σ_1 are positive design parameters, and $\hat{b}_1$ is the estimate of the unknown parameter $\hat{b}_1$.

According to the definition of α_1 as given in (11.42), the term $\zeta\beta g_1\alpha_1$ can be expressed as:

$$\zeta\beta g_1\alpha_1 = -\zeta\beta g_1(c_1z_1 + \hat{b}_1z_1\Phi_1) \leq -\underline{g}_1c_1\zeta^2 - \underline{g}_1\hat{b}_1\zeta^2\Phi_1, \tag{11.43}$$

where the facts that $0 < \underline{g}_1 \leq g_1$ and $\hat{b}_1(t) \geq 0$ are used. Then (11.41) further becomes:

$$\zeta\dot{\zeta} \leq -\underline{g}_1c_1\zeta^2 + \underline{g}_1\tilde{b}_1\zeta^2\Phi_1 + \frac{3}{4\underline{g}_1} + \zeta\beta g_1z_2, \tag{11.44}$$

where $\tilde{b}_1 = b_1 - \hat{b}_1$ denotes the estimate error.

Choosing the Lyapunov function candidate as $V_1 = \frac{1}{2}\zeta^2 + \frac{g_1}{2\gamma_1}\tilde{b}_1^2$, then the derivative of V_1 along (11.44) yields:

$$\dot{V}_1 \leq -\underline{g}_1 c_1 \zeta^2 + \underline{g}_1 \tilde{b}_1 \zeta^2 \Phi_1 + \frac{3}{4\underline{g}_1} + \zeta\beta g_1 z_2 - \frac{g_1}{\gamma_1}\tilde{b}_1\dot{\hat{b}}_1. \tag{11.45}$$

Substituting the adaptive law $\dot{\hat{b}}_1$ as shown in (11.42) into (11.45), one has:

$$\dot{V}_1 \leq -\underline{g}_1 c_1 \zeta^2 + \frac{3}{4\underline{g}_1} + \zeta\beta g_1 z_2 + \frac{g_1 \sigma_1}{\gamma_1}\tilde{b}_1\hat{b}_1$$

$$\leq -\underline{g}_1 c_1 \zeta^2 - \frac{g_1 \sigma_1}{2\gamma_1}\tilde{b}_1^2 + \Pi_1 + \zeta\beta g_1 z_2, \tag{11.46}$$

where $\Pi_1 = \frac{g_1 \sigma_1}{2\gamma_1}b_1^2 + \frac{3}{4\underline{g}_1} > 0$ is an unknown constant, and the term $\zeta\beta g_1 z_2$ will be handled in Step 2.

Step 2. We first clarify the arguments of the function α_1. By examining (11.42) along with the definitions of z_1 and Φ_1, we see that α_1 is a function of x_1, y_d, $\dot{y}_d$, β, $\dot{\beta}$, and $\hat{b}_1$. Differentiating $z_2 = x_2 - \alpha_1$ with the help of $x_3 = z_3 + \alpha_2$ yields:

$$\dot{z}_2 = g_2(z_3 + \alpha_2) + f_2 + \ell_1 - \frac{\partial\alpha_1}{\partial x_1}(g_1 x_2 + f_1),$$

where

$$\ell_1 = -\frac{\partial\alpha_1}{\partial\hat{b}_1}\dot{\hat{b}}_1 - \sum_{k=0}^{1}\frac{\partial\alpha_1}{\partial\beta^{(k)}}\beta^{(k+1)} - \sum_{k=0}^{1}\frac{\partial\alpha_1}{\partial y_d^{(k)}}y_d^{(k+1)}$$

is available for control design. Then the derivative of the quadratic function $V_{21} = V_1 + \frac{1}{2}z_2^2$ is:

$$\dot{V}_{21} \leq -\underline{g}_1 c_1 \zeta^2 - \frac{g_1 \sigma_1}{2\gamma_1}\tilde{b}_1^2 + \Pi_1 + g_2 z_2 \alpha_2 + g_2 z_2 z_3 + \Xi_2, \tag{11.47}$$

where

$$\Xi_2 = \beta\zeta g_1 z_2 + z_2 f_2 + z_2 \ell_1 - z_2\frac{\partial\alpha_1}{\partial x_1}(g_1 x_2 + f_1)$$

is the lumped uncertain function.

Upon using the same procedures in (11.38)–(11.40), we have:

$$\beta\zeta g_1 z_2 \leq \underline{g}_2\beta^2\zeta^2 z_2^2 + \frac{\bar{g}_1^2}{4\underline{g}_2},$$

$$z_2 f_2 \leq \underline{g}_2 a_2^2 z_2^2 \phi_2^2 + \frac{1}{4\underline{g}_2},$$

$$z_2\ell_1 \leq \underline{g}_2 z_2^2 \ell_1^2 + \frac{1}{4\underline{g}_2},$$

$$-z_2\frac{\partial\alpha_1}{\partial x_1}g_1 x_2 \leq \underline{g}_2 z_2^2\left(\frac{\partial\alpha_1}{\partial x_1}x_2\right)^2 + \frac{\bar{g}_1^2}{4\underline{g}_2},$$

$$-z_2\frac{\partial\alpha_1}{\partial x_1}f_1 \leq \underline{g}_2 a_1^2 z_2^2\left(\frac{\partial\alpha_1}{\partial x_1}\phi_1\right)^2 + \frac{1}{4\underline{g}_2}.$$

Therefore, Ξ_2 can be upper bounded by:

$$\Xi_2 \le \underline{g}_2 b_2 z_2^2 \Phi_2 + \frac{3}{4\underline{g}_2} + \frac{\bar{g}_1^2}{2\underline{g}_2},$$

where $b_2 = \max\{1, a_1^2, a_2^2\}$ is an unknown constant parameter, and

$$\Phi_2 = \beta^2 \zeta^2 + \phi_2^2 + \ell_1^2 + \left(\frac{\partial \alpha_1}{\partial x_1} x_2\right)^2 + \left(\frac{\partial \alpha_1}{\partial x_1} \phi_1\right)^2$$

is a computable function. Then (11.47) becomes:

$$\dot{V}_{21} \le -\underline{g}_1 c_1 \zeta^2 - \frac{\underline{g}_1 \sigma_1}{2\gamma_1} \tilde{b}_1^2 + \Pi_1 + g_2 z_2 \alpha_2 + g_2 z_2 z_3 + \underline{g}_2 b_2 z_2^2 \Phi_2 + \frac{3}{4\underline{g}_2} + \frac{\bar{g}_1^2}{2\underline{g}_2}. \quad (11.48)$$

We define the virtual controller α_2 and adaptive law for $\hat{b}_2$ as:

$$\begin{cases} \alpha_2 = -c_2 z_2 - \hat{b}_2 z_2 \Phi_2, \\ \dot{\hat{b}}_2 = \gamma_2 z_2^2 \Phi_2 - \sigma_2 \hat{b}_2, \quad \hat{b}_2(0) \ge 0, \end{cases} \quad (11.49)$$

where c_2, γ_2, and σ_2 are positive design parameters, $\hat{b}_2$ is the estimate of b_2, $\hat{b}_2(0)$ is the arbitrarily given initial value.

Let $V_2 = V_{21} + \frac{1}{2\gamma_2} \tilde{b}_2^2$, then the derivative of V_2 is:

$$\dot{V}_2 \le -\underline{g}_1 c_1 \zeta^2 - \frac{\underline{g}_1 \sigma_1}{2\gamma_1} \tilde{b}_1^2 + \Pi_1 + g_2 z_2 \alpha_2 + g_2 z_2 z_3 + \underline{g}_2 b_2 z_2^2 \Phi_2$$

$$+ \frac{3}{4\underline{g}_2} + \frac{\bar{g}_1^2}{2\underline{g}_2} - \frac{1}{\gamma_2} \tilde{b}_2 \dot{\hat{b}}_2. \quad (11.50)$$

Substituting the virtual controller (11.49) with the adaptive law into (11.50), one has:

$$\dot{V}_2 \le -\underline{g}_1 c_1 \zeta^2 - \underline{g}_2 c_2 z_2^2 - \sum_{k=1}^{2} \frac{\underline{g}_k \sigma_k}{2\gamma_k} \tilde{b}_k^2 + \Pi_2 + g_2 z_2 z_3, \quad (11.51)$$

where $\Pi_2 = \Pi_1 + \frac{3}{4\underline{g}_2} + \frac{\bar{g}_1^2}{2\underline{g}_2} + \frac{\underline{g}_2 \sigma_2}{2\gamma_2} b_2^2 > 0$ is an unknown constant.

Step i $(i = 3, 4, \cdots, n)$. For clarity, let

$$\bar{x}_i = [x_1, \cdots, x_i]^\top, \quad \bar{y}_d^{(i)} = \left[y_d, y_d^{(1)} \cdots, y_d^{(i)}\right]^\top,$$

$$\overline{\beta}^{(i)} = \left[\beta, \beta^{(1)}, \cdots, \beta^{(i)}\right]^\top, \quad \bar{b}_i = [\hat{b}_1, \hat{b}_2, \cdots, \hat{b}_i]^\top,$$

for $i = 1, \cdots, n$, where $\hat{b}_i$ is the estimate of the unknown constant parameter b_i as defined in (11.54).

Note that α_i, $i = 1, 2, \cdots, n$, is a function of $\bar{x}_i$, $\bar{y}_d^{(i)}$, $\bar{\beta}^{(i)}$, and $\hat{b}_i$, then we have:

$$\dot{\alpha}_i = \sum_{k=1}^i \frac{\partial \alpha_i}{\partial x_k}(g_k x_{k+1} + f_k) - \ell_i,$$

with $\ell_i = -\sum_{k=0}^i \frac{\partial \alpha_i}{\partial \beta^{(k)}} \beta^{(k+1)} - \sum_{k=0}^i \frac{\partial \alpha_i}{\partial y_d^{(k)}} y_d^{(k+1)} - \sum_{k=1}^i \frac{\partial \alpha_i}{\partial \hat{b}_k} \dot{\hat{b}}_k$ being computable for controller design.

Since $z_i = x_i - \alpha_{i-1}$, then the derivative of z_i is:

$$\dot{z}_i = g_i z_{i+1} + g_i \alpha_i + f_i + \ell_{i-1} - \sum_{k=1}^{i-1} \frac{\partial \alpha_{i-1}}{\partial x_k}(g_k x_{k+1} + f_k), \tag{11.52}$$

where $z_{n+1} = 0$, $\alpha_n = u$.

Choosing the Lyapunov function candidate as:

$$V_i = \frac{1}{2} \sum_{k=1}^i z_k^2 + \sum_{k=1}^i \frac{g_k}{2\gamma_k} \tilde{b}_k^2,$$

where $\tilde{b}_k = b_k - \hat{b}_k$ is the estimate error and $\gamma_k > 0$ is the design parameter, then the derivative of V_i w.r.t. time along (11.52) yields:

$$\dot{V}_i \leq -\underline{g}_1 c_1 \zeta^2 - \sum_{k=2}^{i-1} \underline{g}_k c_k z_k^2 - \sum_{k=1}^{i-1} \frac{g_k \sigma_k}{2\gamma_k} \tilde{b}_k^2 + \Pi_{i-1} + g_i z_i z_{i+1} + g_i z_i \alpha_i$$

$$+ \Xi_i - \frac{g_i}{\gamma_i} \tilde{b}_i \dot{\hat{b}}_i, \tag{11.53}$$

where

$$\Xi_i = g_{i-1} z_{i-1} z_i + z_i f_i - z_i \sum_{k=1}^{i-1} \frac{\partial \alpha_{i-1}}{\partial x_k}(f_k + g_k x_{k+1}) + z_i \ell_{i-1}$$

is the lumped uncertainty.

Similar to (11.38)–(11.40), we have:

$$\Xi_i \leq \underline{g}_i b_i z_i^2 \Phi_i + \frac{i+1}{4\underline{g}_i} + \sum_{k=1}^{i-1} \frac{\bar{g}_k^2}{4\underline{g}_i} + \frac{\bar{g}_{i-1}^2}{4\underline{g}_i},$$

with

$$b_i = \max\{1, a_1^2, \cdots, a_i^2\}, \tag{11.54}$$

$$\Phi_i = z_{i-1}^2 + \phi_i^2 + \sum_{k=1}^{i-1} \left(\frac{\partial \alpha_{i-1}}{\partial x_k} x_{k+1}\right)^2 + \sum_{k=1}^{i-1} \left(\frac{\partial \alpha_{i-1}}{\partial x_k} \phi_k\right)^2 + \ell_{i-1}^2. \tag{11.55}$$

Therefore, (11.53) becomes:

$$\dot{V}_i \leq -\underline{g}_1 c_1 \zeta^2 - \sum_{k=2}^{i-1} \underline{g}_k c_k z_k^2 - \sum_{k=1}^{i-1} \frac{g_k \sigma_k}{2\gamma_k} \tilde{b}_k^2 + \Pi_{i-1} + g_i z_i z_{i+1} + g_i z_i \alpha_i$$

$$+ \underline{g}_i b_i z_i^2 \Phi_i + \frac{i+1}{4\underline{g}_i} + \sum_{k=1}^{i-1} \frac{\bar{g}_k^2}{4\underline{g}_i} + \frac{\bar{g}_{i-1}^2}{4\underline{g}_i} - \frac{g_i}{\gamma_i} \tilde{b}_i \dot{\hat{b}}_i. \tag{11.56}$$

The virtual controller and actual controller with the adaptive laws are given explicitly as:

$$\begin{cases} \alpha_i = -c_i z_i - \hat{b}_i z_i \Phi_i, \\ u = \alpha_n, \\ \dot{\hat{b}}_i = \gamma_i z_i^2 \Phi_i - \sigma_i \hat{b}_i, \quad \hat{b}(0) \geq 0, \end{cases} \tag{11.57}$$

where c_i, γ_i, and σ_i are positive design parameters.

Now we are ready to present the following theorem and its proof.

11.4.3   THEOREM AND STABILITY ANALYSIS

Theorem 11.4 *Consider the strict-feedback nonlinear systems with non-parametric uncertainties as described in (11.34). Under Assumptions 11.1, 11.4–11.7, if the control algorithm (11.57) is applied, it is ensured that the objectives (i) and (ii) are achieved.*

Proof. Inserting the virtual/actual controller α_i and adaptive law $\hat{b}_i$ as defined in (11.57) into (11.56), we have:

$$\dot{V}_i \leq -\underline{g}_1 c_1 \zeta^2 - \sum_{k=2}^{i} \underline{g}_k c_k z_k^2 - \sum_{k=1}^{i} \frac{\underline{g}_k \sigma_k}{2\gamma_k} \tilde{b}_k^2 + \Pi_i + g_i z_i z_{i+1}, \tag{11.58}$$

where $\Pi_i = \Pi_{i-1} + \frac{i+1}{4\underline{g}_i} + \sum_{k=1}^{i-1} \frac{\bar{g}_k^2}{4\underline{g}_i} + \frac{\bar{g}_{i-1}^2}{4\underline{g}_i} + \frac{\underline{g}_i \sigma_i}{2\gamma_i} b_i^2 > 0$ is an unknown constant. When $i = n$, it holds that $z_{n+1} = 0$, then (11.58) can be rewritten as:

$$\dot{V}_n \leq -\underline{g}_1 c_1 \zeta^2 - \sum_{k=2}^{n} \underline{g}_k c_k z_k^2 - \sum_{k=1}^{n} \frac{\underline{g}_k \sigma_k}{2\gamma_k} \tilde{b}_k^2 + \Pi_n \leq -l_1 V_n + l_2, \tag{11.59}$$

where $l_1 = \min\{2\underline{g}_k c_k, \sigma_k\} > 0$ with $k = 1, 2, \cdots, n$, and $l_2 = \Pi_n > 0$.

Now, we first prove that all signals in the closed-loop systems are bounded. With (11.59) we conclude that $V(t)$ is bounded, which implies that z_k and $\tilde{b}_k$ are bounded for $k = 1, 2, \cdots, n$. As $\tilde{b}_k = b_k - \hat{b}_k$ and b_k is an unknown constant, then $\hat{b}_k$ is bounded. Noting that the reference signal y_d is bounded, then the system state x_1 is bounded, which further indicates that ϕ_1 is bounded. Due to the boundedness of β and $\dot{\beta}$, then it shows that Φ_1 is bounded, then it is seen from (11.42) that $\alpha_1 \in \mathscr{L}_\infty$ and $\hat{b}_1 \in \mathscr{L}_\infty$. In this manner, we can recursively establish that the system states $x_i \in \mathscr{L}_\infty$, $i = 2, 3, \cdots, n$, $\phi_i \in \mathscr{L}_\infty$, $\Phi_i \in \mathscr{L}_\infty$, the virtual control law $\alpha_j \in \mathscr{L}_\infty$, $j = 2, 3, \cdots, n-1$, the actual control law $u \in \mathscr{L}_\infty$, and the adaptive law $\hat{b}_i \in \mathscr{L}_\infty$. Therefore, all signals in the closed-loop systems are bounded.

Next we focus on proving that the tracking error converges to an adjustable region around zero in a prescribed finite time with an assignable decay rate. It is seen from (11.59) that $V(t) \leq (V(0) - \frac{l_2}{l_1}) \exp(-l_1 t) + \frac{l_2}{l_1}$, then one has $\zeta^2(t) \leq 2V(t) \leq 2V(0) + \frac{2l_2}{l_1}$, namely, $|\zeta(t)| \leq \sqrt{2V(0) + \frac{2l_2}{l_1}} := \mathbb{B}_\zeta$. As $z_1 = \beta^{-1}\zeta$, then we have $|z_1(t)| \leq$

$\beta^{-1}|\zeta(t)|$. Combined with the expression of β as defined in (11.2), we obtain:

$$|z_1(t)| < (1-b_f)\left(\frac{t_f-t}{t_f}\right)^{n+2}\varpi\mathbb{B}_\zeta + b_f\mathbb{B}_\zeta, \text{ for } 0 \le t < t_f, \tag{11.60}$$

$$|z_1(t)| < b_f\mathbb{B}_\zeta, \text{ for } t \ge t_f. \tag{11.61}$$

It is seen from (11.60) and (11.61) that, on the one hand, the tracking error converges to an adjustable residual region $\Omega_z := \{z_1(t) : |z_1(t)| < b_f\mathbb{B}_\zeta\}$ in a prescribed finite time $t = t_f$; on the other hand, the tracking error converges to the residual region with the decay rate no less than $\left(\frac{t_f-t}{t_f}\right)^{n+2}\varpi$, which implies that both t_f and ϖ can influence the convergence rate of the tracking error. The proof is completed.

11.4.4 NUMERICAL SIMULATION

To verify the effectiveness of the proposed practical prescribed-time control scheme in Section 11.4.2, here we consider the following strict-feedback nonlinear system:

$$\begin{cases} \dot{x}_1 = g_1(x_1)x_2 + f_1(p_1,x_1), \\ \dot{x}_2 = g_2(\bar{x}_2)u + f_2(p_2,\bar{x}_2), \end{cases}$$

where $f_1 = \sin(\theta_1 x_1)x_1$ and $f_2 = \theta_{21}\sin(\theta_{22}x_1)x_1x_2$ with $\theta_1 = 1$ and $\theta_2 = [\theta_{21},\theta_{22}]^\top = [0.5,1]^\top$. The control coefficients are chosen as $g_1(x_1) = 2+0.2\cos(x_1)$ and $g_2(\bar{x}_2) = 2+0.3\sin(x_1x_2)$. It should be emphasized that the values of θ_i and g_i ($i = 1,2$) are unavailable for control design.

In the simulation, the reference signal is $y_d = \sin(t)$. The initial conditions are given as $x_1(0) = 1$, $x_2(0) = 0.5$, and $\hat{b}_1(0) = \hat{b}_2(0) = 0$. The design parameters are chosen as $c_1 = 2$, $c_2 = 2$, $\gamma_1 = 0.05$, $\gamma_2 = 0.05$, $\sigma_1 = 0.5$, and $\sigma_2 = 0.2$. For the time-varying scaling function, the prescribed time is given as $t_f = 3$, $\varpi(t) = 1$, and $b_f = 0.05$. Under the proposed practical prescribed-time control algorithm, the simulation results are shown in Fig. 11.12–Fig. 11.15. It is seen from Fig. 11.12 and Fig. 11.13 that an adjustable tracking performance is achieved. Furthermore, the evolutions of the control input u and parameter estimates $\hat{b}_k$ ($k = 1,2$) are plotted in Fig. 11.14 and Fig. 11.15, respectively, which are bounded for all time.

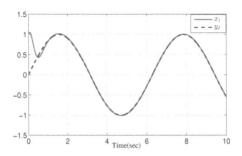

FIGURE 11.12 The evolutions of the system output x_1 and reference signal y_d under the proposed practical prescribed-time control (11.57).

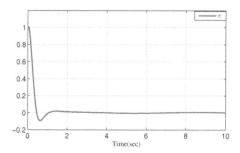

FIGURE 11.13 The evolution of the tracking error z_1 under the proposed practical prescribed-time control (11.57).

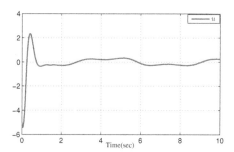

FIGURE 11.14 The evolution of the control input u under the proposed practical prescribed-time control (11.57).

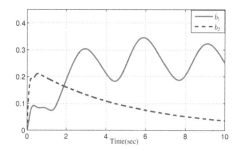

FIGURE 11.15 The evolutions of the parameter estimates $\hat{b}_1$ and $\hat{b}_2$ under the proposed practical prescribed-time control (11.57).

11.5   NOTES

Current fractional power error feedback-based finite-time control methods normally result in non-smooth control action and non-specifiable finite convergence time. In this chapter, we present a new control approach to achieve practical prescribed-time tracking for a class of nonlinear systems. The developed control exhibits several features compared with the fractional power error feedback-based finite-time controls. Firstly, we make use of a prescribed-time function to perform error transformation, with which we build the control scheme upon the transferred error, ensuring that the tracking error converges to an adjustable small residual set within a prescribed time, with a bounded, continuous and $\mathscr{C}^1$ control action. Secondly, as the proposed control is directly built upon the regular feedback of the transferred error, the settling time is independent of the initial conditions and other design parameters, then it can be pre-specified, which is essentially different from the fractional power error feedback-based finite-time controls. Furthermore, we also extend the practical prescribed-time control to strict-feedback nonlinear systems.

12 Prescribed-time Stabilization Control

We illustrated the practical prescribed-time control for nonlinear systems in Chapter 11. For the nonlinear systems with non-vanishing uncertainty, if the corresponding controls in Chapter 11 were used for stabilization control, they ensure that the system state converges to a compact set around zero (rather than the equilibrium) in a prescribed time. Therefore, in this chapter, we focus on introducing a prescribed-time stabilization control algorithm for nonlinear systems in the presence of non-vanishing uncertainty. Notice that the original idea for prescribed-time control is derived from the proportional navigation law in tactical and strategic missile guidance [144], which has been shown to be an optimal feedback control law [72].

To help with the understanding of the fundamental idea of prescribed-time stabilization control, this chapter first introduces a motivating example in Section 12.1, and then we give a systematic comparison among the finite-time control, fixed-time control, and prescribed-time control in Section 12.2. Furthermore, we give the detailed control design about prescribed-time control for normal-form nonlinear systems without/with uncertainty in Section 12.3 and Section 12.4, respectively. The simulation verification is presented in Section 12.5. Finally, some discussions are shown in Section 12.6.

12.1 MOTIVATING EXAMPLE OF PRESCRIBED-TIME CONTROL

To gain insight into the idea of using time-varying scaling to achieve prescribed-time control, we first consider the following single-integrator system:

$$\dot{x} = u, \tag{12.1}$$

where $x \in \mathbb{R}$ and $u \in \mathbb{R}$ are the system state and control input, respectively. It is worth mentioning that, for the prescribed-time control proposed in this section, neither the plant or its solutions nor the control law is expected to exist beyond the prescribed time, and all the stability analysis made for the closed loop is valid only in that time interval.

Before giving the concept of prescribed-time stability, we introduce the following special monotonically increasing function:

$$\mu_1(t) = \frac{t_f}{t_f - t}, \quad t \in [0, t_f), \tag{12.2}$$

where $t_f > 0$ with the properties that $\mu_1(0) = 1$ and $\mu_1(t_f) = +\infty$. Furthermore, the function $\mu_1(t) - 1 = \frac{t}{t_f - t}$ starts from zero at $t = 0$ and grows monotonically to infinity as $t \to t_f$.

DOI: 10.1201/9781003474364-12

Lemma 12.1 *[18] Consider the function*

$$\mu(t) = \frac{t_f^{n+m}}{(t_f - t)^{n+m}} = \mu_1(t)^{n+m}, \quad \mu_1 = \frac{t_f}{t_f - t} \tag{12.3}$$

on $[0, t_f)$, with positive integers m and n (n denotes the order of the considered plant). If a continuously differentiable function $V : [0, t_f) \to [0, +\infty)$ satisfies:

$$\dot{V}(t) \le -2c\mu(t)V(t) + \frac{\mu(t)}{4\lambda}d(t)^2 \tag{12.4}$$

for positive constants c and λ and bounded function $d(t)$, then

$$V(t) \le \zeta(t)^{2c}V(0) + \frac{\|d\|_{[0,t]}^2}{8c\lambda}, \quad \forall t \in [0, t_f), \tag{12.5}$$

where $\|d\|_{[0,t]} := \sup_{\tau \in [0,t]} |d(\tau)|$ and ζ is the monotonically decreasing function with the properties that $\zeta(0) = 1$ and $\zeta(t_f) = 0$, which is expressed as follows:

$$\zeta(t) = \exp\left(\frac{t_f}{m+n-1}\left(1 - \mu_1(t)^{m+n-1}\right)\right).$$

For the first-order system, the time-varying scaling function utilized is of the form:

$$\mu(t) = \frac{t_f^{1+m}}{(t_f - t)^{1+m}}, \quad t \in [0, t_f), \tag{12.6}$$

with $t_f > 0$.

With such μ, we conduct the following simple transformation:

$$w = \mu x$$

on $t \in [0, t_f)$. Then, the prescribed-time controller based on the time-varying gain is of the form:

$$u = -c\mu x, \tag{12.7}$$

with

$$c = c_0 + \frac{1+m}{t_f} > 0, \tag{12.8}$$

with $c_0 > 0$, which leads to the following result.

Theorem 12.1 *For the single-integrator system (12.1), if the prescribed-time controller (12.7) with (12.8) is applied, then*

$$\begin{cases} |x(t)| \le \mu^{-1} \exp\left(-c_0 \frac{t_f}{m}(\mu^{\frac{m}{1+m}} - 1)\right)|x(0)|, & \text{for } t \in [0, t_f), \\ |x(t)| \to 0, & \text{as } t \to t_f^-, \end{cases} \tag{12.9}$$

where

$$\mu^{-1} = \left(1 - \frac{t}{t_f}\right)^{1+m}$$

is a monotonically decreasing function with the properties that $\mu^{-1}(0) = 1$ and $\mu^{-1}(t_f) = 0$, which means that, in particular, the state $x(t)$ and control input $u(t)$ converge to zero within a prescribed time t_f.

Proof. With the transformed variable $w = \mu x$, the system (12.1) is rewritten as:

$$\dot{w} = \mu u + \dot{\mu} x, \tag{12.10}$$

where

$$\dot{\mu} = \frac{(1+m)t_f^{1+m}}{(t_f - t)^{2+m}} = \frac{1+m}{t_f} \mu^{\frac{2+m}{1+m}}.$$

Choose the Lyapunov function candidate as $V = \frac{1}{2}w^2$, whose derivative along (12.10) is:

$$\dot{V} = w\mu u + w\dot{\mu}x.$$

Noting that

$$w\dot{\mu}x = w\frac{1+m}{t_f}\mu^{\frac{2+m}{1+m}}x = w\mu\frac{1+m}{t_f}\mu^{-\frac{m}{1+m}}\mu x \le w\mu\frac{1+m}{t_f}w,$$

then we have:

$$\dot{V} \le w\mu\left(u + \frac{1+m}{t_f}w\right) = w\mu(-c_0 w) = -c_0\mu w^2 = -2c_0\mu V.$$

By using Lemma 12.1, we arrive at:

$$V(t) \le \exp\left(-2c_0\frac{t_f}{m}(\mu^{\frac{m}{1+m}} - 1)\right)V(0),$$

which implies that:

$$x^2(t) = \frac{1}{\mu^2}w^2 = 2\mu^{-2}V(t) \le \mu^{-2}\exp\left(-2c_0\frac{t_f}{m}(\mu^{\frac{m}{1+m}} - 1)\right)x^2(0) \to 0$$

as $t \to t_f$ by using the fact that $\mu^{-2} \to 0$ as $t \to t_f$, and thus, (12.9) holds. Also, the convergence of $u(t)$ can be obtained by the convergence of $x(t)$. This means that the regulation is achieved in a prescribed time t_f, which is independent of the initial conditions and other design parameters and thus can be explicitly pre-specified. The proof is completed.

12.2   DISCUSSIONS ON FINITE-TIME CONTROL AND PRESCRIBED-TIME CONTROL

We first give the definitions for finite-time stability, fixed-time stability, prescribed-time stability, and the related lemmas.

12.2.1 DEFINITIONS

Definition 12.1 *(Finite-time Stability.) Consider the system:*[1]

$$\dot{x} = f(x,t), \quad f(0,t) = 0, \quad x(0) = x_0, \quad t \in \mathbb{R}_+, \tag{12.11}$$

where $x \in \mathbb{R}^n$ denotes the system state, $f : U_0 \times \mathbb{R}_+ \to \mathbb{R}^n$ is a nonlinear vector field locally bounded on an open neighborhood U_0 of the origin and uniformly in time. The equilibrium $x = 0$ of the system is said to be (locally) uniformly finite-time stable if it is uniformly asymptotically stable, and, for any initial condition $x_0 \in U$ ($U \subseteq U_0$), there exists a locally bounded function $T : U \to \mathbb{R}_+$ such that $x(x_0,t) = 0$ for all $t \geq T(x_0)$, where $x(x_0,t)$ is an arbitrary solution of (12.11). The function $T(x_0)$ is called the settling-time function. If $U = U_0 = \mathbb{R}^n$, the origin of the system (12.11) is said to be globally uniformly finite-time stable.

Lemma 12.2 *Consider the system (12.11). If there is a continuously differentiable positive definite Lyapunov function $V(x,t)$ defined on $U \times \mathbb{R}_+$, where $U \subseteq U_0$ is a neighborhood of the origin, real constants $l > 0$ and $0 < \alpha < 1$, such that $\dot{V}(x,t) \leq -lV(x,t)^\alpha$ on U, then the origin of the system (12.11) is finite-time stable. Moreover, the finite settling time T satisfies*

$$T \leq \frac{1}{l(1-\alpha)} V(x_0,0)^{1-\alpha}$$

for any given initial condition $x(0) \in U$. If $U = \mathbb{R}^n$, then the origin of the system (12.11) is globally finite-time stable.

From Definition 12.1, we see that the settling time in traditional finite-time stability is not uniform in the initial condition. For example, the origin of the system

$$\begin{cases} \dot{x} = u, \quad x \in \mathbb{R}, \quad x_0 = x(0), \\ u = -\mathrm{sgn}(x(t))|x(t)|^{\frac{1}{2}}, \end{cases}$$

is globally uniformly finite-time stable because its settling-time function T is locally bounded (i.e., $T(x_0) = 2\sqrt{|x_0|}$) by utilizing Lemma 12.2 and defining $V = \frac{1}{2}x^2$, but it is obvious that the settling time $T(x_0)$ depends on the initial condition.

Definition 12.2 *(Fixed-time Stability.) The origin of the system (12.11) is said to be globally fixed-time stable if it is globally uniformly finite-time stable and the settling-time function T is globally bounded, i.e., $\exists T_{\max} \in \mathbb{R}_+$ such that $T(x_0) \leq T_{\max}, \forall x_0 \in \mathbb{R}^n$.*

According to Definition 12.2, the finite settling time in a fixed-time system, although uniform w.r.t. the initial condition, is still subject to certain restrictions and

---

[1] We assume that the initial time is $t_0 = 0$.

thus can only be pre-specified within a certain range. For example, the origin of the system

$$\begin{cases} \dot{x} = u, \ x \in \mathbb{R}, \\ \dot{x}(t) = -\mathrm{sgn}(x(t))\left(|x(t)|^{\frac{1}{2}} + |x(t)|^{\frac{3}{2}}\right), \end{cases}$$

is globally fixed-time stable since its settling time $T(x_0) = 2\arctan(\sqrt{|x_0|})$ is uniformly bounded by π, from which we see that the upper bound of the finite convergence time is subject to certain constraints, i.e., it cannot be preselected (assigned) to be smaller than π.

In the following, we give two definitions related to the prescribed-time stability.

Definition 12.3 *(Prescribed-time Input-to-state Stable, PT-ISS.) Consider the system:*

$$\dot{x} = f(x,t,d), \ x(0) = x_0, \ t \in \mathbb{R}_+, \tag{12.12}$$

where $x \in \mathbb{R}^n$ denotes the system state, $d \in \mathbb{R}^m$ is the bounded disturbance, and $f : \mathbb{R}^n \times \mathbb{R}_+ \times \mathbb{R}^m \to \mathbb{R}^n$ is a continuous nonlinear vector. Then the system (12.12) is said to be PT-ISS in time t_f if there exist a class $\mathcal{KL}$ function β and a class $\mathcal{K}$ function γ, such that, for all $t \in [0,t_f)$,

$$\|x(t)\| \leq \beta(\|x_0\|, \mu_1(t) - 1) + \gamma(\|d\|_{[0,t]}).$$

Definition 12.4 *(Prescribed-time Input-to-state Stable+Convergent, PT-ISS+C.) The system (12.12) is said to be PT-ISS in time t_f and convergent to zero (PT-ISS+C) if there exist class $\mathcal{KL}$ functions β and β_f, and a class $\mathcal{K}$ function γ, such that, for all $t \in [0,t_f)$,*

$$\|x(t)\| \leq \beta_f\left(\beta(\|x_0\|,t) + \gamma(\|d\|_{[0,t]}), \mu_1(t) - 1\right).$$

Clearly, a system that is PT-ISS+C is also PT-ISS, with the additional property that its state converges to zero in time t_f despite the presence of a disturbance d.

12.2.2 FEATURES

In the following, for the single-integrator system $\dot{x} = u$, we will present some features between the traditional finite-time control and the prescribed-time control, and highlight the major differences between them, with particular attention to the control structure, convergence time, disturbance rejection, and robustness against uncertainties.

⋆ **Control Structure**

• **Traditional Finite-time Controller**

The traditional finite-time controller normally bears the following form:

$$\begin{cases} \dot{x} = u, \\ u = -c(t)\mathrm{sgn}(x)|x|^\alpha, \end{cases} \tag{12.13}$$

where $c(t) > 0$ is the control gain to be designed and $0 \leq \alpha = \frac{p}{q} < 1$ (p and q are positive odd integers). Note that the control law (12.13) covers several different cases:

(1) when $\alpha = 1$, it reduces to the typical asymptotic control $u = -c(t)x$;
(2) when $\alpha = 0$, it reduces to the signum function feedback-based finite-time control, which is discontinuous due to the using of the signum function;
(3) when $0 < \alpha < 1$, it corresponds to the fractional power state feedback-based finite-time control, which is continuous but non-smooth w.r.t. state variables.

It is worth mentioning that with $0 \leq \alpha < 1$, the finite settling time T is determined by $T = \frac{V(x_0)^{1-r}}{l(1-r)}$. According to Lemma 12.2, where $l > 0$ is a constant related to the control gain $c(t)$ and fraction index α, and r is related to α.

• **Prescribed-time Controller**

The prescribed-time controller based on the time-varying gain is of the form:

$$
\begin{cases}
u = -c(t)\mu x, \\
\mu(t) = \dfrac{t_f^{1+m}}{(t_f - t)^{1+m}}, & t \in [0, t_f), \\
c = c_0 + \dfrac{1+m}{t_f} > 0,
\end{cases}
\tag{12.14}
$$

with $t_f > 0$ and $c_0 > 0$.

⋆ **Convergence Time**

• **Traditional Finite-time Controller**

For the traditional finite-time method, applying to the system (12.1), we choose

$$c(t) = c_0, \tag{12.15}$$

with $c_0 > 0$ being a constant gain, then we have the following result. For the system (12.1) with the controllers (12.13) and (12.15), the following result is obtained.

Theorem 12.2 (*i*) *When $0 < \alpha < 1$ (fractional power state feedback-based control), the finite settling time satisfies*

$$T \leq \frac{V(x_0)^{1 - \frac{1+\alpha}{2}}}{2c_0(1 - \frac{1+\alpha}{2})};$$

(*ii*) *When $\alpha = 0$ (signum function feedback-based control), the finite settling time satisfies $T \leq \frac{|x(0)|}{c_0}$;*
(*iii*) *When $\alpha = 1$ (regular state feedback-based control), the settling time satisfies $T = +\infty$.*

260   Control of Nonlinear Systems

Proof. The system (12.1) with (12.13) and (12.15) is rewritten as:

$$\dot{x} = -\rho \operatorname{sgn}(x)|x|^{\alpha}. \tag{12.16}$$

The details of the finite-time stability of (12.16) can be seen in references [49] and [145] for (*i*), in reference [146] for (*ii*), and in reference [9] for (*iii*) and thus are omitted here.

From (*i*) and (*ii*), we see that the finite settling time T under the fractional power state feedback-based finite-time control depends on the initial condition and a number of design parameters, and T by the signum function feedback-based control depends on the initial condition.

• **Prescribed-time Controller**

Theorem 12.3 *The system (12.1) with the controllers (12.7) and (12.8) is PT-ISS+C with a prescribed settling time t_f, and*

$$\begin{cases} |x(t)| \leq \mu^{-1} \exp\left(-c_0 \frac{t_f}{m}(\mu^{\frac{m}{1+m}} - 1)\right) |x(0)|, & \text{for } t \in [0, t_f), \\ |x(t)| \to 0, & \text{as } t \to t_f^-, \end{cases}$$

where

$$\mu^{-1} = \left(1 - \frac{t}{t_f}\right)^{1+m}$$

is a monotonically decreasing linear function with the properties that $\mu^{-1}(0) = 1$ and $\mu^{-1}(t_f) = 0$, which means that, in particular, the state $x(t)$ converges to zero in a prescribed time t_f.

The proof has been proved in Section 12.1, then it is omitted. It is seen from Theorem 12.3 that the regulation is achieved in the prescribed time t_f, which is independent of the initial condition and other design parameters and thus can be explicitly pre-specified. The proof is completed.

⋆ **Disturbance Rejection and Robustness against Uncertainties**
    Of particular interest in this section is the capability for disturbance rejection and robustness against uncertainties. Consider

$$\dot{x} = u + f_d(x, t), \tag{12.17}$$

where $x \in \mathbb{R}$ and $u \in \mathbb{R}$ are the system state and control input, respectively, and $f_d(x, t) \in \mathbb{R}$ denotes the unknown lumped uncertainty that may be non-vanishing.

Assumption 12.1 *(Bound on matched but possibly non-vanishing uncertainty) The nonlinearity f_d in (12.17) obeys $|f_d(x, t)| \leq d(t)\psi(x)$, where $d(t)$ is a disturbance with an unknown bound $\|d\|_{[0,t]} := \sup_{\tau \in [0,t]} |d(\tau)|$, and $\psi(x) \geq 0$ is a known scalar-valued continuous function.*

Lemma 12.3 *Consider the system (12.17). If there is a continuously differentiable positive definite Lyapunov function $V(x,t)$ defined on $U \times \mathbb{R}_+$, where $U \subseteq U_0$ is a neighborhood of the origin, and there are real constants $l > 0$, $0 < \alpha < 1$, and $\|d\|_{[0,t]} := \sup_{\tau \in [0,t]} |d(\tau)|$, such that $\dot{V}(x,t) \leq -lV(x,t)^\alpha + \|d\|_{[0,t]}$, then the system (12.17) is locally finite-time convergent within a finite settling time T, satisfying*

$$T \leq \frac{V(x_0,0)^{1-\alpha}}{l\theta_0(1-\alpha)},$$

where $0 < \theta_0 < 1$ is a real constant, for any given initial condition $x(0) \in U$. Moreover, when $t \geq T$, the state trajectory converges to a compact set

$$\Theta = \left\{ x \middle| V(x,t) \leq \left(\frac{\|d\|_{[0,t]}}{l(1-\theta_0)} \right)^{\frac{1}{\alpha}} \right\}.$$

- **Traditional Finite-time Controller**

For the traditional finite-time method to system (12.17), we choose:

$$c(t) = c_0 + \gamma \psi^2, \tag{12.18}$$

where $\gamma > 0$ is a free design constant parameter, and we have the following result.

Consider the system (12.17) under Assumption 12.1. If the controller (12.13) with (12.18) is applied, the following result is obtained.

Theorem 12.4 (*i*) *When $0 < \alpha < 1$ (fractional power state feedback-based control), the state trajectory is convergent to a compact set*

$$\Theta_1 = \left\{ x \middle| |x(t)| \leq \left(\frac{\|d\|_{[0,t]}^2}{4\gamma c_0(1-\theta_0)} \right)^{\frac{1}{2\alpha}} \right\}$$

within the finite time T, which satisfies

$$T \leq \frac{V(x_0)^{1-\frac{2\alpha}{1+\alpha}}}{\theta_0 c_0 (1+\alpha)^{\frac{2\alpha}{1+\alpha}} \left(1 - \frac{2\alpha}{1+\alpha}\right)},$$

where $V(x_0)$ is the initial value of the Lyapunov function and $0 < \theta_0 < 1$ is defined the same as in Lemma 12.3.

(*ii*) *When $\alpha = 0$ (signum function feedback-based control), the state trajectory is convergent to a compact set*

$$\Theta_2 = \left\{ x \middle| |x(t)| \leq \left(\frac{\|d\|_{[0,t]}^2}{4\gamma c_0(1-\theta_0)} \right)^{\frac{1}{2\times 0}} \right\},$$

which converges to zero if we choose the design parameters such that $\frac{\|d\|^2_{[0,t]}}{4\gamma c_0(1-\theta_0)} < 1$, within the finite time T, which satisfies

$$T \leq \frac{V(x_0)}{\theta_0 c_0}.$$

(iii) When $\alpha = 0$ (regular state feedback-based control), the state trajectory is convergent to a compact set

$$\Theta_3 = \left\{ x \left| |x(t)| \leq \left(\frac{\|d\|^2_{[0,t]}}{4\gamma c_0(1-\theta_0)} \right)^{\frac{1}{2}} \right. \right\}$$

within the infinite time.

Proof. The system (12.17) with (12.13) and (12.18) is rewritten as:

$$\dot{x} = -(c_0 + \gamma\psi^2)\mathrm{sgn}(x)|x|^\alpha + f_d. \qquad (12.19)$$

In the following, we analyze the finite-time stability of (12.19), especially its disturbance rejection and robust ability against uncertainties.

(i) Choose the Lyapunov function candidate as $V = \frac{1}{1+\alpha}x^{1+\alpha}$, whose derivation along (12.19) is:

$$\dot{V} = x^\alpha \dot{x} = -c_0 x^{2\alpha} - \gamma\psi^2 x^{2\alpha} + x^\alpha f_d.$$

Upon using Young's inequality, we have:

$$x^\alpha f_d \leq \gamma\psi^2 x^{2\alpha} + \frac{\|d\|^2_{[0,t]}}{4\gamma}.$$

Then, we arrive at:

$$\dot{V} \leq -c_0 x^{2\alpha} - \gamma\psi^2 x^{2\alpha}\gamma\psi^2 x^{2\alpha} + \frac{\|d\|^2_{[0,t]}}{4\gamma} = -c_0 x^{2\alpha} + \frac{\|d\|^2_{[0,t]}}{4\gamma}. \qquad (12.20)$$

By noting that $V^{\frac{2\alpha}{1+\alpha}} = \left(\frac{1}{1+\alpha}\right)^{\frac{2\alpha}{1+\alpha}} x^{2\alpha}$, we then have:

$$x^{2\alpha} = (1+\alpha)^{\frac{2\alpha}{1+\alpha}} V^{\frac{2\alpha}{1+\alpha}}. \qquad (12.21)$$

According to (12.20) and (12.21), we see that:

$$\dot{V} \leq -c_0(1+\alpha)^{\frac{2\alpha}{1+\alpha}} V^{\frac{2\alpha}{1+\alpha}} + \frac{\|d\|^2_{[0,t]}}{4\gamma}. \qquad (12.22)$$

Let $l = c_0(1+\alpha)^{\frac{2\alpha}{1+\alpha}}$, $\bar{d} = \frac{\|d\|^2_{[0,t]}}{4\gamma}$, and $r = \frac{2\alpha}{1+\alpha}$. According to Lemma 12.3, then we have, for $\forall t \geq T$, that:

$$V(t) \leq \left(\frac{\bar{d}}{(1-\theta_0)l} \right)^{\frac{1}{r}} = \frac{1}{1+\alpha} \left(\frac{\|d\|^2_{[0,t]}}{4\gamma c_0(1-\theta_0)} \right)^{\frac{1+\alpha}{2\alpha}}, \qquad (12.23)$$

with $T \leq \frac{V(x_0)^{1-r}}{I\theta_0(1-r)} = \frac{V(x_0)^{1-\frac{2\alpha}{1+\alpha}}}{\theta_0 c_0 (1+\alpha)^{\frac{2\alpha}{1+\alpha}} (1 - \frac{2\alpha}{1+\alpha})}$.

From (12.23), we then arrive at:

$$|x| = (x^{1+\alpha})^{\frac{1}{1+\alpha}} = [(1+\alpha)V]^{\frac{1}{1+\alpha}} \leq \left(\frac{\|d\|_{[0,t]}^2}{4\gamma c_0 (1 - \theta_0)} \right)^{\frac{1}{2\alpha}}.$$

(*ii*) When $\alpha = 0$, it corresponds to the signum function feedback-based finite-time control:

$$u = -c(t)\mathrm{sgn}(x)|x|^0 = -(c_0 + \gamma \psi^2)\mathrm{sgn}(x). \tag{12.24}$$

The result in (*ii*) is straightforward to be obtained by following the similar line as in the proof of (*i*).

(*iii*) When $\alpha = 1$, the control in (12.13) with (12.18) reduces to the regular state feedback-based nonfinite-time control:

$$u = -c(t)\mathrm{sgn}(x)|x|^1 = -(c_0 + \gamma \psi^2)x, \tag{12.25}$$

and the bound for the state x is:

$$|x(t)| \leq \left(\frac{\|d\|_{[0,t]}^2}{4\gamma c_0 (1 - \theta_0)} \right)^{\frac{1}{2}}.$$

If choosing c_0 and γ such that $\frac{\|d\|_{[0,t]}^2}{4\gamma c_0 (1-\theta_0)} < 1$, and if $0 < \alpha \ll 1$, then

$$\left(\frac{\|d\|_{[0,t]}^2}{4\gamma c_0 (1 - \theta_0)} \right)^{\frac{1}{2\alpha}} \ll \left(\frac{\|d\|_{[0,t]}^2}{4\gamma c_0 (1 - \theta_0)} \right)^{\frac{1}{2}},$$

which implies that the fractional power state feedback-based finite-time control proposed in (12.13) with (12.18) has better disturbance rejection and robustness against uncertainties than the regular state feedback-based nonfinite-time control given in (12.25). Especially, the signum function feedback-based finite-time control (12.24) has better disturbance rejection and robustness against uncertainties than the fractional power state feedback-based finite-time control (12.13) with (12.18) and the regular state feedback-based nonfinite-time control (12.25).

• **Prescribed-time Controller**

In this case, the prescribed-time controller is designed as:

$$\begin{cases} u = -c(t)\mu x, \\ c(t) = c_0 + \dfrac{1+m}{t_f} + \lambda \psi^2. \end{cases} \tag{12.26}$$

Then we have the following result.

Theorem 12.5 *The system (12.17) with the control scheme (12.26) is PT-ISS+C with the pre-specified finite settling time t_f, and*

$$|x(t)| \le \mu^{-1} \left[\exp\left(-c_0 \frac{t_f}{m} \left(\mu^{\frac{m}{1+m}} - 1 \right) \right) |x(0)| + \frac{\|d\|_{[0,t]}}{2\sqrt{c_0 \lambda}} \right] \qquad (12.27)$$

for all $t \in [0, t_f)$, which means that, in particular, the state $x(t)$ and the control input u converge to zero within a prescribed time t_f.

Proof. By using the time-varying scaling function $\mu(t)$ given in (12.6) and the transformed variable $w = \mu x$, we have:

$$\dot{w} = \mu u + \mu f_d + \dot{\mu} x.$$

Choose the Lyapunov function as $V = \frac{1}{2} w^2$, whose derivative is:

$$\dot{V} = w\mu u + w\mu f_d + w\dot{\mu} x.$$

Upon using Young's inequality, we get:

$$w\mu f_d \le w\mu \lambda \psi^2 w + \frac{\mu \|d\|_{[0,t]}^2}{4\lambda}.$$

Therefore, one further has:

$$\dot{V} \le w\mu \left(u + \frac{1+m}{t_f} w \right) + \frac{\mu \|d\|_{[0,t]}^2}{4\lambda}$$

$$= w\mu(-c_0 w) + \frac{\mu \|d\|_{[0,t]}^2}{4\lambda}$$

$$= -c_0 \mu w^2 + \frac{\mu \|d\|_{[0,t]}^2}{4\lambda}$$

$$= -2c_0 \mu V + \frac{\mu \|d\|_{[0,t]}^2}{4\lambda},$$

where the fact that $\dot{\mu} = \frac{(1+m)t_f^{1+m}}{(t_f - t)^{2+m}} = \frac{1+m}{t_f} \mu^{\frac{2+m}{1+m}}$ is utilized.

By using Lemma 12.1, we arrive at:

$$V(t) \le \exp\left(-2c_0 \frac{t_f}{m} (\mu^{\frac{m}{1+m}} - 1) \right) V(0) + \frac{\|d\|_{[0,t]}^2}{8c_0 \lambda},$$

which implies that:

$$x^2 = \mu^{-2} w^2 \le \mu^{-2} \left(\exp\left(-2c_0 \frac{t_f}{m} (\mu^{\frac{m}{1+m}} - 1) \right) x^2(0) + \frac{\|d\|_{[0,t]}^2}{4c_0 \lambda} \right) \to 0$$

as $t \to t_f$, meaning that the state x converges to zero when $t \to t_f$, establishing the same for $u(t)$, and therefore, (12.27) holds naturally.

12.2.3 NUMERICAL SIMULATION

Simulations on two numerical examples are conducted in this section to demonstrate and validate the fast convergence time, good disturbance rejection, and robustness against uncertainties of the prescribed-time control scheme.

Example 1. The simulation model in this example is the single integrator system described by (12.1). The purpose of the numerical simulation is to compare the convergence performance between the traditional finite-time control given in (12.13) and the prescribed-time control given in (12.14). To make a fair comparison, the design parameters for the traditional finite-time control (12.13) are taken as $c_0 = 3.5$, $\alpha = \frac{1}{3}$; the design parameters for the prescribed-time control (12.14) are taken as $c_0 = 0.5$, $t_f = 1$ second, and $m = 2$, and thus, the initial control gain is the same in the two control methods.

The simulation results are represented in Fig. 12.1–Fig. 12.6. From which, we see that the convergence time of the state under the traditional finite-time control depends on the initial state, whereas the convergence time under the prescribed-time control method can be prespecified uniformly, independent of any initial condition and other design parameter. In addition, the control signal under the signum function feedback–based control ($\alpha = 0$) is discontinuous when the state approaches zero.

Example 2. The simulation model in this example is the first-order system subject to non-vanishing uncertainties described by (12.17), in which the lumped uncertainty is taken as $f_d(x,t) = \theta_1 + \theta_2 x \cos(t)$ with $\theta_1 = 1$ and $\theta_2 = 1$.

The purpose of the numerical simulation is to compare the disturbance rejection and robustness against uncertainties between the traditional fractional power state feedback–based finite-time control given in (12.13) with (12.18) (including the signum function feedback–based control as a specific case when $\alpha = 0$) and the prescribed-time control given in (12.26). To make a fair comparison, the design parameters for the traditional finite-time controller (12.13) with (12.18) are taken as $c_0 = 3.5$, $\alpha = \frac{1}{3}$, $\gamma = 0.1$, and $\psi = 1 + |x|$ (the signum function feedback–based control is corresponding to $\alpha = 0$); the design parameters for the prescribed-time

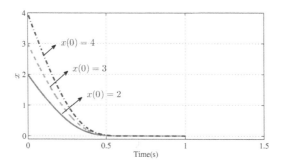

FIGURE 12.1   The evolution of $x(t)$ with the prescribed-time controller (12.14) under $c_0 = 0.5$, $m = 2$, and $t_f = 1$ second.

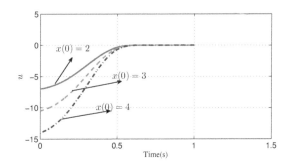

FIGURE 12.2 The evolution of $u(t)$ with the prescribed-time controller (12.14) under $c_0 = 0.5$, $m = 2$, and $t_f = 1$ second.

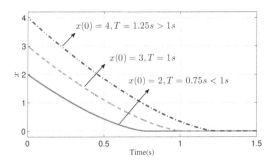

FIGURE 12.3 The evolution of $x(t)$ with the traditional finite-time controller (12.13) under $\alpha = \frac{1}{3}$ and $c_0 = 3.5$.

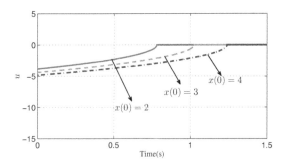

FIGURE 12.4 The evolution of $u(t)$ with the traditional finite-time controller (12.13) under $\alpha = \frac{1}{3}$ and $c_0 = 3.5$.

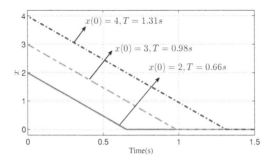

FIGURE 12.5 The evolution of $x(t)$ with the traditional finite-time controller (12.13) under $\alpha = 0$ and $c_0 = 3.5$.

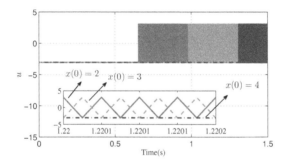

FIGURE 12.6 The evolution of $u(t)$ with the traditional finite-time controller (12.13) under $\alpha = 0$ and $c_0 = 3.5$.

controller (12.26) are taken as $c_0 = 0.5$, $\lambda = 0.1$, $\psi = 1 + |x|$, $t_f = 1$ second, and $m = 2$, and thus, the initial control gain is the same in the two control methods.

The simulation results are represented in Fig. 12.7–Fig. 12.12. The robustness against the uncertainties and the disturbance rejection are reflected by the convergence time in the simulation. From which we see that the convergence time of the state under the traditional finite-time control method does depend on the initial state, whereas the convergence time under the prescribed-time control method can be pre-specified uniformly, which does not depend on the initial conditions and any other design parameter. In particular, the control signal under the signum function feedback–based control ($\alpha = 0$) is discontinuous when the state approaches zero.

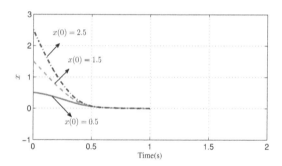

FIGURE 12.7   The evolution of $x(t)$ subject to non-vanishing uncertainties with (12.26) under $c_0 = 0.5$, $\lambda = 0.1$, $m = 2$, and $t_f = 1$ second.

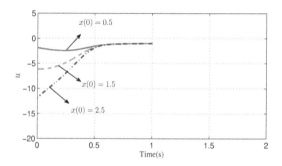

FIGURE 12.8   The evolution of $u(t)$ subject to non-vanishing uncertainties with (12.26) under $c_0 = 0.5$, $\lambda = 0.1$, $m = 2$, and $t_f = 1$ second.

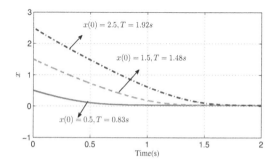

FIGURE 12.9   The evolution of $x(t)$ subject to non-vanishing uncertainties with (12.13) and (12.18) under $\alpha = \frac{1}{3}$, $c_0 = 3.5$, and $\gamma = 0.1$.

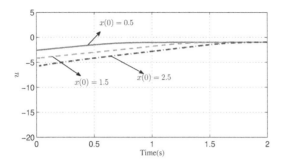

FIGURE 12.10 The evolution of $u(t)$ subject to non-vanishing uncertainties with (12.13) and (12.18) under $\alpha = \frac{1}{3}$, $c_0 = 3.5$, and $\gamma = 0.1$.

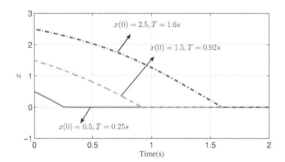

FIGURE 12.11 The evolution of $x(t)$ subject to non-vanishing uncertainties (12.13) and (12.18) under $\alpha = 0$, $c_0 = 3.5$, and $\gamma = 0.1$.

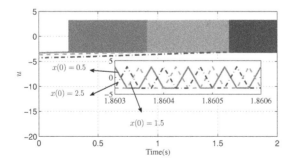

FIGURE 12.12 The evolution of $u(t)$ subject to non-vanishing uncertainties with (12.13) and (12.18) under $\alpha = 0$, $c_0 = 3.5$, and $\gamma = 0.1$.

12.3   MODEL-DEPENDENT PRESCRIBED-TIME CONTROL

In this section, we extend the comparison results in Sections 12.1 and 12.2 to the following n-order normal-form nonlinear systems:

$$\begin{cases} \dot{x}_k = x_{k+1}, \ k = 1, 2, \cdots, n-1, \\ \dot{x}_n = f(x,t) + g(x,t)u, \end{cases} \tag{12.28}$$

where $x = [x_1, x_2, \cdots, x_n]^\top \in \mathbb{R}^n$ is the system state, $u \in \mathbb{R}$ is the control input, $b(x,t)$ represents the control gain of the system, and $f(x,t)$ denotes the nonlinear function. Here, $g(x,t) \neq 0$ and $f(x,t)$ are assumed to be known and bounded.

In this section, the control objective is to design a robust controller for the system (12.28) such that:

(i)   all the closed-loop signals are uniformly bounded;
(ii) the system states converge to zero within a prescribed time.

12.3.1   SCALED STATE AND SYSTEM

To achieve prescribed-time stabilization for the normal-form nonlinear system (12.28), we use the following scaling function:

$$\begin{cases} \omega_1(t) = \mu(t) x_1(t), \\ \omega_j(t) = \frac{d\omega_{j-1}}{dt}, \quad j = 2, \cdots, n+1. \end{cases} \tag{12.29}$$

Denote $\omega_{n+1} = \dot{\omega}_n$ and $x_{n+1} = \dot{x}_n$. We present the following two lemmas to facilitate analysis.

Lemma 12.4 [18] The scaling transformation $x(t) \mapsto \omega(t)$ given by

$$\omega = \mu_1^{m+1} P(\mu_1) x \tag{12.30}$$

yields the system (12.29), where $\omega = [\omega_1, \omega_2, \cdots, \omega_n]^\top$, $x = [x_1, x_2, \cdots, x_n]^\top$, and the matrix $P(\mu_1)_{n \times n}$ is a lower triangular matrix in the following form:

$$P(\mu_1)_{n \times n} = \begin{bmatrix} \mu_1^{n-1} & 0 & \cdots & 0 \\ \frac{n+m}{t_f} \mu_1^n & \mu_1^{n-1} & \cdots & 0 \\ \vdots & \vdots & \cdots & \vdots \\ \frac{(2n+m-2)!}{t_f^{n-1}(n+m-1)!} \mu_1^n & \frac{(n-1)\cdot(2n+m-2)!}{t_f^{n-2}(n-m+1)!} & \cdots & \mu_1^{n-1} \end{bmatrix}_{n \times n} ,$$

$$p_{ij}(\mu_1) = \bar{p}_{ij} \mu_1^{n+i-j-1}, \quad 1 \le j \le i \le n,$$

$$\bar{p}_{ij} = \binom{i-1}{i-j} \frac{(n+m+i-j-1)!}{t_f^{i-j}(n+m-1)!}, \quad \binom{i}{j} = \frac{i!}{j!(i-j)!}.$$

Proof. From (12.29), we have:

$$\omega_1 = \mu x_1, \tag{12.31}$$

$$\omega_2 = \dot{\omega}_1 = \mu^{(1)} x_1 + \mu x_1^{(1)} = \sum_{k=0}^{1} \binom{i}{k} \mu^{(k)} x_1^{(i-k)}, \tag{12.32}$$

$$\omega_3 = \dot{\omega}_2 = \mu^{(2)} x_1 + 2\mu^{(1)} x_1^{(1)} + \mu x_1^{(2)} = \sum_{k=0}^{2} \binom{i}{k} \mu^{(k)} x_1^{(i-k)}. \tag{12.33}$$

From (12.31) to (12.33), we find that $\omega_i = \sum_{k=0}^{i-1} \binom{i-1}{k} \mu^{(k)} x_1^{(i-k)}$, whose derivative w.r.t. time is:

$$\omega_{i+1} = d\left(\sum_{k=0}^{i-1} \binom{i-1}{k} \mu^{(k)} x_1^{(i-k)} \right) \bigg/ dt = \sum_{k=0}^{i-1} \binom{i-1}{k} d\left(\mu^{(k)} x_1^{(i-k)} \right) \bigg/ dt$$

$$= \sum_{k=0}^{i-1} \binom{i-1}{k} \left[\mu^{(k+1)} x_1^{(i-k)} + \mu^{(k)} x_1^{(i-k+1)} \right]$$

$$= \mu x_1^{(i)} + \sum_{k=0}^{i} \left[\binom{i-1}{k} + \binom{i-1}{k+1} \right] \left(\mu^{(i-k)} x_1^{(k)} \right) + \mu^{(i)} x_1$$

$$= \sum_{k=0}^{i} \binom{i}{k} \mu^{(i-k)} x_1^{(k)}.$$

Using the principle of mathematical induction, it follows that:

$$\omega_i = \sum_{k=0}^{i-1} \binom{i-1}{k} \mu^{(k)} \times x_1^{(i-k)} = \sum_{k=0}^{i-1} \binom{i-1}{k} \mu^{(k)} x_{i-k},$$

for $i = 1, 2, \cdots, n$, namely,

$$\omega = [\omega_1, \omega_2, \omega_3 \cdots, \omega_n]^\top$$

$$= \begin{bmatrix} \mu & 0 & 0 & \cdots & 0 \\ \mu^{(1)} & \mu & 0 & \cdots & 0 \\ \mu^{(2)} & 2\mu^{(1)} & \mu & \cdots & 0 \\ \vdots & \vdots & \vdots & \ddots & 0 \\ \mu^{(n-1)} & (n-1)\mu^{(n-2)} & \binom{n-1}{2}\mu^{(n-3)} & \cdots & \mu \end{bmatrix} \begin{bmatrix} x_1 \\ x_2 \\ x_3 \\ \vdots \\ x_n \end{bmatrix}.$$

Let $j = i - k$ $(j = 1, 2, \cdots, i)$, then

$$\omega_i = \sum_{j=1}^{i} \binom{i-1}{j-1} \mu^{(i-j)} x_j, \quad i = 1, 2, \cdots, n. \tag{12.34}$$

Noting that

$$\mu^{(k)} = \frac{(n+m+k-1)!}{t_f^k (n+m-1)!} \mu_1^{n+m+k} \quad (k=1,2,\cdots,n), \tag{12.35}$$

then (12.34) can be further written as:

$$\omega_i = \sum_{j=1}^{i} \binom{i-1}{j-1} \mu^{(i-j)} x_j = \sum_{j=1}^{i} \left[\binom{i-1}{j-1} \frac{(n+m+i-j-1)!}{t_f^k (n+m-1)!} \mu_1^{n+m+i-j} \right] x_j$$

$$= \mu_1^{n+m} \sum_{j=1}^{i} \left[\binom{i-1}{j-1} \frac{(n+m+i-j-1)!}{t_f^k (n+m-1)!} \mu_1^{i-j} \right] x_j.$$

Therefore, we have $\omega = [\omega_1, \omega_2, \cdots, \omega_n]^\top = \mu_1^{m+1} P(\mu_1) x$, where the matrix $P(\mu_1)$ is a lower triangular matrix, and each element $\{p_{ij}\}$ satisfies:

$$p_{ij}(\mu_1) = \bar{p}_{ij} \mu_1^{n+i-j-1}, \quad p_{ij}(\mu_1) = \binom{i-1}{i-j} \frac{(n+m+i-j-1)!}{t_f^{i-j}(n+m-1)!},$$

for $1 \leq j \leq i \leq n$. The proof is completed.

Lemma 12.5 *[18] Given the transformation* $x(t) \mapsto \omega(t)$ *defined by* $\omega(t) = \mu_1^{1+m} P(\mu_1) x(t)$ *in (12.30), the inverse transformation* $\omega(t) \mapsto x(t)$ *is given by:*

$$x(t) = \upsilon^{m+1} Q(\upsilon) \omega, \tag{12.36}$$

where $\upsilon(t) = \mu_1(t)^{-1} = 1 - \frac{t}{t_f}$ *and the inverse matrix* $Q(\upsilon) = P(\mu)^{-1}$ *is a lower triangular matrix having elements* $\{q_{ij}\}$ *given by:*

$$q_{ij}(\upsilon) = \bar{q}_{ij} \upsilon^{n+j-i-1}, \quad \bar{q}_{ij} = \binom{i-1}{i-j} \frac{(-1)^{i-j}(n+m)!}{t_f^{i-j}(n+m+j-i)!},$$

for $1 \leq j \leq i \leq n$. *Furthermore,* $\bar{q} = \sup_{\upsilon \in (0,1]} |Q(\upsilon)|$ *is finite.*

Proof. Since $x_1 = \frac{1}{\mu} \omega_1$, one has:

$$x_{i+1} = x_1^{(i)} = \left(\frac{1}{\mu} w_1\right)^{(i)} = \sum_{k=0}^{i} \binom{i}{k} \left(\frac{1}{\mu}\right)^{(k)} \omega_1^{(i-k)} = \sum_{k=0}^{i} \binom{i}{k} \left(\frac{1}{\mu}\right)^{(k)} \omega_{i+1-k}, \tag{12.37}$$

for $i = 0, 1, \cdots, n$.

Taking the k-th derivative of $\frac{1}{\mu}$, we have:

$$\left(\frac{1}{\mu}\right)^{(k)} = \frac{(-1)^k (n+m)!}{t_f^k (n+m-k)!} \upsilon^{n+m-k}. \tag{12.38}$$

Inserting (12.38) into (12.37), one obtains:

$$x_{i+1} = v^{n+m} \sum_{k=0}^{i} \binom{i}{k} \frac{(-1)^k (n+m)!}{t_f^{\ k} (n+m-k)!} v^{-k} \omega_{i+1-k}, \quad i = 0, 1, \cdots, n, \qquad (12.39)$$

which implies that x_i can be expressed by:

$$x_i = v^{n+m} \sum_{k=0}^{i-1} \binom{i-1}{k} \frac{(-1)^{-k} (n+m)!}{t_f^{\ k} (n+m-k)} v^{-k} \omega_{i-k}. \qquad (12.40)$$

Let $i := j - k$ and $j = 1, \cdots, i$, then (12.40) can be rewritten as:

$$x_i = v^{n+m} \sum_{j=1}^{i} \binom{i-1}{i-j} \frac{(-1)^{i-j} (n+m)!}{t_f^{\ i-j} (n+m+j-i)!} v^{j-i} \omega_j, \qquad (12.41)$$

then it is not difficult to get the expression of $\{q_{ij}\}$. The boundedness of $\bar{q}_{ij}$ indicates that $Q(v)$ is bounded. The proof is completed.

12.3.2 CONTROL DESIGN

We introduce the following transformation:

$$\begin{cases} r_1 = [\omega_1, \omega_2, \cdots, \omega_{n-1}]^\top = J_1 \omega \in \mathbb{R}^{n-1}, \\ r_2 = [\omega_2, \omega_3, \cdots, \omega_n]^\top = J_2 \omega \in \mathbb{R}^{n-1}, \end{cases} \qquad (12.42)$$

where $\omega = [\omega_1, \omega_2, \cdots, \omega_n]^\top \in \mathbb{R}^n$, and

$$J_1 = \begin{bmatrix} 1 & \cdots & 0 & 0 \\ \vdots & \ddots & \vdots & \vdots \\ 0 & \cdots & 1 & 0 \end{bmatrix} \in \mathbb{R}^{(n-1) \times n}, \quad J_2 = \begin{bmatrix} 0 & 1 & \cdots & 0 \\ \vdots & \vdots & \ddots & \vdots \\ 0 & 0 & \cdots & 1 \end{bmatrix} \in \mathbb{R}^{(n-1) \times n}.$$

In addition, let

$$\Lambda = \begin{bmatrix} 0 & 1 & \cdots & 0 \\ \vdots & \vdots & \ddots & \vdots \\ 0 & 0 & \cdots & 1 \\ -k_1 & -k_2 & \cdots & -k_{n-1} \end{bmatrix} \in \mathbb{R}^{(n-1) \times (n-1)},$$

and

$$z = \omega_n + k_1 \omega_1 + k_2 \omega_2 + \cdots + k_{n-1} \omega_{n-1}, \quad K_{n-1} = [k_1, k_2, \cdots, k_{n-1}]^\top, \qquad (12.43)$$

where K_{n-1} is the design parameter that makes the matrix Λ Hurwitz.

According to (12.42), the derivative of r_1 w.r.t. time is:

$$\dot{r}_1 = [\omega_2, \omega_3, \cdots, \omega_n]^\top = \Lambda r_1 + e_{n-1} z, \qquad (12.44)$$

where $e_{n-1} = [0, \cdots, 0, 1]^{\top}$, then the derivative of z is:

$$\dot{z} = \dot{\omega}_n + k_1 \dot{\omega}_1 + k_2 \dot{\omega}_2 + \cdots + k_{n-1} \dot{\omega}_{n-1} = \dot{\omega}_n + K_{n-1}^{\top} J_2 \omega. \qquad (12.45)$$

Since

$$\dot{\omega}_n = \omega_{n+1} = \sum_{k=0}^{n} \binom{n}{k} \mu^k x_{n+1-k} = \mu x_{n+1} + \sum_{k=1}^{n} \binom{n}{k} \mu^k x_{n+1-k}$$

$$= \mu \dot{x}_n + \sum_{k=1}^{n} \binom{n}{k} \mu^k x_{n+1-k},$$

and $\dot{\omega}_n = \omega_{n+1}$, $\dot{x}_n = x_{n+1}$, then (12.45) can be further rewritten as:

$$\dot{z} = \dot{\omega}_n + K_{n-1}^{\top} J_2 \omega = \mu x_{n+1} + \sum_{k=1}^{n} \binom{n}{k} \mu^k x_{n+1-k} + K_{n-1}^{\top} J_2 \omega$$

$$= \mu x_{n+1} + \mu \sum_{k=1}^{n} \binom{n}{k} \frac{\mu^k}{\mu} x_{n+1-k} + \mu \cdot \upsilon^{n+m} K_{n-1}^{\top} J_2 \omega$$

$$= \mu \left(\dot{x}_n + \sum_{k=1}^{n} \binom{n}{k} \frac{\mu^k}{\mu} x_{n+1-k} + \upsilon^{n+m} K_{n-1}^{\top} J_2 \omega \right)$$

$$= \mu (\dot{x}_n + L_0 + L_1) = \mu (bu + f + L_0 + L_1), \qquad (12.46)$$

where

$$L_0 = \sum_{k=1}^{n} \binom{n}{k} \frac{\mu^{(k)}}{\mu} x_{n+1-k}, \quad \text{and} \quad L_1 = \upsilon^{n+m} K_{n-1}^{\top} J_2 \omega \qquad (12.47)$$

are computable functions.

In the following lemma, the quantity L_0 is expressed in terms of ω.

Lemma 12.6 [18] *The quantity L_0 in (12.47) can be expressed as:*

$$L_0 = \upsilon^m l_0 (\upsilon) \omega, \qquad (12.48)$$

where $l_0(\upsilon) = [l_{0,1}, l_{0,2}, \cdots, l_{0,n}]$, $l_{0,j}(\upsilon) = \bar{l}_{0,j} \upsilon^{j-1}$, $(j = 1, 2, \cdots, n)$, $\upsilon = \mu_1^{-1} = 1 - \frac{t}{t_f}$, $\omega = [\omega_1, \cdots, \omega_n]^{\top}$, and

$$\bar{l}_{0,j} = \frac{n+m}{t_f^{n+1-j}} \sum_{i=0}^{n-j} \binom{n}{n-i-j+1} \binom{i+j-1}{i} \frac{(-1)^i (2n+m-i-j)!}{(n+m-i)!}.$$

Furthermore, $l_0(\upsilon)$ is bounded.

Proof. According to the proof in Lemma 12.5, inserting the expression of x_{n+1-k} as shown in (12.39) into (12.47), with the aid of (12.34), we obtain:

$$L_0 = \sum_{k=1}^{n} \binom{n}{k} \frac{\mu^{(k)}}{\mu} \left[\sum_{i=0}^{n-k} \binom{n-k}{i} \left(\frac{1}{\mu} \right)^{(i)} \omega_{n-k+1-i} \right]$$

$$= \sum_{k=1}^{n} \sum_{i=0}^{n-k} \binom{n}{k} \binom{n-k}{i} \frac{\mu^{(k)}}{\mu} \left(\frac{1}{\mu} \right)^{(i)} \omega_{n-k+1-i},$$

which can be viewed as the sum of function evaluations $f_{ki}(n)$ defined on a triangle in (k,i) space. Therefore, $\sum\limits_{k=1}^{n}\sum\limits_{i=0}^{n-k} f_{ki}(n) = \sum\limits_{i=0}^{n-1}\sum\limits_{k=1}^{n-i} f_{ki}(n)$, which leads to

$$L_0 = \sum_{i=0}^{n-1}\sum_{k=1}^{n-i} \binom{n}{k}\binom{n-k}{i} \frac{\mu^{(k)}}{\mu}\left(\frac{1}{\mu}\right)^{(i)} \omega_{n-k+1-i}.$$

Let $j := n-k+1-i$, $j = n-i, n-i-1, \cdots 1$, then

$$L_0 = \sum_{i=0}^{n-1}\sum_{j=1}^{n-i} \binom{n}{n-i-j+1}\binom{i+j-1}{i} \frac{\mu^{(n-i-j+1)}}{\mu}\left(\frac{1}{\mu}\right)^{(i)} \omega_j,$$

which can be rewritten in another form:

$$L_0 = \sum_{j=1}^{n}\sum_{i=0}^{n-j} \binom{n}{n-i-j+1}\binom{i+j-1}{i} \frac{\mu^{(n-i-j+1)}}{\mu}\left(\frac{1}{\mu}\right)^{(i)} \omega_j.$$

Upon using (12.35) for $\mu^{(n-i-j+1)}$, together with $\frac{1}{\mu} = \upsilon^{n+m}$, we arrive at:

$$L_0 = \sum_{j=1}^{n}\sum_{i=0}^{n-j} \binom{n}{n-i-j+1}\binom{i+j-1}{i} \upsilon^{n+m}$$
$$\times \left[\frac{(2n+m-i-j)!}{t_f^{n-i-j+1}(n+m-1)!}\mu_1^{2n+m-i-j+1}\right]\left[\frac{(-1)^i(n+m)!}{t_f^i(n+m-i)!}\upsilon^{n+m-i}\right]\omega_j,$$

it can be rewritten as:

$$L_0 = \sum_{j=1}^{n} \frac{n+m}{t_f^{n-j+1}} \times$$
$$\sum_{i=0}^{n-j} \binom{n}{n-i-j+1}\binom{i+j-1}{i} \frac{(-1)^i(2n+m-i-j)!}{(n+m-i)!}\upsilon^{m+j-1}\omega_j,$$

which proves the first part of the lemma. The boundedness of $l_{o,j}(\upsilon)$ follows by inspection, since $\upsilon(t) \le 1$ for $\forall t \in [0,t_f)$, and $\bar{l}_{0,j}$ are bounded since they are finite sums of real numbers. The proof is completed.

Now, for the nonlinear system (12.28), we give the following theorem.

12.3.3 THEOREM AND STABILITY ANALYSIS

Theorem 12.6 *If $g(x,t)$ and $f(x,t)$ are known, then the system (12.28) with the controller*

$$u = -\frac{1}{g}(f + L_0 + L_1 + cz), \tag{12.49}$$

where $c > 0$, $L_0 = \sum\limits_{k=1}^{n} \binom{n}{k}\frac{\mu^{(k)}}{\mu}x_{n+1-k}$, $L_1 = \upsilon^{n+m}K_{n-1}^{\top}J_2\omega$, has a globally prescribed-time asymptotically stable equilibrium at the origin, with a prescribed convergence time t_f, and there exist $\tilde{M}, \tilde{\delta} > 0$, for all $t \in [0,t_f)$, such that:

$$\|x(t)\| \le \upsilon(t)^{m+1}\tilde{M}\exp\left(-\tilde{\delta}t\right)\|x(0)\|. \tag{12.50}$$

Furthermore, the control u remains uniformly bounded over $t \in [0, t_f)$, and if $f(x,t)$ is vanishing at $x = 0$, $u(t)$ also converges to zero as $t \to t_f$.

Proof. Inserting the controller (12.49) into (12.46), we have:

$$\dot{z} = -c\mu z. \tag{12.51}$$

According to Lemma 12.1, one has:

$$|z(t)| \leq \varsigma(t)^c |z_0|, \quad \forall t \in [0, t_f). \tag{12.52}$$

Furthermore, (12.51) is a linear system that is ISS w.r.t. z, then it follows from (12.44) that there exist some positive constants M_1, δ_1, and γ_1 such that:

$$\|r_1(t)\| \leq M_1 \exp(-\delta_1 t) \|r_1(0)\| + \gamma_1 \|z\|_{[0,t]}, \quad \forall t \in [0, t_f). \tag{12.53}$$

Define the following variable:

$$\overline{\omega}(t) = \begin{bmatrix} r_1 \\ z \end{bmatrix} = [\omega_1, \omega_2, \cdots, \omega_{n-1}, z]^\top. \tag{12.54}$$

According to (12.52)–(12.54) and Theorem 2.14, we have:

$$
\begin{aligned}
\|\overline{\omega}\| &= \sqrt{\|r_1(t)\|^2 + |z(t)|^2} \leq \|r_1(t)\| + |z(t)| \\
&\leq M_1 \exp(-\delta_1 t) \|r_1(0)\| + \exp\left(\frac{ct_f}{m+n-1}\left(1 - \mu_1(t)^{m+n-1}\right)\right) |z_0| + \gamma_1 \|z\|_{[0,t]} \\
&\leq M_1 \exp(-\delta_1 t) \|r_1(0)\| + \gamma_1 |z_0| \\
&\quad + \exp\left(\frac{ct_f}{m+n-1}\right) \cdot \exp\left(\frac{kt_f}{m+n-1}\left(-\mu_1(t)^{m+n-1}\right)\right) |z_0| \\
&\leq M_1 \exp(-\delta_1 t) \|r_1(0)\| + \left(\gamma_1 + \exp\left(\frac{ct_f}{m+n-1}\right) \cdot \exp(-\delta_2 t)\right) |z_0| \\
&\leq M_1 \exp(-\delta_1 t) \|r_1(0)\| + \left(\gamma_1 \exp(\delta_2 t_f) + \exp\left(\frac{ct_f}{m+n-1}\right)\right) \exp(-\delta_2 t) |z_0| \\
&\leq \overline{M}_1 \exp(-\overline{\delta} t) (\|r_1(0)\| + |z_0|) \\
&\leq \overline{M} \exp(-\overline{\delta} t) \|\overline{\omega}(0)\|,
\end{aligned}
\tag{12.55}
$$

where $\overline{M}_1 = \max\left\{M_1, \gamma_1 \exp(\delta_2 t_f) + \exp\left(\frac{ct_f}{m+n-1}\right)\right\}$, $\overline{\delta} = \min\{\delta_1, \delta_2\}$, $\overline{M} = \sqrt{2}\overline{M}_1$, which implies that there exist positive constants $\overline{M}$ and $\overline{\delta}$ such that:

$$\|\overline{\omega}\| \leq \overline{M} \exp(-\overline{\delta} t) \|\overline{\omega}(0)\|, \quad t \in [0, t_f). \tag{12.56}$$

As $\overline{\omega}$ can be rewritten as:

$$\overline{\omega}(t) = \begin{bmatrix} r_1 \\ z \end{bmatrix} = \begin{bmatrix} \omega_1 \\ \omega_2 \\ \vdots \\ \omega_{n-1} \\ \omega_n + \sum\limits_{i=1}^{n-1} k_i \omega_i \end{bmatrix} = \begin{bmatrix} 1 & 0 & \cdots & 0 & 0 \\ 0 & 1 & \cdots & 0 & 0 \\ \vdots & \vdots & \ddots & \vdots & \vdots \\ 0 & 0 & \cdots & 1 & 0 \\ k_1 & k_2 & \cdots & k_{n-1} & 1 \end{bmatrix} \begin{bmatrix} \omega_1 \\ \omega_2 \\ \vdots \\ \omega_{n-1} \\ \omega_n \end{bmatrix}$$

$$= \Re\omega, \tag{12.57}$$

where

$$\Re = \begin{bmatrix} 1 & 0 & \cdots & 0 & 0 \\ 0 & 1 & \cdots & 0 & 0 \\ \vdots & \vdots & \ddots & \vdots & \vdots \\ 0 & 0 & \cdots & 1 & 0 \\ k_1 & k_2 & \cdots & k_{n-1} & 1 \end{bmatrix}, \; \Re^{-1} = \begin{bmatrix} 1 & 0 & \cdots & 0 & 0 \\ 0 & 1 & \cdots & 0 & 0 \\ \vdots & \vdots & \ddots & \vdots & \vdots \\ 0 & 0 & \cdots & 1 & 0 \\ -k_1 & -k_2 & \cdots & -k_{n-1} & 1 \end{bmatrix}, \tag{12.58}$$

then it is deduced from (12.36) that:

$$x = v^{m+1} Q(v)\omega = v^{m+1} Q(v)\Re^{-1}\overline{\omega}, \; \overline{\omega}_0 = \Re P(0) x_0. \tag{12.59}$$

In addition, (12.57) and (12.58) can be rewritten as:

$$\overline{\omega}(t) = \left(J_1^\top + \mathbf{e}_n K_{n-1}^\top \right) r_1 + \mathbf{e}_n \omega_n; \; \Re = I + \mathbf{e}_n K_{n-1}^\top J_1, \; \Re^{-1} = I - \mathbf{e}_n K_{n-1}^\top J_1.$$

By utilizing Lemma 12.5, i.e., $x(t) = v^{m+1} Q(v) \omega$, for all $t \in [0, t_f)$, one has:

$$\|x(t)\| \le \left| v^{m+1} \right| \|Q(v)\| \|\omega\| \le \left| v^{m+1} \right| \|Q(v)\| \|\Re^{-1}\| \|\overline{\omega}\|. \tag{12.60}$$

Inserting (12.55) into (12.60), it is checked that:

$$\|x(t)\| \le v(t)^{m+1} \overline{q} \|\Re^{-1}\| \|\Re P(0)\| \overline{M} \exp\left(-\bar{\delta}t\right) \|x(0)\|,$$

therefore (12.50) holds, i.e.,

$$\|x(t)\| \le v(t)^{m+1} \tilde{M} \exp\left(-\tilde{\delta}t\right) \|x(0)\|,$$

where $\tilde{M} = \overline{q} \|\Re^{-1}\| \|\Re P(0)\| \overline{M}$, $\tilde{\delta} = \bar{\delta}$.

Now we turn our attention to proving the claims regarding the input u given in (12.49). Since $L_0 + L_1 = v^m \left(l_0(v) + v^n K_{n-1}^\top J_2 \right) \omega$, then it is bounded and converges to zero as $t \to t_f$. According to (12.52), the term cz in (12.49) is bounded, and converges to zero. Finally, f is bounded and, if $f(0, t) = 0$, then $f(x, t)$ also goes to zero as $t \to t_f$, establishing the same for u. The proof is completed.

12.4   MODEL-INDEPENDENT PRESCRIBED-TIME CONTROL

In this section, we consider a model-independent prescribed-time stabilization control algorithm for normal-form nonlinear systems (12.28), i.e., the nonlinear function
f and control gain g are unknown for control design. Meanwhile, the control objective is the same as that in Section 12.3, namely, the closed-loop signals are uniformly
bounded over the finite time interval and the system states converge to zero within a
prescribed time.

To this end, the following assumptions are imposed.

Assumption 12.2 *(Global controllability) There exists a known constant $\underline{g} \neq 0$, such
that, for all $x \in \mathbb{R}^n$, $t \in \mathbb{R}_+$, $0 < \underline{g} \le |g(x,t)| < \infty$.*

Assumption 12.3 *The nonlinearity f obeys $|f(x,t)| \le d(t)\psi(x)$, where $d(t)$ is a
disturbance with an unknown bound $\|d\|_{[0,t]} := \sup_{\tau \in [0,t]} |d(\tau)|$, and $\psi(x) \ge 0$ is a
known scalar-valued continuous function.*

Theorem 12.7 *Under Assumptions 12.2 and 12.3, consider the system (12.28) with
the controller*

$$u = -\frac{1}{g}\left(c + \theta + \lambda \psi(x)^2\right) z, \tag{12.61}$$

where

$$z = \mu_1(t)^{m+1} K_+^\top P(\mu_1) x, \quad K_+^\top = [k_1, k_2, \cdots, k_{n-1}, 1]^\top, \tag{12.62}$$

and $\theta > \theta_*$ with

$$\theta_* = k_{n-1} + \bar{l}_{0,n} + \rho \max_{\upsilon \in [0,1]} \left|\left(\upsilon^n K_{n-1}^\top J_2 + l_0(\upsilon)\right)\left(J_1^\top - e_n K_{n-1}^\top\right)\right|^2.$$

*If the control gains satisfy ρ, c, $\lambda > 0$, $\rho c > \frac{\gamma_1}{4}$, where γ_1 depends on the choice of
the gain vector K_{n-1} in (12.43), then the closed-loop system (12.28) with the control
law (12.61) is PT-ISS+C and there exist $\breve{M}$, $\breve{\delta}$, $\breve{\gamma}$ for any $t \in [0,t_f)$ such that:*

$$\|x(t)\| \le \upsilon(t)^{m+1}\left(\breve{M}\exp\left(-\breve{\delta}t\right)\|x(0)\| + \breve{\gamma}\|d\|_{[0,t]}\right). \tag{12.63}$$

Furthermore, the control u remains uniformly bounded over $t \in [0,t_f)$.

Proof. According to the definition of the filter variable z, the derivative of the
Lyapunov function $V = \frac{1}{2}z^2$ along (12.46) yields,

$$\dot{V} = \mu z b u + \mu z f + \mu z (L_0 + L_1). \tag{12.64}$$

By using Young's inequality, we have:

$$\mu z f \le \mu |z| d\psi \le \mu \lambda z^2 \psi^2 + \frac{\mu}{4\lambda} d^2 \le z \mu \underline{g} \frac{1}{\underline{g}} \lambda \psi^2 + \frac{\mu}{4\lambda} d^2.$$

Since $\omega = \left(J_1^\top - \mathbf{e}_n K_{n-1}^\top \right) r_1 + \mathbf{e}_n z$, $L_0 + L_1$ can be rewritten as:

$$
\begin{aligned}
L_0 + L_1 &= v^m \left(l_0\left(v\right) + v^n K_{n-1}^\top J_2 \right) \left[\left(J_1^\top - \mathbf{e}_n K_{n-1}^\top \right) r_1 + \mathbf{e}_n z \right] \\
&= v^m \left[\left(l_0\left(v\right) + v^n K_{n-1}^\top J_2 \right) \mathbf{e}_n z + \left(l_0\left(v\right) + v^n K_{n-1}^\top J_2 \right) \left(J_1^\top - \mathbf{e}_n K_{n-1}^\top \right) r_1 \right] \\
&= v^m \left[v^{n-1} \left(\bar{l}_{0,n}\left(v\right) + k_{n-1} v \right) z + \left(l_0\left(v\right) + v^n K_{n-1}^\top J_2 \right) \left(J_1^\top - \mathbf{e}_n K_{n-1}^\top \right) r_1 \right].
\end{aligned}
$$

Thus,

$$
\begin{aligned}
&\mu z \left(L_0 + L_1 \right) \\
&= \mu z \left(v^m \left[v^{n-1} \left(\bar{l}_{0,n}\left(v\right) + k_{n-1} v \right) z + \left(l_0\left(v\right) + v^n K_{n-1}^\top J_2 \right) \left(J_1^\top - \mathbf{e}_n K_{n-1}^\top \right) r_1 \right] \right),
\end{aligned}
$$

where $\bar{l}_{0,n} = \frac{n(n+m)}{t_f} \frac{(n+m)!}{(n+m)!} > 0$.

Upon using Young's inequality, we get:

$$
\begin{aligned}
\mu z \left(L_0 + L_1 \right) \leq{} & \mu v^{m+n-1} \left| k_{n-1} v + \bar{l}_{0,n} \right| z^2 + \mu \frac{\|r_1\|^2}{4\rho} \\
&+ \mu \rho v^{2m} \left| \left(v^n K_{n-1}^\top J_2 + l_0\left(v\right) \right) \left(J_1^\top - \mathbf{e}_n K_{n-1}^\top \right) \right|^2 z^2,
\end{aligned}
$$

where $\rho > 0$, define $\rho_1\left(\rho\right) = \rho \max_{v \in [0,1]} \left| \left(v^n K_{n-1}^\top J_2 + l_0\left(v\right) \right) \left(J_1^\top - \mathbf{e}_n K_{n-1}^\top \right) \right|^2$.

Since $k_{n-1} > 0$, $\bar{l}_{0,n} > 0$, one has:

$$
\mu z \left(L_0 + L_1 \right) \leq \mu \left[\left(k_{n-1} + \bar{l}_{0,n} + \rho_1\left(\rho\right) \right) z^2 + \frac{\|r_1\|^2}{4\rho} \right],
$$

then (12.64) can be expressed as:

$$
\dot{V} \leq \mu z g \left[u + \frac{1}{\underline{g}} \left(k_{n-1} + \bar{l}_{0,n} + \rho_1(\rho) + \lambda \psi^2 \right) z \right] + \frac{\mu}{4} \left(\frac{\|r_1\|^2}{\rho} + \frac{d^2}{\lambda} \right). \quad (12.65)
$$

Inserting the control law $u = -\frac{1}{g} \left(c + \theta + \lambda \psi(x)^2 \right) z$ into (12.65), we have:

$$
\begin{aligned}
\dot{V} \leq{} & \mu z g \left[u + \frac{1}{\underline{g}} \left(k_{n-1} + \bar{l}_{0,n} + \rho_1(\rho) + \lambda \psi^2 \right) z \right] + \frac{\mu}{4} \left(\frac{\|r_1\|^2}{\rho} + \frac{d^2}{\lambda} \right) \\
\leq{} & -c \mu z^2 + \frac{\mu}{4} \left(\frac{\|r_1\|^2}{\rho} + \frac{d^2}{\lambda} \right) \\
\leq{} & -2c \mu V + \frac{\mu}{4} \left(\frac{\|r_1\|^2}{\rho} + \frac{d^2}{\lambda} \right).
\end{aligned}
$$

According to Lemma 12.1, we obtain:

$$
V\left(t\right) \leq \zeta\left(t\right)^{2c} V\left(0\right) + \frac{\left\| \lambda \|r_1\|^2 / \rho + d^2 \right\|}{8c\lambda}.
$$

Since $V = \frac{1}{2}z^2$, one has:

$$z^2 \leq \zeta(t)^{2c}z_0^2 + \frac{\left\| \lambda \|r_1\|^2 \big/ \rho + d^2 \right\|}{4c\lambda},$$

then it follows that:

$$|z| \leq \zeta(t)^c |z_0| + \frac{1}{2\sqrt{c}}\left(\frac{\|r_1\|_{[0,t]}}{\sqrt{\rho}} + \frac{\|d\|_{[0,t]}}{\sqrt{\lambda}} \right).$$

Hence, the z-system is PT-ISS w.r.t. the r_1-input with the gain of $\frac{1}{2\sqrt{c\rho}}$, and is, additionally, PT-ISS w.r.t. the d-input. By being PT-ISS, the z-system is, in particular, ISS, w.r.t. the same inputs (r_1, d). From (12.53) we recall that the r_1-system is ISS (though not PT-ISS) w.r.t. the z-input with the gain of γ_1 (see 12.53). Hence, if $\frac{\gamma_1}{2\sqrt{c\rho}} < 1$, namely, if $c\rho > \gamma_1/4$, the (r_1, z)-system is ISS w.r.t. d. Note that we cannot conclude the PT-ISS property for the overall (r_1, z)-system w.r.t. d, but only the ISS property, because the r_1-subsystem is merely ISS. Now, since, from (12.54), $(r_1, z) = \bar{\omega}$, from the small-gain and ISS argument that we have just completed it follows that there exist positive constants $\hat{M}$, $\hat{\delta}$, and $\hat{\gamma}$ such that:

$$\|\bar{\omega}(t)\| \leq \hat{M}\exp\left(-\hat{\delta}t\right)\|\bar{\omega}(0)\| + \hat{\gamma}\|d\|_{[0,t]}, \quad \forall t \in [0, t_f).$$

Following the same argument as in (12.59), we obtain:

$$\|x(t)\| \leq \upsilon(t)^{m+1}\bar{q}\|\mathfrak{R}^{-1}\|\left(\|\mathfrak{R}P(0)\|\hat{M}\exp\left(-\hat{\delta}t\right)\|x(0)\| + \hat{\gamma}\|d\|_{[0,t]}\right),$$

and arrive at (12.63) with $\breve{M} = \bar{q}\|\mathfrak{R}^{-1}\|\|\mathfrak{R}P(0)\|\hat{M}$, $\breve{\gamma} = \bar{q}|\mathfrak{R}^{-1}|\hat{\gamma}$, $\breve{\delta} = \hat{\delta}$, establishing that the x-system is PT-ISS+C w.r.t. input d. The boundedness of u is proved as in Theorem 12.6. The proof is completed.

12.5 NUMERICAL SIMULATION

12.5.1 EXAMPLE 1

In this section, we use a simple example to verify the effectiveness of the finite-time control and prescribed-time control, and compare the differences between them. A second-order integral system is considered:

$$\begin{cases} \dot{x}_1 = x_2, \\ \dot{x}_2 = u, \end{cases} \tag{12.66}$$

where $x_1(t)$, $x_2(t)$ are the system states, and $u(t)$ is the control input.

The finite-time controller and prescribed-time controller are given as follows:

$$\text{Finite-time control:} \quad u = -5\left(x_2^{\frac{5}{3}} + \frac{3}{20}x_1\right)^{\frac{1}{5}},$$

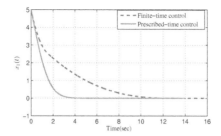

FIGURE 12.13 The evolution of the system state x_1 under the finite-time control and prescribed-time control.

$$
\text{Prescribed-time control:} \quad
\begin{cases}
u_{PT} = -\mu z_2 - \mu^{-1}\dot{\mu} z_2 - x_1 + \dot{\alpha}_1, \\
z_2 = x_2 - \alpha_1, \\
\alpha_1 = -\mu x_1 - \mu^{-1}\dot{\mu} x_1,
\end{cases}
$$

where $\mu = \left(t_f/(t_f - t)\right)^{2+m}$ is a monotonically increasing function and grows to infinity at t_f. The simulation results are shown in Fig. 12.13–Fig. 12.15. It is seen from Fig. 12.13 and Fig. 12.14 that the system states converge to zero within a finite time under the finite-time control and prescribed-time control. However, due to the fractional power term contained in the finite-time controller, the chattering phenomenon may happen, which brings a great burden to the actuator operation in practical engineering applications. It is worth mentioning that the control signal in the prescribed-time control is smooth, and the chattering phenomenon is avoided.

12.5.2 EXAMPLE 2

We use the "wing rock" unstable motion model for high performance aircraft introduced in reference [147] to verify the effectiveness of the proposed method. The

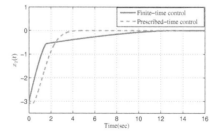

FIGURE 12.14 The evolution of the system state x_2 under the finite-time control and prescribed-time control.

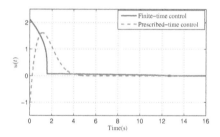

FIGURE 12.15 The evolution of the control signal u under the finite-time control and prescribed-time control.

mathematical expression of the model is given as:

$$\begin{cases} \dot{x}_1 = x_2, \\ \dot{x}_2 = g(\cdot)u(t) + f(\cdot), \end{cases} \tag{12.67}$$

where $g(\cdot) = 2 + 0.4\sin(t)$, $f(\cdot) = 1 + \cos(t)x_1 + 2\sin(t)x_2 + 2|x_1|x_2 + 3|x_2|x_2 + x_1^3$. According to Assumption 12.3, we have $\psi(x) = 1 + |x_1| + |x_2| + |x_1x_2| + x_2^2 + |x_1|^3$.

In the simulation, the specific expression form of the prescribed-time controller is:

$$u = -\frac{1}{g}\left(c + \theta + \lambda \psi(x)^2\right)z,$$

where the design parameters are chosen as $g = 1.6$, $c = 4$, $\theta = 8$, $\lambda = 0.1$, $z = 0.1\mu_1^3 x_1 + \mu_1^5 \frac{4}{t_f} x_1 + \mu_1^4 x_2$, and $t_f = 1s$. Furthermore, in order to verify the prescribed-time method is independent of initial states, we choose the following different initial conditions:

$$[x_1(0), x_2(0)]^\top = [0.1, 0]^\top, [x_1(0), x_2(0)]^\top = [0.2, 0]^\top, [x_1(0), x_2(0)]^\top = [0.3, 0]^\top.$$

The simulation results are shown in Fig. 12.16–Fig. 12.18, in which the evolutions of the system states under different conditions are shown in Fig. 12.16 and Fig. 12.17, respectively, from which it is seen that the system states converge to the equilibrium point within a finite time independent of initial conditions. The evolution of the prescribed-time control signal is plotted in Fig. 12.18, which is bounded for $t \in [0, t_f)$.

12.6 NOTES

A prescribed-time stabilization control scheme has been proposed in this chapter. By constructing a time-varying function $\mu(t)$ that grows to infinity at the prescribed

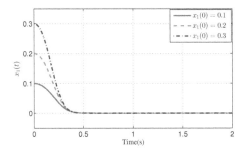

FIGURE 12.16 The evolution of x_1 under different initial conditions.

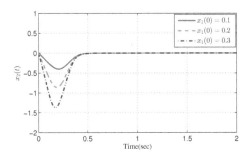

FIGURE 12.17 The evolution of x_2 under different initial conditions.

time, the developed control can not only effectively offset the uncertain disturbances in the system, but also make the system state converge to zero within a prescribed time.

Although the control signal is proved to be bounded as $t \to t_f$, there exists a "$\infty \times 0$" term in the prescribed-time controller. For the simulation, the system operation time cannot exceed t_f, this is because: (1) infinite numbers cannot be stored in the

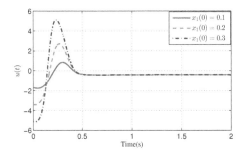

FIGURE 12.18 The evolution of the control input u under different initial conditions.

computer; and (2) $\mu(t)$ can only be defined on $[0,t_f)$ and it is not defined over the interval $[t_f,\infty)$. To solve this problem, $\mu(t)$ and its derivative must be some constants greater than 0 after $t_f - \varepsilon$, where ε is an extremely small positive number, such as $\varepsilon = 0.01$. This technique not only avoids the occurrence of an infinite number, but also allows $\mu(t)$ to be defined for all time.

13 Prescribed-time Tracking Control

We illustrated the prescribed-time stabilization control of normal-form nonlinear systems in Chapter 12. The basic assumption therein is that the system is in SISO form as well as the lower bound of the control gain is available for control design. In this chapter, we relax this assumption and develop a prescribed-time tracking control method for MIMO systems involving unknown control gain matrices and non-vanishing disturbances/uncertainties (part of which stems from time-varying reference signals), rendering the model and underlying problem setting more general than those considered in Chapter 12.

To help with the understanding of the fundamental idea of prescribed-time tracking control, this chapter first introduces the control design and stability analysis for MIMO normal-form systems in Section 13.1. We then discuss how to apply such method to the Euler-Lagrange (EL) systems in Section 13.2, and also perform a simulation verification for a robotic manipulator in Section 13.3. Finally, some discussions are shown in Section 13.4.

13.1 PRESCRIBED-TIME TRACKING CONTROL OF MIMO SYSTEMS

13.1.1 SYSTEM MODEL AND ASSUMPTIONS

Consider the following MIMO normal-form nonlinear systems:

$$\begin{cases} \dot{x}_k = x_{k+1}, & k = 1, \cdots, n-1, \\ \dot{x}_n = F(\bar{x}_n, \theta) + G(\bar{x}_n)u, \end{cases} \tag{13.1}$$

where $x_i = [x_{i1}, \cdots, x_{im}]^\top \in \mathbb{R}^m$ with $i = 1, \cdots, n$, and $\bar{x}_n = [x_1^\top, \cdots, x_n^\top]^\top \in \mathbb{R}^{mn}$ is the state vector, $u = [u_1, \cdots, u_m]^\top \in \mathbb{R}^m$ is the control input vector, $G(\bar{x}_n) \in \mathbb{R}^{m \times m}$ represents the control gain matrix, $F(\bar{x}_n, \theta) \in \mathbb{R}^m$ denotes an unknown smooth nonlinear vector, and $\theta \in \mathbb{R}^r$ denotes a parameter vector.

Assumption 13.1 *[148] The reference trajectory $y_d(t)$ and its derivatives up to n-th are known, bounded, and piecewise continuous.*

Assumption 13.2 *[148] Certain crude structural information on $F(\cdot)$ is available to allow an unknown constant a and a known scalar-valued smooth function $\phi(\bar{x}_n)$ to be extracted, such that:*

$$\|F(\bar{x}_n, \theta)\| \leq a\phi(\bar{x}_n),$$

where $\phi(\bar{x}_n)$ is bounded if $\bar{x}_n$ is bounded.

DOI: 10.1201/9781003474364-13

Assumption 13.3 *The control gain matrix $G(\cdot)$ is symmetric and positive definite, and there exists a positive constant $\underline{g}$ such that $0 < \underline{g} \leq \lambda_{\min}(G) < +\infty$, where $\lambda_{\min}(\bullet)$ represents the minimum eigenvalue of the matrix $\bullet$.*

Remark 13.1 *It is worth mentioning that even for EL systems as considered in reference [149], the control gains are required to be known exactly. In contrast, as stated in Assumption 13.3, here in this work the control gain is allowed to be unknown, and as stated in Assumption 13.2, only certain crude structural information on nonlinear functions is needed. Thus, a more practical and less restrictive condition is imposed.*

With t_f being a user-designed constant representing the prescribed settling time, the control objective is to design a bounded and continuous $u(t)$ for the MIMO normal-form nonlinear system (13.1) to make the trajectories of the closed-loop system bounded, while achieving precise full state tracking such that the state tracking errors $E^{(i)}$ $(i = 0, \cdots, n-1)$ converge to zero as $t \to t_f$.

13.1.2   CONTROL DESIGN

Scaling of Virtual Errors: In order to facilitate later controller design, we define two kinds of variables: $z_i \in \mathbb{R}^m$ and $\eta_i \in \mathbb{R}^m$ $(i = 1, \cdots, n)$, which represent the virtual error and the scaling of the virtual error, respectively,

$$\begin{cases} z_1 = E, \\ z_k = x_k - \alpha_{k-1}, \ k = 2, \cdots, n, \\ \zeta_i = \mu(t)z_i, \ i = 1, \cdots, n, \end{cases} \tag{13.2}$$

where $\alpha_k \in \mathbb{R}^m$, $k = 1, 2, \cdots, n-1$, is the virtual control, $E = x_1 - y_d$ is the tracking error, and $\mu(t) = \left(\frac{t_f}{t_f - t}\right)^{m+n}$ with positive integers m and n.

Now, we introduce the backstepping-based Lyapunov design.

Step 1. From the first equation of (13.1) and the definition of z_i $(i = 1, \cdots, n)$ in (13.2), we have:

$$\dot{x}_1 = x_2 = z_2 + \alpha_1.$$

By using the notion of the scaling of the virtual errors as mentioned in Section 13.1.2, we have:

$$\zeta_1 = \mu z_1 = \mu E = \mu(x_1 - y_d),$$

then the time derivative of ζ_1 becomes:

$$\dot{\zeta}_1 = \mu \dot{E} + \dot{\mu} E = \mu(z_2 + \alpha_1) - \mu \dot{y}_d + \dot{\mu} \mu^{-1}(\mu E)$$
$$= \mu \alpha_1 + \dot{\mu} \mu^{-1} \zeta_1 + \zeta_2 - \mu \dot{y}_d. \tag{13.3}$$

The derivation of the quadratic function $V_1 = \frac{1}{2} \zeta_1^\top \zeta_1$ is:

$$\dot{V}_1 = \mu \zeta_1^\top \alpha_1 + \dot{\mu} \mu^{-1} \|\zeta_1\|^2 + \zeta_1^\top \zeta_2 - \mu \zeta_1^\top \dot{y}_d, \tag{13.4}$$

hence the virtual control α_1 is developed as:

$$\alpha_1(x_1, y_d, \dot{y}_d, \mu, \dot{\mu}) = -\left(c_1\zeta_1 + \frac{z_1}{2} + \dot{\mu}\mu^{-1}z_1 - \dot{y}_d\right), \tag{13.5}$$

where $c_1 > 0$, and $\frac{z_1}{2}$ can be regarded as the static damping term, then the first term on the right-hand side of (13.4) becomes:

$$\mu\zeta_1^\top \alpha_1 = -c_1\mu\|\eta_1\|^2 - \frac{1}{2}\|\zeta_1\|^2 - \dot{\mu}\mu^{-1}\|\eta_1\|^2 + \mu\zeta_1^\top \dot{y}_d. \tag{13.6}$$

Substituting (13.6) into (13.4) yields:

$$\dot{V}_1 = -c_1\mu\|\zeta_1\|^2 - \frac{1}{2}\|\zeta_1\|^2 + \zeta_1^\top \zeta_2. \tag{13.7}$$

Step i ($i = 2, 3, \cdots, n-1$). From (13.1) and the definitions of α_i and z_{i+1} in (13.2), the dynamics of x_i can be written as:

$$\dot{x}_i = x_{i+1} = z_{i+1} + \alpha_i.$$

Again, performing scaling on the virtual error $z_i = x_i - \alpha_{i-1}$, we have:

$$\zeta_i = \mu z_i = \mu(x_i - \alpha_{i-1}),$$

and

$$\dot{\zeta}_i = \mu\alpha_i + \dot{\mu}\mu^{-1}\zeta_i + \zeta_{i+1} - \mu\dot{\alpha}_{i-1}. \tag{13.8}$$

Choose a quadratic function V_i as $V_i = \frac{1}{2}\zeta_i^\top \zeta_i$, it is easy to get the time derivative of V_i along (13.8) as:

$$\dot{V}_i = \mu\zeta_i^\top \alpha_i + \dot{\mu}\mu^{-1}\|\zeta_i\|^2 + \zeta_i^\top \zeta_{i+1} - \mu\zeta_i^\top \dot{\alpha}_{i-1}, \tag{13.9}$$

where

$$\dot{\alpha}_{i-1} = \sum_{k=1}^{i-1}\frac{\partial\alpha_{i-1}}{\partial x_k}x_{k+1} + \sum_{k=0}^{i-1}\left(\frac{\partial\alpha_{i-1}}{\partial\mu^{(k)}}\mu^{(k+1)} + \frac{\partial\alpha_{i-1}}{\partial y_d^{(k)}}y_d^{(k+1)}\right).$$

The virtual controller α_i is designed as:

$$\alpha_i = -\left(c_i\zeta_i + z_i + \dot{\mu}\mu^{-1}z_i - \dot{\alpha}_{i-1}\right), \tag{13.10}$$

where $c_i > 0$.

Recall that $\zeta_i = \mu z_i$, it is thus derived from (13.10) that:

$$\mu\zeta_i^\top \alpha_i = -c_i\mu\|\zeta_i\|^2 - \|\zeta_i\|^2 - \dot{\mu}\mu^{-1}\|\zeta_i\|^2 + \mu\zeta_i^\top \dot{\alpha}_{i-1}. \tag{13.11}$$

Substituting (13.11) into (13.9) yields,

$$\dot{V}_i = -k_i\mu\|\zeta_i\|^2 - \|\zeta_i\|^2 + \zeta_i^\top \zeta_{i+1}. \tag{13.12}$$

Step n. Following the same procedure as in previous steps, we utilize $\mu(t)$ to scale the virtual error $z_n = x_n - \alpha_{n-1}$ to get $\zeta_n = \mu z_n = \mu(x_n - \alpha_{n-1})$, then it follows that:

$$\dot{\zeta}_n = \mu G u + \dot{\mu}\mu^{-1}\zeta_n + \mu(F - \dot{\alpha}_{n-1}). \tag{13.13}$$

Choose a quadratic function V_n as:

$$V_n = \frac{1}{2\underline{g}}\zeta_n^\top \zeta_n$$

with $\underline{g}$ being a positive constant as defined in Assumption 13.3, then the derivative of V_n along the trajectory of (13.13) is evaluated as:

$$\dot{V}_n = \frac{1}{\underline{g}}\mu\zeta_n^\top G u + \frac{1}{\underline{g}}\mu\zeta_n^\top \left(\dot{\mu}\mu^{-2}\zeta_n + F - \dot{\alpha}_{n-1}\right). \tag{13.14}$$

With the help of Assumption 13.2, we have $\|F(\cdot)\| \leq a\phi(\cdot)$, and using Young's inequality with $\lambda > 0$, it follows that:

$$\frac{1}{\underline{g}}\mu\zeta_n^\top \left(\dot{\mu}\mu^{-2}\zeta_n + F - \dot{\alpha}_{n-1}\right) \leq \Delta\mu\|\zeta_n\|\Phi \leq \lambda\mu\|\zeta_n\|^2\Phi^2 + \frac{\mu\Delta^2}{4\lambda},$$

with $\Delta = 1/\underline{g} \times \max\{1, a\}$ being an unknown constant, and

$$\Phi = \dot{\mu}\mu^{-2}\|\zeta_n\| + \phi + \|\dot{\alpha}_{n-1}\|$$

being a computable scalar-valued function, in which

$$\dot{\alpha}_{n-1} = \sum_{k=0}^{n-1}\left(\frac{\partial\alpha_{n-1}}{\partial\mu^{(k)}}\mu^{(k+1)} + \frac{\partial\alpha_{n-1}}{\partial y_d^{(k)}}y_d^{(k+1)}\right) + \sum_{k=1}^{n-1}\frac{\partial\alpha_{n-1}}{\partial x_k}x_{k+1}.$$

Thus, (13.14) can be further rewritten as:

$$\dot{V}_n \leq \frac{\mu\zeta_n^\top G u}{\underline{g}} + \lambda\mu\|\zeta_n\|^2\Phi^2 + \frac{\mu\Delta^2}{4\lambda}. \tag{13.15}$$

The actual controller u is designed as:

$$u = -\left(c_n\zeta_n + \frac{1}{2}z_n + \lambda\zeta_n\Phi^2\right), \quad c_n > 0. \tag{13.16}$$

Now we state the following theorem according to the proposed controller (13.16).

Theorem 13.1 *Consider the MIMO normal-form nonlinear system (13.1) under Assumptions 13.1–13.3, if the robust controller (13.16) is applied, then for $t \in [0, t_f)$,*

(i)   *all the internal signals as well as the control action are bounded;*
(ii)   *each state tracking errors $E^{(j)}$, $j = 0, 1, \cdots, n-1$, converge to zero as $t \to t_f$.*

13.1.3 STABILITY ANALYSIS

To establish the closed-loop system stability under the control of (13.16), we first need to further process the first term in (13.15). Under Assumption 13.3, it can be shown that:

$$\zeta_n^\top G \zeta_n \geq \underline{g} \|\zeta_n\|^2.$$

Hence, one can obtain:

$$\frac{\mu \zeta_n^\top G u}{\underline{g}} \leq -k_n \mu \|\zeta_n\|^2 - \frac{1}{2}\|\zeta_n\|^2 - \mu \lambda \|\zeta_n\|^2 \Phi^2. \qquad (13.17)$$

Substituting (13.17) into (13.15) yields:

$$\dot{V}_n \leq -c_n \mu \|\zeta_n\|^2 - \frac{1}{2}\|\zeta_n\|^2 + \frac{\mu \Delta^2}{4\lambda}. \qquad (13.18)$$

To proceed, we choose the Lyapunov function as:

$$V = \sum_{k=1}^{n} V_k.$$

By combining (13.7), (13.12), and (13.18), it is readily shown that:

$$\dot{V} = \sum_{k=1}^{n} \dot{V}_k \leq - \sum_{k=1}^{n} c\mu \|\zeta_k\|^2 + \frac{\mu \Delta^2}{4\lambda} + L_1, \qquad (13.19)$$

with $c = \min\{c_1 \cdots, c_n\}$, and

$$L_1 = \sum_{k=1}^{n-1} \zeta_k^\top \zeta_{k+1} - \frac{1}{2}\|\zeta_1\|^2 - \sum_{k=2}^{n-1} \|\zeta_k\|^2 - \frac{1}{2}\|\zeta_n\|^2 \leq 0. \qquad (13.20)$$

Note that (13.20) shows that the perturbations caused by the interconnected z_i-subsystems can be compensated by the damping terms added to the control inputs. Consequently, (13.19) becomes:

$$\dot{V} \leq - \sum_{k=1}^{n} c\mu \|\zeta_k\|^2 + \frac{\mu \Delta^2}{4\lambda} \leq -2\gamma\mu V + \mu \chi, \qquad (13.21)$$

where $\gamma = c\min\{1,\underline{g}\}$ and $\chi = \frac{\Delta^2}{4\lambda}$.

The following proofs discuss the boundedness and convergence of signals in detail. We first show that all the internal signals as well as $u(t)$ are bounded for $t \in [0,t_f)$. In view of Lemma 12.1, we have from (13.21) that $V \in \mathscr{L}_\infty[0,t_f)$, which means that z_i and ζ_i $(i=1,\cdots,n) \in \mathscr{L}_\infty[0,t_f)$. It follows from (13.2) that $z_1 = \mu^{-1}\zeta_1$ and $\dot{y}_d$ are bounded, thus $\alpha_1 \in \mathscr{L}_\infty[0,t_f)$ is ensured since all the signals in α_1 are bounded. Furthermore, as $x_2 = z_2 + \alpha_1$, then x_2 is also bounded. To show the boundedness of α_2, let us note that $\dot{\alpha}_1$ remains finite as $t \to t_f$. Since from (13.3) and (13.5) we know that $\dot{\zeta}_1 = -c_1 \mu \zeta_1 + \zeta_2 - \frac{\zeta_1}{2}$, which yields:

$$\lim_{t \to t_f} \dot{z}_1 = \lim_{t \to t_f} \frac{d}{dt}(\mu^{-1}\zeta_1) = \lim_{t \to t_f}(-\mu^{-2}\dot{\mu}\zeta_1 + \mu^{-1}\dot{\zeta}_1) = \lim_{t \to t_f} -\check{c}\zeta_1. \qquad (13.22)$$

where $\check{c} = c_1 + \lim_{t \to t_f} \dot{\mu}/\mu^2$ is a constant. Furthermore, from the dynamics of ζ_1 and the boundedness of $\zeta_i(i = 1,2)$, we know that ζ_1 converges to zero as $t \to t_f$. Specifically, we have:

$$\zeta_1(t) = e^{-c_1 \int_0^t \mu(v)dv} \zeta_1(0) + e^{-c_1 \int_0^t \mu(v)dv} \int_0^t e^{c_1 \int_0^s \mu(v)dv} \left(\zeta_2 - \frac{\zeta_1}{2}\right) ds, \qquad (13.23)$$

and

$$\begin{aligned}
\lim_{t \to t_f} \zeta_1(t) &= \lim_{t \to t_f} \frac{\int_0^t e^{c_1 \int_0^s \mu(v)dv} \left(\zeta_2 - \frac{\zeta_1}{2}\right) ds}{e^{c_1 \int_0^t \mu(v)dv}} \\
&= \lim_{t \to t_f} \frac{e^{c_1 \int_0^t \mu(v)dv} \left(\zeta_2 - \frac{\zeta_1}{2}\right)}{c_1 \mu(t) e^{c_1 \int_0^t \mu(v)dv}} \qquad (13.24) \\
&= \lim_{t \to t_f} \frac{1}{c_1} \mu^{-1} \left(\zeta_2 - \frac{\zeta_1}{2}\right) = 0.
\end{aligned}$$

It follows that $\lim_{t \to t_f} \frac{t_f^{m+n} w_1}{(t_f - t)^{m+n}} = 0$. Using L'Hôpital's rule and recalling (13.22), we have:

$$\lim_{t \to t_f} \frac{\frac{\check{c}}{m+n} t_f^{m+n} \zeta_1}{(t_f - t)^{m+n-1}} = \lim_{t \to t_f} \frac{\frac{\check{c}}{m+n} t_f^{2m+2n} w_1}{(t_f - t)^{2m+2n-1}} = 0, \qquad (13.25)$$

which shows that ζ_1 tends to zero no slower than $(m+n-1)$-th power of $(t_f - t)$ as $t \to t_f$, and w_1 tends to zero no slower than $(2m+2n-1)$-th power of $(t_f - t)$ as $t \to t_f$. Continue using L'Hôpital's rule for the first term of (13.25), it holds that:

$$\lim_{t \to t_f} \frac{-t_f^{m+n} \check{c} \dot{\zeta}_1}{(m+n)(m+n-1)(t_f - t)^{m+n-2}} = 0, \qquad (13.26)$$

which shows that $\dot{\zeta}_1$ tends to zero no slower than $(m+n-2)$-th power of $(t_f - t)$ as $t \to t_f$. Now one can conclude that:

$$\dot{\alpha}_1 = -\left(c_1 \dot{\zeta}_1 + \frac{\dot{z}_1}{2} + \frac{d}{dt}(\dot{\mu}\mu^{-2}\zeta_1) - \ddot{y}_d\right) \in \mathscr{L}_\infty[0, t_f). \qquad (13.27)$$

Therefore, the boundedness of $\alpha_2 = -(c_2\zeta_2 + z_2 + \dot{\mu}\mu^{-2}\zeta_2) + \dot{\alpha}_1$ can be guaranteed. Since $x_3 = z_3 + \alpha_2$, then $x_3 \in \mathscr{L}_\infty[0, t_f)$. A careful examination of (13.10) reveals that the quantity μ always appears multiplied by ζ_i, and by repetitively taking the derivatives of (13.25) and using L'Hôpital's Rule, it can be concluded that $\mu \zeta_i$ are always bounded. In the same manner, it follows that, for $i = 1, \ldots, n$, x_i, $\mu^{n+1-i}\zeta_i$, $\dot{\eta}_i$, $\dot{\alpha}_{i-1}$, α_{i-1}, and $u(t)$ are bounded on $[0, t_f)$. The continuity of these signals follows from the fact that $\mu(t)$ and x_i are continuous functions on $[0, t_f)$.

To proceed, we prove that precise full state tracking is achieved. Note that $E = z_1$, we can rewrite (13.22) as $\lim_{t \to t_f} \dot{E} = \lim_{t \to t_f} -\check{c}\mu E$. By using L'Hôpital's rule j-th times ($j = 1, \cdots, n$), it can be derived that:

$$\lim_{t \to t_f} \dot{E} = \lim_{t \to t_f} \frac{\check{c} t_f^{n+m} \dot{E}}{(n+m)(t_f - t)^{n+m-1}}$$

$$= \cdots = \lim_{t \to t_f} \frac{\check{c} t_f^{n+m} E^{(j)}}{(-1)^{j+1} \frac{(n+m)!}{(n+m-j)!} (t_f - t)^{n+m-j}},$$

Note that $\lim_{t \to t_f} \dot{E} = \lim_{t \to t_f} -\check{c}\zeta_1 = 0$, the following recurrence relation can be established:

$$\lim_{t \to t_f} E^{(j)} = \lim_{t \to t_f} \frac{(-1)^{j+1} \frac{(n+m)!}{(n+m-j)!} (t_f - t)^{n+m-j}}{\check{c} t_f^{n+m}} \dot{E} = 0. \qquad (13.28)$$

Therefore, the full state tracking errors are bounded and converge to zero as $t \to t_f$. The proof is completed.

Remark 13.2 *Note that reference [18] solves the regulation problem for SISO nonlinear systems, whereas Theorem 13.1 solves the prescribed-time tracking for MIMO systems, covering reference [18] as a special case. This chapter eliminates the restrictive assumption in reference [18] that the lower bound of the control gain is known, and uses a unified virtual error variable instead of two different but interconnected variables in reference [18], avoiding the ISS small-gain analysis and resulting in the parameter selection rule that is only based upon the Lyapunov theorem. The prescribed-time tracking control for MIMO systems with mismatched uncertainties is not covered in this chapter, and the related content can be found in reference [148].*

13.2   APPLICATION TO ROBOTIC SYSTEMS

Recall the dynamic model of the nonlinear robotic systems in Chapter 7, it can be described by:

$$H(q,p)\ddot{q} + N_g(q,\dot{q},p)\dot{q} + G_g(q,p) + \tau_d(\dot{q},p) = u, \qquad (13.29)$$

where $q = [q_1, \cdots, q_m]^\top \in \mathbb{R}^m$ denotes the link position, $H(q,p) \in \mathbb{R}^{m \times m}$ denotes the inertia matrix, $p \in \mathbb{R}^l$ denotes the unknown parameter vector, $N_g(q,\dot{q},p) \in \mathbb{R}^{m \times m}$ denotes the centripetal-Coriolis matrix, $G_g(q,p) \in \mathbb{R}^m$ represents the gravitation vector, $\tau_d(\dot{q},p) \in \mathbb{R}^m$ denotes the frictional and disturbing force, and $u \in \mathbb{R}^m$ is the control input. It is seen from the discussion in Chapter 7 that Assumptions 13.2 and 13.3 are applicable to robotic systems.

Define $q = x_1 \in \mathbb{R}^m$ and $\dot{q} = x_2 \in \mathbb{R}^m$, then (13.29) can be expressed as the following form:

$$\begin{cases} \dot{x}_1 = x_2, \\ \dot{x}_2 = G(\cdot)u + F(\cdot), \end{cases} \tag{13.30}$$

where $F(\cdot) = H^{-1}(-N_g x_2 - G_g - \tau_d)$ and $G(\cdot) = H^{-1}$ are completely unknown. Using the structural property of EL systems, it is easy to show that $\|F(\cdot)\| \le a\phi(\cdot)$ with a being an unknown constant and $\phi(\cdot) = 1 + \|x_1\| + \|x_2\| + \|x_2\|^2$.

The prescribed-time controller can be derived from (13.16) as:

$$u = -\left(c\zeta_2 + \frac{1}{2}z_2 + \lambda\zeta_2\Phi^2\right), \tag{13.31}$$

with $c > 0$, and

$$\begin{cases} \Phi = \|\dot{\mu}\mu^{-2}\zeta_2\| + \phi + \|\dot{\alpha}_1\|, \\ \alpha_1 = -\left(c\zeta_1 + \frac{1}{2}E + \dot{\mu}\mu^{-1}E - \dot{y}_d\right), \\ E = x_1 - y_d, \\ z_2 = x_2 - \alpha_1, \\ \zeta_1 = \mu(t)E, \\ \zeta_2 = \mu(t)z_2, \\ \dot{\alpha}_1 = \frac{\partial\alpha_1}{\partial\mu}\dot{\mu} + \frac{\partial\alpha_1}{\partial\dot{\mu}}\ddot{\mu} + \frac{\partial\alpha_1}{\partial y_d}\dot{y}_d + \frac{\partial\alpha_1}{\partial\dot{y}_d}\ddot{y}_d + \frac{\partial\alpha_1}{\partial x_1}x_2. \end{cases} \tag{13.32}$$

Theorem 13.2 *Consider the nonlinear robotic system (13.29). If the time-varying function is selected as $\mu = (t_f/(t_f - t))^3$ and the prescribed-time controller (13.31) is applied, then for all $t \in [0, t_f)$:*

(i)   *the internal signals including the control action are bounded;*
(ii) *both the position tracking error $E(t)$ and velocity tracking error $\dot{E}(t)$ converge to zero within a prescribed time t_f.*

Proof: By using the notion of the scaling of the virtual errors, we have that $\zeta_1 = \mu E = \mu(x_1 - y_d)$ and $\zeta_2 = \mu z_2 = \mu(x_2 - \alpha_1)$. Take the time derivative of ζ_i, then system (13.30) is transformed to the following form:

$$\begin{cases} \dot{\zeta}_1 = \mu\alpha_1 + \dot{\mu}\mu^{-1}\zeta_1 + \zeta_2 - \mu\dot{y}_d, \\ \dot{\zeta}_2 = \mu Gu + \mu\left(\dot{\mu}\mu^{-2}\zeta_2 + \mu(F - \dot{\alpha}_1)\right). \end{cases} \tag{13.33}$$

Choose the Lyapunov function candidate as $V = \frac{1}{2}\zeta_1^\top\zeta_1 + \frac{1}{2g}\zeta_2^\top\zeta_2$, with $\underline{g}$ being a positive constant. The derivative of V along the trajectory of (13.33) is evaluated as:

$$\dot{V} = \zeta_1^\top \dot{\zeta}_1 + \frac{1}{g}\zeta_2^\top \dot{\zeta}_2$$

$$= \zeta_1^\top \left(\dot{\mu}\mu^{-1}\zeta_1 + \mu(x_2 - \dot{y}_d) \right) + \frac{1}{g}\mu \zeta_2^\top \left(\dot{\mu}\mu^{-2}\zeta_2 + Gu + F - \dot{\alpha}_1 \right)$$

$$\leq \zeta_1^\top \left(\dot{\mu}\mu^{-1}\zeta_1 + \zeta_2 + \mu(\alpha_1 - \dot{y}_d) \right) + \frac{1}{g}\mu \zeta_2^\top Gu + \frac{1}{g}\mu \zeta_2^\top \left(\dot{\mu}\mu^{-2}\zeta_2 + F - \dot{\alpha}_1 \right).$$

$$\tag{13.34}$$

With the help of the fact $\|F\| \leq a\phi$, it follows that:

$$\frac{1}{g}\mu \zeta_2^\top \left(\dot{\mu}\mu^{-2}\zeta_2 + F - \dot{\alpha}_1 \right) \leq \Delta\mu\|\zeta_2\|\Phi \leq \lambda\mu\|\zeta_2\|^2\Phi^2 + \frac{\mu\Delta^2}{4\lambda},$$

where $\Delta = 1/g \times \max\{1,a\}$ is an unknown constant, and $\Phi = \dot{\mu}\mu^{-2}\|\zeta_2\| + \phi + \|\dot{\alpha}_1\|$ is a computable scalar-valued function.

Inserting (13.31) and (13.32) into (13.34), we have:

$$\dot{V}_n \leq -c\mu\|\zeta_1\|^2 + \zeta_1^\top \zeta_2 - \frac{1}{2}\|\zeta_1\|^2 - \frac{1}{g}\mu\zeta_2^\top G\left(c\zeta_2 + \frac{1}{2}z_2 + \theta\zeta_2\Phi^2 \right)$$

$$+ \lambda\mu\|\zeta_2\|^2\Phi^2 + \frac{\mu\Delta^2}{4\lambda} \tag{13.35}$$

$$\leq -c\mu\|\zeta_1\|^2 + \zeta_1^\top \zeta_2 - \frac{1}{2}\|\zeta_1\|^2 - c\mu\|\zeta_2\|^2 - \frac{1}{2}\|\zeta_2\|^2 + \frac{\mu\Delta^2}{4\lambda}$$

$$\leq -2\gamma\mu V + \mu\chi,$$

where $\gamma = c\min\{1, 1/g\}$ and $\chi = \frac{\Delta^2}{4\lambda}$.

Now we are ready to establish the results of the theorem as follows.

Using (13.35) and Lemma 12.1, $V \in \mathscr{L}_\infty[0, t_f)$ is established, from which it holds that $\zeta_i \in \mathscr{L}_\infty[0, t_f)$. Inserting α_1 into (13.33) yields:

$$\dot{\zeta}_1 = -c\mu\zeta_1 + \zeta_2 - \frac{\zeta_1}{2}.$$

Then, it follows that:

$$\lim_{t \to t_f} \dot{z}_1 = \lim_{t \to t_f} \frac{d}{dt}(\mu^{-1}\zeta_1) = \lim_{t \to t_f} (-\mu^{-2}\dot{\mu}\zeta_1 + \mu^{-1}\dot{\zeta}_1) = \lim_{t \to t_f} -\check{c}\zeta_1,$$

where $\check{c} = c + \lim_{t \to t_f} \dot{\mu}/\mu^2$ is a constant. It follows that $\lim_{t \to t_f} \frac{t_f^3 z_1}{(t_f - t)^3} = 0$. Employing L'Hôpital's rule and recalling (13.22), we have:

$$\lim_{t \to t_f} \frac{\check{c}t_f^3\zeta_1}{3(t_f - t)^2} = \lim_{t \to t_f} \frac{\check{c}t_f^6 z_1}{3(t_f - t)^5} = 0. \tag{13.36}$$

Continue using L'Hôpital's rule for the first term of (13.25), it holds that

$$\lim_{t \to t_f} \frac{-t_f^3 \check{c} \dot{\zeta}_1}{6(t_f - t)} = 0, \tag{13.37}$$

which shows that $\dot{\zeta}_1$ tends to zero no slower than $(t_f - t)$ as $t \to t_f$. Now one can conclude that:

$$\dot{\alpha}_1 = -\left(c \dot{\zeta}_1 + \frac{\dot{z}_1}{2} + \frac{d}{dt} (\dot{\mu} \mu^{-2} \zeta_1) - \ddot{y}_d \right) \in \mathcal{L}_\infty[0, t_f). \tag{13.38}$$

Therefore, the boundedness of $\Phi = \dot{\mu} \mu^{-2} \|\zeta_2\| + \phi + \|\dot{\alpha}_1\|$ can be guaranteed, and the boundedness of $u = -(c\zeta_2 + \frac{1}{2} z_2 + \theta \zeta_2 \Phi^2)$ can also be guaranteed. Furthermore, using L'Hôpital's Rule, we have E and $\dot{E} \in \mathcal{L}_\infty[0, t_f)$, and

$$\lim_{t \to t_f} E = \lim_{t \to t_f} \mu^{-1}(t) \zeta_1 = 0,$$

$$\lim_{t \to t_f} \dot{E} = \lim_{t \to t_f} -\frac{ct_f^3 E}{(t_f - t)^3} = \lim_{t \to t_f} \frac{ct_f^2 \dot{E}}{3(t_f - t)^2}. \tag{13.39}$$

It follows that both the position tracking error $E(t)$ and velocity tracking error $\dot{E}(t)$ converge to zero within the prescribed time t_f. The proof is completed.

13.3 NUMERICAL SIMULATION

To verify the effectiveness of the proposed control method, we conduct a numerical simulation on a 2-DOF robotic manipulator, the detailed expressions of the model can be found in Section 7.2.3. For simulation, the initial condition is given as: $q(0) = [q_1(0), q_2(0)]^\top = [0.1, 0.1]^\top$ and $\dot{q}(0) = [\dot{q}_1(0), \dot{q}_2(0)]^\top = [0, 0]^\top$. The reference trajectories of joint angles are $y_d = [y_{d1}, y_{d2}]^\top = [0.7 + 0.1 \cos(2t), 0.3 + 0.1 \sin(2t)]^\top$; the design parameters are set as $\lambda = 0.1$ and $c = 2$. The position tracking error is $E = q - y_d$ and the velocity tracking error is $\dot{E} = \dot{q} - \dot{y}_d$. The prescribed convergence time is set as $t_f = 2$ seconds. For comparison, we test and compare the performance of the proposed control (13.31) with a well known algorithm developed in reference [1] (refer to example 9.3 of reference [1]). The tracking process is shown in Fig. 13.1 and the tracking errors and control torques are shown in Fig. 13.2 and Fig. 13.3, respectively, from which, one can find a counterintuitive phenomenon, that is, the proposed method achieves faster system response and higher tracking accuracy with relatively small control efforts.

We also test different finite-time functions $\mu(t)$ [with $\mu_1 = (t_f/(t_f - t))^3$ and $\mu_2 = \tan(\pi t/(2t_f)) + 1$], different initial conditions, and different prescribed settling time t_f. The results are depicted in Fig. 13.4–Fig. 13.6, from which it is seen that, with the proposed control without using system model information, precise position and velocity tracking are achieved within a given time t_f, irrespective of initial conditions, confirming the theoretical prediction.

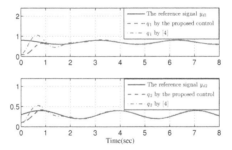

FIGURE 13.1   The tracking process under the proposed prescribed-time controller and the controller in reference [1].

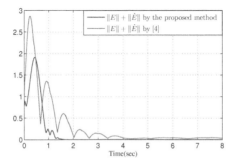

FIGURE 13.2   The evolution of the tracking errror $\|E\| + \|\dot{E}\|$ under the proposed prescribed-time controller and the controller in reference [1].

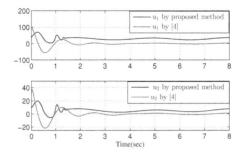

FIGURE 13.3   The evolutions of the control inputs u_1 and u_2 under the proposed prescribed-time controller and the controller in reference [1].

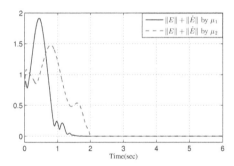

FIGURE 13.4   The evolution of the tracking error under different $\mu(t)$.

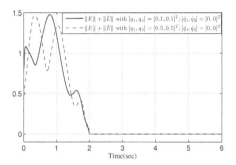

FIGURE 13.5   The evolution of the tracking error under different initial conditions.

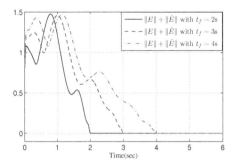

FIGURE 13.6   The evolution of the tracking error under different settling time.

13.4   NOTES

This chapter illustrates the prescribed-time tracking control design methods for MIMO normal-form nonlinear systems with unknown control gain matrices. Different from the existing related results (see reference [149]) where the control gain matrix is assumed to be known, the approach presented in this chapter can be applied to a broader class of systems, and therefore has a wide range of application prospects. Although the actual control gain matrix $G(\cdot)$ of the system is unknown and time-varying, the proposed prescribed-time control (13.16) does not require any information on it. When the control gain matrix $G(\cdot)$ is nonsquare and partially known, satisfying $G = AM$ with $A \in \mathbb{R}^{m \times q}$ being a known bounded matrix with full row rank and $M \in \mathbb{R}^{q \times q}$ being completely unknown and unnecessarily symmetric, we can also design a prescribed-time controller to achieve full state zero-error tracking. An additional assumption is needed: $A(M + M^\top)A^\top$ is symmetric and positive definite. Based upon this assumption, we can use the same ideas as introduced in this chapter for controller design and stability analysis. The only difference is that we should use $\zeta_n^\top \frac{A(M+M^\top)A^\top}{2\|A\|} \zeta_n \geq \underline{m} \|\zeta_n\|^2$ to replace the inequality $\zeta_n^\top G \zeta_n \geq \underline{g} \|\zeta_n\|^2$ in stability analysis, where $\underline{m}$ is an unknown constant and satisfies $\underline{m} \leq \frac{1}{2\|A\|} \lambda_{\min}\{A(M + M^\top)A^\top\}$.

In addition to the introduced results in prescribed-time control, some important yet challenging directions deserving further research are listed as follows.

(1)   *Prescribed-time adaptive control:* Although adaptive prescribed-time control is investigated in reference [150], the system unknown parameter is assumed to be a constant, and the control coefficient is assumed to be known. In some practical applications, such assumptions may not be satisfied since we know that systems with changing structures usually have time-varying parameters and that the inertia matrix (which can be viewed as the control coefficient) of a robotic system is usually unknown [151]. It is necessary and challenging to develop more powerful solutions to meet such scenarios.

(2)   *Robust prescribed-time control to measurement noise:* Even though prescribed-time designs have been proven robust w.r.t. structural uncertainties and external input perturbations, some simulation studies have revealed that this is not the case in the presence of measurement noise, which may be non-constant and unknown. In this direction, new tools specifically for existing frameworks are needed.

(3)   *Output feedback-based prescribed-time control:* Output feedback schemes often imply low cost, which is very attractive in practical applications (especially for large-scaled/networked/multi-agent systems). However, the existing results can only achieve prescribed-time output feedback stabilization for some special systems (e.g., linear systems), it is therefore important to explore the output feedback-based prescribed-time control for more general nonlinear systems.

(4) *Prescribed-time distributed control:* As for prescribed-time control for multi-agent systems, it is interesting to generalize the simple framework on first- or second-order integrators to agents having high-order uncertain nonlinear dynamics and to investigate prescribed-time decentralized control algorithms under complex communication topologies, as well as to study how to achieve consensus with as little information interaction between agents as possible without losing controllability.

References

1. Slotine J J E, Li W P. Applied Nonlinear Control [M]. Englewood Cliffs: Prentice Hall, 1991.
2. Galicki M. Finite-time control of robotic manipulators [J]. Automatica, 2015, 51: 49–54.
3. Lewis F L, Jagannathan S, Yesildirak A. Neural Network Control of Robot Manipulators and Nonlinear Systems [M]. London: Taylor & Francis Ltd, 1999.
4. Craig J J, Hsu P, Sastry S. Adaptive control of technical manipulators [C]. Proceedings, 1986 IEEE International Conference on Robotics and Automation, 1986: 190–195.
5. Song Q, Song Y D, Tang T, et al. Computationally inexpensive tracking control of high-speed trains with traction/braking saturation [J]. IEEE Transactions on Intelligent Transportation Systems, 2011, 12(4): 1116–1125.
6. Li S, Yang L, Gao Z. Distributed optimal control for multiple high-speed train movement: An alternating direction method of multipliers [J]. Automatica, 2020, 112: 108646.
7. Tao G, Chen S H, Tang X D, et al. Adaptive Control of Systems with Actuator Failures [M]. London: Springer-Verlag Ltd, 2004.
8. Tang X D, Tao G, Joshi S M. Adaptive actuator failure compensation for parametric strict feedback systems and an aircraft application [J]. Automatica, 2003, 39(11): 1975–1982.
9. Krstić M, Kanellakopoulos I, Kokotović P. Nonlinear and Adaptive Control Design [M]. New York: John Wiley & Sons, Inc., 1995.
10. Astolfi A, Karagiannis D, Ortega R. Nonlinear and Adaptive Control with Applications [M]. London: Springer, 2008.
11. Goodwin G C, Graebe S F, Salgado M E. Control System Design [M]. Upper Saddle River: Prentice Hall, 2001.
12. Åström K J, Wittenmark B. Adaptive Control [M]. Mineola: Dover Publications, Inc., 2008.
13. Utkin V I. Sliding Modes in Control and Optimization [M]. Berlin: Springer-Verlag, 1992.
14. Khalil H K. Nonlinear Systems [M]. Upper Saddle River: Prentice Hall, 2002.
15. Chen Z Y, Huang J. Stabilization and Regulation of Nonlinear Systems: A Robust and Adaptive Approach [M]. Cham: Springer, 2015.
16. Zhao K, Song Y D, Ma T D, et al. Prescribed performance control of uncertain Euler-Lagrange systems subject to full-state constraints [J]. IEEE Transactions on Neural Networks and Learning Systems, 2018, 29(8): 3478–3489.
17. Song Y D. Adaptive motion tracking control of robot manipulators–non-regressor based approach [J]. International Journal of Control, 1996, 63(1): 41–54.
18. Song Y D, Wang Y J, Holloway J, et al. Time-varying feedback for regulation of normal-form nonlinear systems in prescribed finite time [J]. Automatica, 2017, 83: 243–251.
19. Wang M, Ge S S, Hong K S. Approximation-based adaptive tracking control of pure-feedback nonlinear systems with multiple unknown time-varying delays [J]. IEEE Transactions on Neural Networks, 2010, 21(11): 1804–1816.
20. Zhao K, Song Y D, Meng W C, et al. Low-cost approximation-based adaptive tracking control of output-constrained nonlinear systems [J]. IEEE Transactions on Neural Networks and Learning Systems, 2021, 32(11): 4890–4900.
21. LaSalle J, Lefschetz S. Stability by Lyapunov's Direct Method [M]. London: Academic Press Inc., 1961.

22. Ioannou P A, Sun J. Robust Adaptive Control [M]. Upper Saddle River: Prentice Hall, 2012.
23. Rudin W. Real and Complex Analysis [M]. McGraw Hill Inc., 1987.
24. Zuo Z Y, Wang C L. Adaptive trajectory tracking control of output constrained multi-rotors systems [J]. IET Control Theory and Applications, 2015, 8(13): 1163–1174.
25. Wang M, Zhang S Y, Chen B, et al. Direct adaptive neural control for stabilization of nonlinear time-delay systems [J]. Science China Information Sciences, 2010, 53(4): 800–812.
26. Qian C J, Lin W. Non-lipschitz continuous stabilizers for nonlinear systems with uncontrollable unstable linearization [J]. Systems & Control Letters, 2001, 42(3): 185–200.
27. Hardy G H, Littlewood J E, Pólya G. Inequalities [M]. Cambridge: Cambridge University Press, 1952.
28. Glover K. All optimal hankel-norm approximations of linear multi-variable systems and their L^{∞}-error bounds [J]. International Journal of Control, 1984, 39(6): 1115–1193.
29. Kokotović P, Khalil H K, Reilly J. Singular Perturbation Methods in Control: Analysis and Design [M]. New York: Academic Press, 1986.
30. Zhou K, Doyle J C. Essentials of Robust Control [M]. Englewood Cliffs: Prentice Hall, 1998.
31. Haimo V T. Finite time controllers [J]. SIAM Journal on Control and Optimization, 1986, 24(4): 760–770.
32. Slotine J J E, Li W P. Composite adaptive control of robot manipulators [J]. Automatica, 1989, 25(4): 509–519.
33. Nguyen N, Krishnakumar K, Kaneshige J, et al. Flight dynamics and hybrid adaptive control of damaged aircraft [J]. Journal of Guidance, Control, and Dynamics, 2008, 31(3): 751–764.
34. Guo B J, Jiang A P, Hua X M, et al. Nonlinear adaptive control for multivariable chemical processes [J]. Chemical Engineering Science, 2001, 56(23): 6781–6791.
35. Pierre D A. A perspective on adaptive control of power systems [J]. IEEE Transactions on Power Systems, 1987, 2(2): 387–395.
36. Kahveci N E, Ioannou P A. Adaptive steering control for uncertain ship dynamics and stability analysis [J]. Automatica, 2013, 49(3): 685–697.
37. Bastin G, Dochain D. On-line Estimation and Adaptive Control of Bioreactors [M]. Netherlands: Elsevier, 1990.
38. Kanellakopoulos I, Kokotović P, Morse A. Systematic design of adaptive controllers for feedback linearizable systems [J]. IEEE Transactions on Automatic Control, 1991, 36(11): 1241–1253.
39. Song Y D, Zhao K, Krstić M. Adaptive control with exponential regulation in the absence of persistent excitation [J]. IEEE Transactions on Automatic Control, 2017, 62(5): 2589–2596.
40. Qu Z H, Dawson D M, Dorsey J F. Exponentially stable trajectory following of robotic manipulators under a class of adaptive controls [J]. Automatica, 1992, 28(3): 579–586.
41. Wen C Y, Zhou J, Liu Z T, et al. Robust adaptive control of uncertain nonlinear systems in the presence of input saturation and external disturbance [J]. IEEE Transactions on Automatic Control, 2011, 56(7): 1672–1678.
42. Song Y D, Huang X C, Wen C Y. Tracking control for a class of unknown nonsquare MIMO nonaffine systems: A deep-rooted information based robust adaptive approach [J]. IEEE Transactions on Automatic Control, 2016, 61(10): 3227–3233.

43. Bhat S P, Bernstein D S. Finite-time stability of continuous autonomous systems [J]. SIAM Journal on Control and Optimization, 2000, 38(3): 751–766.

44. Weiss L, Infante E F. On the stability of systems defined over a finite time interval [J]. Proceeding of the National Academy of Sciences, 1965, 54(1): 44–48.

45. Weiss L, Infante E F. Finite time stability under perturbing forces and on product spaces [J]. IEEE Transactions on Automatic Control, 1967, 12(1): 54–59.

46. Emelyanov S V, Taranm V A. On a class of variable structure control systems [R]. Proceeding of USSR Academy of Sciences, Energy and Automation, Moskov, 1962, 3: 94.

47. Yu X, Man Z. Model reference adaptive control systems with terminal sliding modes [J]. International Journal of Control, 1996, 64(6): 1165–1176.

48. Xu X, Wang J Z. Finite-time consensus tracking for second-order multi-agent systems [J]. Asian Journal of Control, 2013, 15(4): 1246–1250.

49. Bhat S P, Bernstein D S. Continuous finite-time stabilization of the translational and rotational double integrators [J]. IEEE Transactions on Automatic Control, 1998, 43(5): 678–682.

50. Bhat S P, Bernstein D S. Finite time stability of homogeneous systems [C]. Proceedings of the American Control Conference, 1997, 4: 2513–2514.

51. Hong Y G, Huang J, Xu Y S. On an output feedback finite-time stabilization problem [J]. IEEE Transactions on Automatic Control, 2001, 46(2): 305–309.

52. Hong Y G, Wang J, Cheng D Z. Adaptive finite-time control of nonlinear systems with parametric uncertainty [J]. IEEE Transactions on Automatic Control, 2006, 51(5): 858–862.

53. Hui Q, Haddad W M, Bhat S P. Finite-time semi-stability and consensus for nonlinear dynamical networks [J]. IEEE Transactions on Automatic Control, 2008, 53(8): 1887–1990.

54. Du H B, He Y G, Cheng Y Y. Finite-time synchronization of a class of second-order nonlinear multi-agent systems using output feedback control [J]. IEEE Transactions on Circuits and Systems I: Regular Papers, 2014, 61(6): 1778–1788.

55. Guan Z H, Sun F L, Wang Y W, et al. Finite-time consensus for leader-following second-order multi-agent networks [J]. IEEE Transactions on Circuits and Systems I: Regular Papers, 2012, 59(11): 2646–2654.

56. Lu X Q, Lu R Q, Chen S H, et al. Finite-time distributed tracking control for multi-agent systems with a virtual leader [J]. IEEE Transactions on Circuits and Systems I: Regular Papers, 2013, 60(2): 352–362.

57. Li S H, Wang X Y. Finite-time consensus and collision avoidance control algorithms for multiple AUVs [J]. Automatica, 2013, 49(11): 3359–3367.

58. Zhang B, Jia Y M, Matsno F. Finite-time observers for multi-agent systems without velocity measurements and with input saturations [J]. Systems & Control Letters, 2014, 68: 86–94.

59. Yu H, Shen Y J, Xia X H. Adaptive finite-time consensus in multi-agent networks [J]. Systems & Control Letters, 2013, 62(10): 880–889.

60. Wang L, Xiao F. Finite-time consensus problems for networks of dynamic agents [J]. IEEE Transactions on Automatic Control, 2010, 55(4): 950–955.

61. Xiao F, Wang L, Chen J. Finite-time formation control for multi-agent systems [J]. Automatica, 2009, 45(11): 2605–2611.

62. Xiao F, Wang L, Chen T W. Finite-time consensus in networks of integrator-like dynamic agents with directional link failure [J]. IEEE Transactions on Automatic Control, 2014, 59(3): 756–762.

63. Cao Y C, Ren W. Finite-time consensus for multi-agent networks with unknown inherent nonlinear dynamics [J]. Automatica, 2014, 50(10): 2648–2656.

64. Zuo Z Y, Tie L. A new class of finite-time nonlinear consensus protocols for multi-agent systems [J]. International Journal of Control, 2014, 87(2): 363–370.

65. Caron J M, Praly L. Adding an integrator for the stabilization problem [J]. Systems & Control Letters, 1991, 17: 89–104.

66. Lin W, Qian C J. Adding one power integrator: A tool for global stabilization of high-order lower-triangular systems [J]. Systems & Control Letters, 2000, 39(5): 339–351.

67. Lin W, Qian C J. Adaptive control of nonlinearly parameterized systems: The smooth feedback case [J]. IEEE Transactions on Automatic Control, 2002, 47(8): 1249–1266.

68. Lin W, Qian C J. Adaptive control of nonlinearly parameterized systems: A non-smooth feedback framework [J]. IEEE Transactions on Automatic Control, 2002, 47(5): 757–774.

69. Song Y D, Wang Y J, Holloway J, et al. Time-varying feedback for finite-time robust regulation of normal-form nonlinear systems [C]. 2016 IEEE 55th Conference on Decision and Control, 2016: 3837–3842.

70. Wang Y J, Song Y D. Leader-following control of high-order multi-agent systems under directed graphs: Pre-specified finite time approach [J]. Automatica, 2018, 87: 113–120.

71. Slater G L, Wells W R. Optimal evasive tactics against a proportional navigation missile with time delay [J]. Journal of Spacecraft and Rockets, 1973, 10(5): 309–313.

72. Ho Y, Bryson A, Baron S. Differential games and optimal pursuit-evasion strategies [J]. IEEE Transactions on Automatic Control, 1965, 10(4): 385–389.

73. Rekasius Z. An alternate approach to the fixed terminal point regulator problem [J]. IEEE Transactions on Automatic Control, 1964, 9(3): 290–292.

74. Sidar M. On closed-loop optimal control [J]. IEEE Transactions on Automatic Control, 1964, 9(3): 292–293.

75. Polyakov A. Nonlinear feedback design for fixed-time stabilization of linear control systems [J]. IEEE Transactions on Automatic Control, 2011, 57(8): 2106–2110.

76. Song Y D, Wang Y J. Cooperative Control of Nonlinear Networked Systems: Infinite-time and Finite-time Design Methods [M]. Cham: Springer-Verlag, 2019.

77. Ge S S, Wang C. Adaptive NN control of uncertain nonlinear pure-feedback systems [J]. Automatica, 2002, 38(4): 671–682.

78. Song Y D, Huang X C, Jia Z J. Dealing with the issues crucially related to the functionality and reliability of NN-associated control for nonlinear uncertain systems [J]. IEEE Transactions on Neural Networks and Learning Systems, 2017, 28(11): 2614–2625.

79. Ge S S, Wang J. Robust adaptive tracking for time-varying uncertain nonlinear systems with unknown control coefficients [J]. IEEE Transactions on Automatic Control, 2003, 48(8): 1463–1469.

80. Ye X D, Jiang J P. Adaptive nonlinear design without a prior knowledge of control directions [J]. IEEE Transactions on Automatic Control, 1998, 43(11): 1617–1621.

81. Ye X D. Asymptotic regulation of time-varying uncertain nonlinear systems with unknown control directions [J]. Automatica, 1999, 35(5): 929–935.

82. Psillakis H E. Further results on the use of Nussbaum gains in adaptive neural network control [J]. IEEE Transactions on Automatic Control, 2010, 55(12): 2841–2846.

83. Nussbaum R D. Some remarks on a conjecture in parameter adaptive control [J]. Systems & Control Letters, 1983, 3(5): 243–246.

84. Song Y D, Wang Y J, Wen C Y. Adaptive fault-tolerant PI tracking control with guaranteed transient and steady-state performance [J]. IEEE Transactions on Automatic Control, 2016, 62(1):481–487.

85. Zhao K, Song Y D, Qian J Y, et al. Zero-error tracking control with pre-assignable convergence mode for nonlinear systems under nonvanishing uncertainties and unknown control direction [J]. Systems & Control Letters, 2018, 115(5): 34–40.

86. Krstić M, Kanellakopoulos I, Kokotović P. Adaptive nonlinear control without overparametrization [J]. Systems & Control Letters, 1992, 19(3): 177–185.

87. Zhou J, Wen C Y, Zhang Y. Adaptive backstepping control of a class of uncertain nonlinear systems with unknown backlash-like hysteresis [J]. IEEE Transactions on Automatic Control, 2004, 49(10): 1751–1759.

88. Deng H, Krstić M. Stochastic nonlinear stabilization–I: A backstepping design [J]. Systems & Control Letters, 1997, 32(3): 143–150.

89. Zhou J, Wen C Y. Adaptive Backstepping Control of Uncertain Systems: Nonsmooth Nonlinearities, Interactions or Time-variations [M]. Berlin Heidelberg: Springer, 2008.

90. Liu Y J, Gong M Z, Liu L, et al. Fuzzy observer constraint based on adaptive control for uncertain nonlinear MIMO systems with time-varying state constraints [J]. IEEE Transactions on Cybernetics, 2021, 51(3): 1380–1389.

91. Boulkroune A, Tadjine M, Saad M M, et al. Fuzzy adaptive controller for MIMO nonlinear systems with known and unknown control direction [J]. Fuzzy Sets and Systems, 2010, 161(6): 797–820.

92. Chen M, Ge S S, Hou B. Robust adaptive neural network control for a class of uncertain MIMO nonlinear systems with input nonlinearities [J]. IEEE Transactions on Neural Networks, 2010, 21(5): 796–812.

93. Shi W X. Adaptive fuzzy control for MIMO nonlinear systems with nonsymmetric control gainmatrix and unknown control direction [J]. IEEE Transactions on Fuzzy Systems, 2014, 22(5): 1288–1300.

94. Jin X. Fault tolerant nonrepetitive trajectory tracking for MIMO output constrained nonlinear systems using iterative learning control [J]. IEEE Transactions on Cybernetics, 2019, 49(8): 3180–3190.

95. Zhao K, Song Y D, Zhang Z R. Tracking control of MIMO nonlinear systems under full state constraints: A single-parameter adaptation approach free from feasibility conditions. Automatica, 2019, 107: 52–60.

96. Zhao K, Chen J W. Adaptive neural quantized control of MIMO nonlinear systems under actuation faults and time-varying output constraints [J]. IEEE Transactions on Neural Networks and Learning Systems, 2020, 31(9): 3471–3481.

97. Song Y D, Zhang B, Zhao K. Indirect neuroadaptive control of unknown MIMO systems tracking uncertain target under sensor failures [J]. Automatica, 2017, 77: 103–111.

98. He W, Chen Y, Yin Z. Adaptive neural network control of an uncertain robot with full-state constraints [J]. IEEE Transactions on Cybernetics, 2016, 46(3): 620–629.

99. He W, Huang H F, Ge S S. Adaptive neural network control of a robotic manipulator with time-varying output constraints [J]. IEEE Transactions on Cybernetics, 2017, 47(10): 3136–3147.

100. Makavita C D, Jayasinghe S G, Nguyen H D, et al. Experimental study of command governor adaptive control for unmanned underwater vehicles [J]. IEEE Transactions on Control Systems Technology, 2017, 27(1): 332–345.

101. Faramin M, Goudarzi R H, Maleki A. Track-keeping observer-based robust adaptive control of an unmanned surface vessel by applying a 4-DOF maneuvering model [J]. Ocean Engineering, 2019, 183: 11–23.

102. Sanner R M, Slotine J J E. Structurally dynamic wavelet networks for adaptive control of robotic systems [J]. International Journal of Control, 1998, 70(3): 405–421.

103. Song Y D. Neuro-adaptive control with application to robotic systems [J]. Journal of Robotic Systems, 1997, 14(6): 433–447.
104. Wang C L, Wen C Y, Zhang X Y, et al. Output-feedback adaptive control for a class of mimo nonlinear systems with actuator and sensor faults [J]. Journal of the Franklin Institute, 2020, 357(12): 7962–7982.
105. Gao R Z, Wang H Q, Zhao L, et al. Exponential tracking control of MIMO nonlinear systems with actuator failures and nonparametric uncertainties [J]. International Journal of Robust and Nonlinear Control, 2022, 32(3): 1604–1617.
106. Zheng Z W, Sun L, Xie L H. Error-constrained los path following of a surface vessel with actuator saturation and faults [J]. IEEE Transactions on Systems, Man, and Cybernetics: Systems, 2018, 48(10): 1794–1805.
107. Yang Q M, Ge S S, Sun Y X. Adaptive actuator fault tolerant control for uncertain nonlinear systems with multiple actuators [J]. Automatica, 2015, 60: 92–99.
108. Song Y D, Guo J X, Huang X C. Smooth neuroadaptive PI tracking control of nonlinear systems with unknown and nonsmooth actuation characteristics [J]. IEEE Transactions on Neural Networks and Learning Systems, 2017, 28(9): 2183–2195.
109. Zhang Y, Tao G, Chen M, et al. A matrix decomposition based adaptive control scheme for a class of mimo non-canonical approximation systems [J]. Automatica, 2019, 103: 490–502.
110. Costa R R, Hsu L, Imai A K, et al. Lyapunov-based adaptive control of MIMO systems [J]. Automatica, 2003, 39(7): 1251–1257.
111. Ye X D, Jiang J P. Adaptive nonlinear design without a priori knowledge of control directions [J]. IEEE Transactions on Automatic Control, 1998, 43(11): 1617–1621.
112. Chen Z Y. Nussbaum functions in adaptive control with time-varying unknown control coefficients [J]. Automatica, 2019, 102: 72–79.
113. Tang X D, Tao G, Joshi S M. Adaptive actuator failure compensation for nonlinear mimo systems with an aircraft control application [J]. Automatica, 2007, 43(11): 1869–1883.
114. Xu H, Ioannou P A. Robust adaptive control for a class of MIMO nonlinear systems with guaranteed error bounds [J]. IEEE Transactions on Automatic Control, 2003, 48(5): 728–742.
115. Jin X. Adaptive finite-time fault-tolerant tracking control for a class of MIMO nonlinear systems with output constraints [J]. International Journal of Robust and Nonlinear Control, 2017, 27(5): 722–741.
116. Ge S S, Zhang J. Neural-network control of nonaffine nonlinear system with zero dynamics by state and output feedback [J]. IEEE Transactions on Neural Networks and Learning Systems, 2003, 14(4): 900–918.
117. Khalil H K. Universal integral controllers for minimum-phase nonlinear systems [J]. IEEE Transactions on Automatic Control, 2000, 45(3): 490–494.
118. Jiang Z P. Global tracking control of underactuated ships by Lyapunov's direct method [J]. Automatica. 2002, 38(2): 301–309.
119. Ilchmann A, Ryan E P, Sangwin C J. Tracking with prescribed transient behavior [J]. ESAIM: Control, Optimisation and Calculus of Variations, 2002, 7: 471–493.
120. Berger T, Reis T. Zero dynamics and funnel control for linear electrical circuits [J]. Journal of the Franklin Institute, 2014, 351(11): 5099–5132.
121. Berger T, Otto S, Reis T, et al. Combined open-loop and funnel control for underactuated multibody systems [J]. Nonlinear Dynamics, 2019, 95: 1977–1998.

122. Berger T, Rauert A. L. Funnel cruise control [J]. Automatica, 2020, 119: 109061.

123. Hopfe N, Ilchmann A, Ryan E P. Funnel control with saturation: Nonlinear SISO systems [J]. IEEE Transactions on Automatic Control, 2010, 55(9): 2177–2182.

124. Hackl C M. Non-identifier Based Adaptive Control in Mechatronics: Theory and Application [M]. Berlin: Springer International Publishing, 2017.

125. Ilchmann A, Schuster H. PI-funnel control for two mass systems [J]. IEEE Transactions on Automatic Control, 2009, 54(4): 918–923.

126. Ryan E P, Sangwin C J, Townsend P. Controlled functional differential equations: Approximate and exact asymptotic tracking with prescribed transient performance [J]. ESAIM: Control, Optimisation and Calculus of Variations, 2019, 15: 745–762.

127. Berger T, Reis T. Funnel control via funnel precompensator for minimum phase systems with relative degree two [J]. IEEE Transactions on Automatic Control, 2018, 63(7): 2264–2271.

128. Hackl M C, Hopfe N, Ilchmann A, et al. Funnel control for systems with relative degree two [J]. SIAM Journal on Control and Optimization, 2013, 51(2): 965–995.

129. Chowdhury D, Khalil H K. Funnel control for nonlinear systems with arbitrary relative degree using high-gain observers [J]. Automatica, 2019, 105: 107–116.

130. Ilchmann A, Ryan E P, Townsend P. Tracking with prescribed transient behavior for nonlinear systems of known relative degree [J]. SIAM Journal on Control and Optimization, 2007, 46(1): 210–230.

131. Liberzon D, Trenn S. The bang-bang funnel controller for uncertain nonlinear systems with arbitrary relative degree [J]. IEEE Transactions on Automatic Control, 2013, 58(12): 3126–3141.

132. Song Y D, Zhao K. Accelerated adaptive control of nonlinear uncertain systems [C]. Proceedings of the American Control Conference, Seattle, 2017: 2471–2476.

133. Wang W, Wen C Y. Adaptive actuator failure compensation control of uncertain nonlinear systems with guaranteed transient performance [J]. Automatica, 2010, 46(12): 2082–2091.

134. Wang Y J, Song Y D. Fraction dynamic-surface-based neuroadaptive finite-time containment control of multiagent systems in nonaffine pure-feedback form [J]. IEEE Transactions on Neural Networks and Learning Systems, 2017, 28(3): 678–689.

135. Ding S H, Li S H. A survey for finite-time control problems [J]. Control and Decision, 2011, 26(2): 161–169.

136. Jiang B Y. Research on Finite-time Control of Second-order Systems [D]. Harbin: Harbin Institute of Technology, 2018.

137. Liu Y, Jing Y W, Liu X P, et al. An Overview of research on finite-time control of nonlinear system [J]. Control Theory & Applications, 2020, 37(1): 1–12.

138. Bhat S P, Bernstein D S. Lyapunov analysis of finite-time differential equations [C]. the Proceedings of the American Control Conference, Seattle, 1995: 1831–1832.

139. Bhat S P, Bernstein D S. Finite-time stability of homogeneous systems [C]. Proceedings of the American Control Conference, Albuquerque, 1997: 2513–2514.

140. Wang Y J, Song Y D, Krstic M, et al. Fault-tolerant finite time consensus for multiple uncertain nonlinear mechanical systems under single-way directed communication interactions and actuation failures [J]. Automatica, 2016, 63: 374–383.

141. Song Y D, Ye H F, and Lewis F L. Prescribed-time control and its latest developments [J]. IEEE Transactions on Systems, Man, and Cybernetics: Systems, 2023, 53(7): 4102–4116.

142. Zhu Z, Xia Y Q, Fu M Y. Attitude stabilization of rigid spacecraft with finite-time convergence [J]. International Journal of Robust and Nonlinear Control, 2011, 21(6): 686–702.

143. Zuo Z Y. Nonsingular fixed-time consensus tracking for second-order multi-agent networks [J]. Automatica, 2015, 54: 305–309.

144. Puckett A, Ramo S. Guided Missile Engineering [M]. New York: McGraw-Hill, 1959.

145. Poznyak A, Polyakov A, Strygin V. Analysis of finite-time convergence by the method of Lyapunov functions in systems with second-order sliding modes [J]. Journal of Applied Mathematics and Mechanics, 2011, 75(3): 289–303.

146. Polyakov A, Poznyak A. Lyapunov function design for finite-time convergence analysis: "twisting" controller for second-order sliding mode realization [J]. Automatica, 2009, 45(2): 444–448.

147. Monahemi M, Krstić M. Control of wing rock motion using adaptive feedback linearization [J]. Journal of Guidance, Control, and Dynamics, 1996, 19(4): 905–912.

148. Ye H F, Song Y D. Prescribed-time tracking control of MIMO nonlinear systems with non-vanishing uncertainties [J]. IEEE Transactions on Automatic Control, 2023, 68(6), 3664–3671.

149. Shakouri A, Assadian N. Prescribed-time control for perturbed Euler-Lagrange systems with obstacle avoidance [J]. IEEE Transactions on Automatic Control, 2022, 67(7): 3745–3761.

150. Hua C C, Ning P J, Li K. Adaptive prescribed-time control for a class of uncertain nonlinear systems [J]. IEEE Transactions on Automatic Control, 2022, 67(11): 6159–6166.

151. Ye H F, Song Y D. Prescribed-time control for time-varying nonlinear systems: A temporal scaling based robust adaptive approach [J]. Systems & Control Letters, 2023, 181: 105602.

Index

Printed in the USA
CPSIA information can be obtained
at www.ICGtesting.com
LVHW011824041124
795688LV00003B/354